Performance All the Way Down

science · culture
A series edited by Adrian Johns and Joanna Radin

Performance All the Way Down

Genes, Development, and Sexual Difference

RICHARD O. PRUM

The University of Chicago Press
Chicago and London

The University of Chicago Press, Chicago 60637
The University of Chicago Press, Ltd., London
© 2023 by Richard O. Prum
All rights reserved. No part of this book may be used or reproduced in any manner whatsoever without written permission, except in the case of brief quotations in critical articles and reviews. For more information, contact the University of Chicago Press, 1427 E. 60th St., Chicago, IL 60637.
Published 2023
Printed in the United States of America

32 31 30 29 28 27 26 25 24 23   1 2 3 4 5

ISBN-13: 978-0-226-77175-5 (cloth)
ISBN-13: 978-0-226-82978-4 (paper)
ISBN-13: 978-0-226-82977-7 (e-book)
DOI: https://doi.org/10.7208/chicago/9780226829777.001.0001

Library of Congress Cataloging-in-Publication Data

Names: Prum, Richard O., author.
Title: Performance all the way down : genes, development, and sexual difference / Richard O. Prum.
Other titles: Science.culture.
Description: Chicago : The University of Chicago Press, 2023. | Series: Science.culture | Includes bibliographical references and index.
Identifiers: LCCN 2023008521 | ISBN 9780226771755 (cloth) | ISBN 9780226829784 (paperback) | ISBN 9780226829777 (ebook)
Subjects: LCSH: Sex differentiation. | Sex (Psychology) | Gender identity.
Classification: LCC QP278 .P78 2023 | DDC 612.6—dc23/eng/20230524
LC record available at https://lccn.loc.gov/2023008521

Contents

List of Illustrations ix

Prologue / 1

Taking Birds Seriously *2*
A Humanistic Turn *9*
An Ornithologist for Intersectionality *13*

CHAPTER ONE
Performance All the Way Down / 17

Material Feminisms *20*
A Performative Continuum *22*
What Is the Role of Metaphor in Biology? *25*
What Is at Stake? *27*
The Stakes for Evolutionary Biology *30*
Mind the Gap *32*
Why Queer Biology? *34*
Where Are We Going? *35*

CHAPTER TWO
Critical Concepts / 37

What Are Male and Female? *37*
Historical Ontology *38*
Sex *Is* a History *43*
Sex Difference versus Sexual Difference *48*
Sex and Race *51*
Sexual Development and Differentiation *54*

The Sexual Phenotype *55*
Sex Determination and Sex Reversal *56*
Discourse *58*
Agency *59*
Queer and Queering *60*

CHAPTER THREE
Gender Performativity / 63

What Is Performativity? *64*
Elements of Performativity *67*
Performativity and Trans Experience *72*
Between Butler and Barad *73*

CHAPTER FOUR
The Enactment of the Biological Self / 77

Genes and Development *79*
What Is Molecular Discourse? *83*
Want to Go to the Movies on Friday? *89*
How Discourse Becomes Genetic Action within Cells *91*
The Choreography of Gene Expression *94*
The Performative Phenotypic Landscape *96*
Canonical versus Performative Pathways *98*
How Does the Body Regulate Growth over Space? *100*
What Is the Role of Physical Forces in Development? *102*
Performativity of Cellular Discourse *106*
Are Genes Causes? *108*
Agency in Developmental Biology *111*
Citationality and Homology *114*
Posthuman Power *117*
Physiology and Immunity *119*
Neurobiology and Psychology *120*
Sexual Selection *122*
What Is Not Performative in Biology? *124*
Why Performative Biology Now? *125*

CHAPTER FIVE
How Do Our Sexual Bodies Develop? / 127

The Role of Chromosomes *130*
On Gene Nomenclature *132*
How Do Gonads Differentiate? *134*
Reproductive Tract Development *142*
Genital Development *145*

Post-embryonic Sexual Development *151*
Sexual Development Summary *154*

CHAPTER SIX
Variations in Our Sexual Development / 157

Terminology and the Framing of Embodied Sexual Variation *160*
Moving beyond Pathology *162*
Chromosomal Contributions to Differences in Sexual Development *166*
Genetic Variations in Gonad Development *167*
X Chromosome Inactivation *171*
Genital and Reproductive Tract Development *172*
Noncoding Genetic Variation *177*
How Does the Environment Affect Sexual Development? *178*
Queer Science *181*

CHAPTER SEVEN
How Evolution Generates Sexual Variability / 189

The Evolution of Sex *191*
Evolutionary Variability of Sexual Development Initiation *195*
Why Sexual Differentiation Mechanisms Are Generatively Queering *201*
Sexually Disruptive Selection *207*
Evolution of the Molecular Discourse of Sexual Development *209*
Evolution of Sexual Transition *212*
Evolution Is Incompatible with Sexual Essences *215*
Norms and Innovation *216*
Placental Performativity *219*
Limits of the Binary Bottleneck *224*

CHAPTER EIGHT
The Future of Performative Biology / 227

Performative Scientific Hypotheses *227*
Performativity of Illness and Disability *231*
Recalibrating Causality *236*
Biology Is Ready to Think Performatively *240*
Pluralism and the Phenotype *242*
What Is Evolutionary Biology About? *242*

CHAPTER NINE
Performance All the Way Up / 245

Sexual Reproduction Is an Intra-action *246*
A Posthuman Genealogy of Performative Discourse *247*

Is There Gender in Nature? *249*
"What Is Sex?" Revisited *252*
Toward a Scientific/Cultural Concept of Gender/Sex *256*
Performative Perspectives on Transsexual Experience *258*
Sex and Race as Scientific Apparatuses *259*
Sex and Race Categories in Biomedical Research *264*
Post-disciplinary Material Feminisms *269*
An Intellectually Queer Space in Science *272*

Acknowledgments 277

Appendixes 281

APPENDIX ONE
Material Feminisms *283*

APPENDIX TWO
Acquired Immunity *287*

APPENDIX THREE
Current Models of the Genotype-Phenotype Relationship *293*

APPENDIX FOUR
Modularity *299*

APPENDIX FIVE
Genetic Assimilation *301*

APPENDIX SIX
Why Gene-Level Selection Is Insufficient *305*

APPENDIX SEVEN
Internal Selection *311*

Notes 315
References 345
Index 365

Illustrations

1. Structure of DNA *81*
2. Endocrine signaling *87*
3. Paracrine signaling *88*
4. Developmental landscapes *97*
5. African Grey Parrot (*Psittacus erithacus*) *104*
6. Molecular signaling pathways in gonad development *135*
7. SRY transcription factor function *137*
8. SRY mRNA stabilization *139*
9. Human reproductive tract development *143*
10. Müllerian duct fusion *144*
11. Development of human external genitalia *146*
12. Descent of the testis *150*
13. SRY binding to SOX9 enhancer *169*
14. Developmental variations in reproductive tract anatomy *174*
15. Simplified phylogeny of vertebrate animals *192*
16. An embryonic Male mouse *220*

Prologue

This book is simultaneously an exploration of the definition and meaning of sex, a feminist material philosophy of the body, a scientific exercise in rethinking causality in genetics and development, an invitation to take a queer perspective on organismal biology and evolution, and a nonreductive model for the relationship between science and culture. The book is a scientific exploration of ideas and approaches that have been pursued in queer and feminist philosophy and science studies for decades, but from a different, distinctly biological perspective. I ask what sex is, and what it means, scientifically, to question the essentialist, binary concept of sex. This book connects queer feminist thought broadly to the fields of genetics, developmental biology, and evolutionary biology in new and, I think, unexpected and productive ways.

In the book, I will openly question whether certain foundational scientific categories—like male and female—have been defined in specific ways to preclude the results of their investigation, to prevent even the perception of their scientific shortcomings and weaknesses, and to maintain a patriarchal, heteronormative status quo. To pursue this idea, we will delve into details of the genetics and developmental biology of sex that pose real stakes for society at large. The goal will be to investigate phenomena and identify new commonalities that would be difficult to perceive without interdisciplinary conversations, and substantial give and take between biology and queer feminist studies.

As a lifelong ornithologist, I have focused my own scientific research and writing on issues in avian biology and evolution that lie far from questions of human gender and sex. You may rightly wonder how a bird-watching evolutionary biologist ends up writing a book about a queer scientific perspective on human gender and sex. What expertise can I bring to these well-studied issues? I don't have prior, formal education in feminist theory or developmental genetics, which are, of course, vast, diverse, and rich fields of research populated by many dedicated, creative, and influential scholars. Nor do I have the lived, personal experience of a sexual minority that could otherwise inform my views of these topics. What can I—pale, male, and Yale—bring to the discussion of the profound questions of sex and gender?

How did a life's work in ornithology and evolutionary biology lead me to write this book? The full story would involve delving deep into ornithology, which I would love, but this is not a book about birds. (Nevertheless, bird fans can take heart that a few examples from avian biology will pop up to briefly peer out of the intellectual foliage in these pages from time to time.) Rather, this book presents a queer feminist theory of the organism. Here, I want to describe how my research in ornithology contributed specifically to the scientific perspectives and intellectual goals that led directly to this work.

Taking Birds Seriously

Having started bird-watching as a child, I always felt certain that I would pursue a life of birds, even though I did not have any concrete ideas about what that could actually be. Before college, I was intrigued by ecology and vitally interested in conservation. I imagined myself working as a park ranger or refuge manager, which were the only jobs I thought might involve birds. But from my very first undergraduate classes, I realized that evolution was the area of biology that addressed the issues that I found so fascinating about birds—their astounding diversity, complicated history, and patchy distributions around the world. I soon became involved in early efforts to reconstruct the history of avian diversification through phylogenies—explicit hypotheses of the historical relationships, or genealogies, of species and higher groups of birds. In the early 1980s, the concepts and practice of producing phylogenies were considered by many to be controversial, disruptive, and even revolutionary. But I found the new focus on reconstructing organismal history irresistible.

The fascinating thing about the concept of phylogeny is that it transforms the traditional Linnaean classification—a system of nested taxonomic groups—from a human hierarchy imposed upon biodiversity into an empirical search to uncover the actual history of evolutionary descent and diversification. Organismal classification is transformed from an act of intellectual colonization, an imposition of control, and anthropocentric power into a (still human!) exploration of the genealogical history and individuality of life.

Every one of the tens of millions of living species and every ancestral lineage shared by subsets of those species is a product of the four-billion-year history of Earth's biodiversity. Within the massive intellectual effort to reconstruct that history, my focus has been on one very diverse branch of the vertebrates—the extant, feathered dinosaurs known as birds. Exploring the avian branch of the so-called Tree of Life—from its species-twigs to its ancient Mesozoic trunk, and back even earlier to its Jurassic origin among the bipedal, carnivorous theropod dinosaurs—remains an important part of my current research interests. For our purposes, my interest in avian phylogeny introduced me to the unique intellectual qualities of studying history for its own sake, and the constant tension between making broader generalizations and studying individual instances.

In my early work, I combined my bird-watching roots with my newfound fascination with phylogeny through the study of the evolution of courtship display behavior in a family of polygynous, Neotropical birds, called manakins (Pipridae). Within the fifty or so species of manakins, the females build the nest, incubate the eggs, and raise the young entirely on their own. However, by conducting all the parental work, female manakins have also gained complete freedom of choice over whom they will mate with. Consequently, female manakins have evolved mating preferences for strikingly patterned and colorful male plumages, and elaborate courtship displays that include acrobatic movements, vocalizations, and even mechanical wing sounds.[1] (We will return to discuss sexual selection briefly in chapter 4.)

This research required classic observations and descriptions of wild animal behavior—a field of research called ethology. Ideally for me, however, this behavioral fieldwork required trips to remote rainforests and cloud forests of South America, which perfectly suited my open-ended interests in bird-watching across the continent. To produce a phylogeny of the manakin family, I dissected, described, and analyzed the anatomy of their vocal organs, called syringes. (Molecular phylogenetic tools were still rudimentary and unproductive at that time.) My goal was to investigate the evolution and homology of behavior—that is, those shared similarities in manakin display behavior and syringeal anatomy that were due to common ancestry.

The result was an explicitly historical account of the evolution of the courtship display behavior repertoires of manakin species.[2] The display repertoire of each manakin species included behaviors that had evolved in more ancient common ancestors shared with other species, and other, unique elements that had originated in that species alone. I was able to trace the patterns of origin and conservation of display elements; identify historical instances of behavioral novelty and innovation; and document the expansion of display diversity through the addition or insertion of new behavioral elements at the beginning, middle, or end of a complex sequence of display behaviors.

My findings were an intensive history of the specific instances of sameness, change, difference, and innovation in an elaborate, evolutionary radiation in behavior and anatomy. Instead of broad, lawlike generalizations, I documented and described detailed evolutionary events and possibilities. I was able to use new phylogenetic tools to provide the strongest confirmation to date of a fundamental tenet of ethology, proposed in the 1930s by Konrad Lorenz and Niko Tinbergen, that behavioral variation among species should reveal evidence of phylogenetic history just like anatomical features of the animal body, and the molecular details of the organismal genome.[3]

This deep engagement with the evolutionary history of manakins demonstrated to me the scientific productivity of focusing on *individuality* in biology—the individuality of species and clades (i.e., the bigger, more inclusive branches on a phylogenetic tree), the individuality of behavioral and anatomical homologs, and the individuality of avian lives. At its broadest, the research was about how to use "tree thinking" to propose and test hypotheses of homology—homologies in behavior and anatomy—all of which require a deep exploration of anatomical and behavioral similarity and difference, including what they are, how to explain them, and their histories. The work also established the vital role of contingency in evolutionary history. This research involved the deepest possible engagement with the evolutionary radiation of the manakins, yet the concept of adaptation by natural selection was almost entirely irrelevant.[4]

In later years, my research expanded to include two unexpected topics (unexpected, most of all, to me) about feathers—the development and evolution of feathers, and the physics and material science of structural coloration.

Feathers are the most complex structures to grow out of the skin of any animal, and they pose a genuine challenge to evolutionary understanding. Feathers are branched like a tree, yet their branches zipper together to form a coherent plane, or vane, that allows birds to fly. Along with the origin of birds and the origin of avian flight, the origin of feathers was among the most fundamental and consequential questions in ornithology. For most of the twentieth century, the evolutionary origin of feathers had been a classic but intractable problem. Decades of efforts to explain feathers as adaptations for flight derived from elongate scales had failed to yield any significant empirical support. Meanwhile, the fossil record had not yet revealed any evidence about what ancestral, or primitive, feathers might have looked like.

My approach to the question involved using our understanding of how feathers *grow* on extant birds today to construct a model of the stages of the evolution of feathers from simple to more complex. Without getting into the feathery weeds, I predicted a series of ancestral feather morphologies that feathers would have passed through to reach contemporary feather complexity. Each stage in feather evolution was predicted to have involved the origin of a specific innovation in the mechanisms of feather development. Each stage would generate a new class of anatomical diversity in feather morphology. The model would allow us to extrapolate backwards from modern feather diversity into their deep evolutionary past and back to their initial origin.[5]

This developmental theory of feather evolution was made possible by the fortuitous details of feather development itself. To co-opt surfer slang, feathers are "totally tubular"—tubes of epidermis growing out of the skin. Like the first tubular body plan evolved in the most recent, wormy shared ancestor of humans and insects, the origin of a tube creates a host of entirely new spatial dimensions over which anatomical development and differentiation can occur. A tube of epidermis has a tip and a base, an inside and an outside, a front side and a back side, a left side and a right side, and a myriad of possible radial sections. Feathers have achieved their remarkable anatomical complexity and diversity by developmentally and evolutionarily differentiating along all of these dimensions. In short, feathers evolved hierarchical modularity, by which I mean that feather complexity is built from numerous replicate parts, or modules, and that these replicate modules are anatomically *nested* within each other. (In biology, hierarchy refers *not* to the structures or relations of power and control, but to spatial, anatomical, historical, or functional nestedness.)

When discoveries of non-avian dinosaurs with fossil feathers began to explode out of northeastern China in the late 1990s and 2000s, these radically new feather fossils confirmed both the specific morphologies and evolutionary sequence predicted by the developmental model of feather evolution. My colleagues and I further tested the developmental theory of feather evolution with experimental molecular-developmental biology. We will learn more about molecular genetic mechanisms of animal development in chapter 4, but, in brief, we investigated the intercellular molecular signaling pathways that are used by naive, developing feather cells to organize themselves into developmental modules, and to control their growth in ways that generate mature feather structures, complexity, and diversity. This was my introduction to the molecular cacophony of the developmental environment of the cells of the body—an idea that will be a major component of our exploration of the process of individual human sexual becoming.[6]

In contrast to previous, adaptationist theories of feather evolution, this developmental theory made no mention of the adaptive value or mechanisms for natural selection for any of the predicted stages of feather evolution. It is not that adaptation and selection were irrelevant to the history of feather evolution; rather, it was only by temporarily shelving questions about mechanisms of selection that it was possible to reconstruct a detailed and accurate account of the actual historical stages of feather evolution. Thinking about adaptation was an intellectual hindrance to progress on the question of the origin of evolutionary innovations, like feathers.

About the same time that I was working on the evolution of feathers, I also became accidentally intrigued in the phenomenon of structural coloration in bird feathers and skin, especially the non-iridescent colors of bluebirds, Blue Jay, and parrot plumages, and the blue skin of a cassowary or the blue beak of a Ruddy Duck. Such structural colors are produced by optical interactions between ambient white light and the physical material of the feathers or skin.

This basic research on avian structural coloration focused on questions that were fundamental to optics and materials science, respectively. *How* do these optical nanostructures produce their structural colors? And *how* do these optical nanostructures grow? In a classic bluebird feather, light is scattered by a spongy mixture of air bubbles in a matrix of solid β-keratin protein in the feather barbs. Imagine a jar filled with marbles of similar size. The marbles are all closely packed and touching their nearest neighbors, but they are otherwise a great jumble and not organized in any other way. Now, to add more realism, imagine that the marbles are *really* small— around 150 nanometers in diameter (about one-third of the size of the wavelength of blue light)—and that the "marbles" are actually *air bubbles* embedded in an otherwise solid block of protein, or feather β-keratin.

How would ambient white light respond to such a nanoscale jumble of closely packed air bubbles? Before we started our research on blue bird feathers, physicists had never really asked, let alone answered, this question. The answer is that such an aggregation works optically like the sheen of an oil slick, rather than like the blue sky, but it produces a non-iridescent color (i.e., a color that looks the same from multiple angles) under multi-directional natural light. While the details are fascinating (to me at least), the bigger issue here is that thinking deeply about bird feathers can lead to fundamental new contributions to physics.[7] Indeed, amorphous or quasi-ordered nanostructures, like the air bubbles in blue bird feathers, are being investigated for new optical technologies that could produce reflective colors without actually producing light, like a color Kindle or "electronic paper."

Physicists had never asked about the optical properties of a quasi-ordered nanostructure before because they hadn't yet imagined that it could be important. They were too busy investigating other major and important questions of the day with obvious broad impacts. However, our structural color research demonstrated that pursuing a deep curiosity for birds themselves—or really any other specific organism—can actually lead to fundamentally new insights that one might never encounter in the regular pursuit of science as usual.

How do birds control the development of their optical nanostructures that produce their colorful social and sexual signals so precisely? Organisms have evolved fantastically detailed systems to engineer biochemical structures at the molecular, or angstrom, scale. Likewise, organisms have also evolved detailed mechanisms for the growth of structure and pattern at the cellular-size scales. Interestingly, however, organisms have no specially evolved or biologically specific tools for creating pattern and spatial structure at the intermediate, nanoscale that is necessary for optical function. For this reason, the question of optical nanostructure development provides a window into a profound issue in molecular and cellular biology.

In brief, feather cells create the conditions under which the nanostructures *assemble themselves*. Molecular self-assembly exploits particular properties of soft matter—which we can think of generally as the squishy stuff that lies between the better understood material phases of solids and liquids.[8] Specifically, the spongy air and protein nanostructures in bird feathers grow by phase separation—a process of molecular unmixing, like bubbles in champagne, oil and vinegar, or cooling miso soup. In the feather cells, unmixing proceeds as the β-keratin protein polymerizes—or gloms together—to form a solid out of the solution of the cell's liquid cytoplasm. The phase separation hypothesis is supported by the observation that the nanomorphologies of these structurally colored feathers match the highly specific spherical shapes of bubbles in beer and the tortuously swirling forms of miso soup.[9]

This research was life changing for me—and not merely because every mug of beer, glass of champagne, or bowl of miso soup now reminds me of blue birds. Engaging with the material properties of molecular mixtures exposed me to thinking about self-organization, difference within mixed materials, and the emergence of pattern and self-assembly of structure in contexts beyond biology and human culture.

All during this science journey, I kept my interest in the evolution of bird behavior, song, and courtship display. There has been a long tradition in biology to conceive of sexual selection by mate choice as simply a kind of adaptation by natural selection. Going back to my first exposure to the idea, I found this view to be simply inadequate to explain the diversity and complexity of avian sexual communication and display. Rather, I pursued an alternative, authentically Darwinian view of sexual selection by mate choice as a nonadaptive (even maladaptive), arbitrary, and aesthetic evolutionary process, which sometimes interacts with natural selection to produce adaptive sexual ornaments. In brief, this research means that birds are beautiful because they are beautiful to themselves. Through their sexual and social choices, birds are active agents in their own evolution.

This view implies that the evolutionary history, diversity, and complexity of avian biology has been deeply shaped by the subjectivities of the birds themselves—by the sensory/cognitive experiences of aesthetic attraction of individual birds. The heart of the process of aesthetic evolution by mate choice is the *coevolution* of mating preferences and sexual ornaments and displays. In this context, beauty can be defined as a coevolved attraction in which the form of the preference and the perceivable qualities of display have shaped one another through evolutionary time. The aesthetic view of mate choice and sexual selection reframes reproduction as a downstream consequence of animal subjectivity—that is, sexual desire—and focuses scientific investigation on those subjectivities and away from exclusively adaptive (and normative) conceptions of sex and reproduction. Thinking about the arbitrary coevolution of sexual displays and mating preferences—which have no biological functions other than their specific, coevolved correspondences with each other—also presaged philosopher Karen Barad's concept of *intra-action*, which is central to the themes of this book (see chapter 3).[10]

My research on avian beauty established for me the importance of the *subjective agency* of animals—the cognitive capacities necessary for their complex sensory impressions, autonomous preferences, and the realization of their individual social and sexual preferences as choices. This research program conflicts directly with adaptationism—the prevalent idea that natural selection is a strong force that dominates all other evolutionary mechanisms in nature. And it highlighted yet again the intellectual costs of reductionism in biology. If all female mate choices are adaptive, then you already have an explanation of them, and any further curiosity about the subjective agency of female birds is foreclosed. I am pleased to say that there is a growing appreciation among professional biologists of the cognitive complexity of animals and an increasing acknowledgment of the subjective, aesthetic agency of birds and many other nonhuman animals. Surprisingly, perhaps, my research on aesthetic evolution of birds prepared me to understand how contemporary biology works to obscure and deny the numerous agencies involved in the development and function of organisms.

A Humanistic Turn

Ducks (Anatidae) are peculiar among birds because ducks still have a penis. I say "still have" because, although most birds have evolutionary lost this structure, ducks are among the few living groups of birds that have retained it. Although the avian penis is homologous with the mammalian penis—that is, it originated in the most recent common ancestor between birds and mammals—the duck penis has a number of unusual features, at least unusual to us. It has a counterclockwise (or right-handed) corkscrew shape, uses an explosive lymphatic erection mechanism, is highly flexible rather than stiff when erect, and is covered with tough, ribbed, ridged, barbed, or even thorny surface projections. In high-density populations of some species of ducks, the presence of the penis allows unpaired males to forcibly copulate with females that have already chosen a male partner. These violent forced copulations are vigorously resisted by female ducks, and can cause physical harm or even death.[11]

During the nineteenth and most of the twentieth century, the term *rape* was commonly used in ornithology and biology to refer to such forced or coerced copulations. However, in response to the second-wave feminist arguments by Susan Brownmiller in *Against Our Will*, evolutionary biologist Patty Gowaty, and others, the word *rape* was reserved to refer to the special social and political role of sexual violence, coercion, and social control in the subjugation of women and girls in human societies. At that point, biologists began to use the drier and less pointed term *forced copulation* to refer to these acts of sexual violence in nonhuman animals.[12]

An unfortunate ancillary consequence of the cultural/political substitution of "forced copulation" for "rape" in nonhuman biology was to facilitate the elimination of individual agency of nonhuman animals from consideration in biology. Since the 1990s, the study of sexual conflict and sexual coercion in biology was intellectually framed to view female resistance to male sexual violence and coercion as an adaptive strategy to limit the costs of their *own* overly promiscuous sexual preferences (the so-called chase-away model), or as an adaptation by females to obtain the most successful coercers as mates so that their male offspring will inherit genes to be successful through further sexual violence (the so-called resistance-as-choice hypothesis). (A much longer analysis would be required to fully unpack the destructive social and cultural implications of this scientific framework, but these ideas have many theoretical and empirical weaknesses from a scientific perspective.[13]) The term *forced copulation* in biology has allowed scientists to avoid the recognition that sexual violence is, to paraphrase Brownmiller, *against the will* of the ducks. In order to reinstate the recognition of female agency in the scientific literature and research on sexual coercion, I have suggested that biologists should consider using the term *rape* in nonhuman biology.[14]

How do ducks respond evolutionarily to persistent sexual violence—the equivalent of rape in the animal world? In 2005, Patricia Brennan came to my lab at Yale as a postdoc to work on this question. To our complete surprise, Brennan discovered that, in response to male sexual coercion, females of many species of ducks have *coevolved* complicated vaginal morphologies—including side-pocket cul-de-sacs and clockwise (left-handed) coiling—that function defensively to disrupt intromission and prevent successful fertilization during forced copulation. Although female ducks have not been able to completely eliminate the physical risk and harm of sexual coercion and violence, they have evolved the capacity to maintain extensive individual control over *which* males will fertilize their eggs—their chosen sexual partner or violently coercive males.[15] In short, we discovered that freedom of sexual choice *matters* to ducks, and that there are evolutionary consequences to infringing upon that individual sexual freedom. Although they cannot avoid the direct harms of sexual violence and coercion, the capacity of female ducks to resist forced fertilization has evolved to expand and reinforce their sexual autonomy in the face of persistent sexual violence.[16]

I considered our findings on the sexual autonomy of ducks to be a feminist discovery (or insight if you prefer) in the natural sciences. It is not feminist in assuming or accommodating any particular political theory, ideology, or framework (unless you count our conscious recognition of the obvious terror of a female duck struggling for its life as an unnecessarily political stance). Nor is it feminist science because the scientists themselves would personally identify as feminists. Rather, we found that fundamental features of feminist analyses of the dynamics of sexual conflict, coercion, and sexual violence within patriarchal human cultures are present in the social and sexual biology of many nonhuman, nonliterate, animal species. Sexual autonomy is not merely a political ideology invented by human suffragettes and feminists in the nineteenth and twentieth centuries, but an evolved feature of the evolutionary histories and lived experiences of many social, sexual animals.

Trying to understand what it meant to make a "feminist discovery" *in* science led me to start reading in feminist science studies and queer theory, which opened up a whole new direction of research possibilities. In a similar way, my interests in aesthetic evolution led me to begin reading, and ultimately to do research in, aesthetic philosophy.[17] Thus, various ornithological interests contributed directly to a humanistic turn in my work.

One of the critical commitments of all my science has been to take birds seriously—as worthy subjects themselves of scientific investigation, as cognitively complex individuals with subjective agency, and as the result of inherently fascinating evolutionary processes and complex histories of contingent events. Taking birds seriously has contributed to my personal identification with the now old-fashioned professional label of *ornithologist*, which has been largely abandoned by several academic generations of bird-studying scientists in favor of disciplinary categories based on scientific phenomena, like behavioral ecologist, population biologist, geneticist, sensory biologist, conservation biologist, and so on. To me, however, there were no limits to what I would be willing to study, the data I would need to gather, or the intellectual tools I would deploy in order to better understand birds—whether that meant studying genetics, anatomy, developmental biology, behavior, chemistry, physics, game theory, or even aesthetics and queer theory.

In a highly influential 1988 essay on feminist objectivity and science, biologist-turned-philosopher Donna Haraway writes, "Situated knowledges require that the object of knowledge be pictured as an actor and agent, not as a screen or ground or resource."[18] Ultimately, I came to understand that my own bird-watching roots, my personal commitment to natural history itself as science, and my lifelong engagement with birds had fostered a *situated style* of science—situated in the lives, natural history, behavior, development, phylogeny, and evolutionary history of specific lineages of birds. This book is really a product of that situated style of science, my curiosity to learn more, and a deep dissatisfaction with the conceptual state of much contemporary biology.

As my curiosity drew me to into the humanities and science studies, I found that a situated ornithology framework contributed naturally to a particularly *posthuman* perspective in my interdisciplinary research. Specifically, taking birds seriously as agents in their own lives and evolution means that one is already prepared to view the traditional topics of the humanities and social sciences as reframed without human agency, needs, and concerns at the organizing center of these disciplines. This posthuman perspective has been critical to the views developed throughout this entire book.

My personal scientific path has shaped my intellectual concerns and values in distinct ways: toward a fundamental intellectual commitment to the study of history itself; to the intellectual value of individuality and individual instances—events—over lawlike generalization; to emergence over reductionism; to the recognition of the agency of organisms, including the sexual and aesthetic agencies of animals; and to a situated style of inquiry. Through my research on both aesthetic evolution and sexual coercion in ducks, I experienced how doing science can lead you directly *beyond* the traditional boundaries of science. How the practice of science can blur the boundaries of science itself. How science can uncover new stakes in the natural world. However, instead of becoming anxious about this, I found it invigorating and productive to work across, and beyond, disciplinary boundaries, indeed, without disciplinary boundaries at all.

This book grew out of my curiosity about the intellectual opportunities that lie outside of the traditional boundaries of science—at the interfaces of the biology of sex and feminist and queer studies. Through my reading and explorations of feminist and queer theory, feminist history of science, and queer feminist science studies, I became fascinated by commonalities and contrasts between biological and queer/feminist thought on the nature of sex and gender. At first, I thought I had simply identified a new shared vocabulary for these fields to interact and expand their shared conversations. But ultimately I reached the point at which queer theory began to transform my understanding of biology and science—where concepts from queer feminist theory became productive new tools *in* genetics, developmental biology, and evolutionary biology. That was when I realized that communicating this perspective needed more than a brief commentary or interdisciplinary paper, but required the in-depth treatment of a book.

An Ornithologist for Intersectionality

By pursuing my curiosity about birds, I have been repeatedly drawn outside the core discipline of ornithology to explore other areas of science and ultimately the humanities. In all this work, I have applied my ornithological and bird-watching methods of close observation and critical thinking as intellectual tools in broader contexts. As in my core scientific work, I was not interested in the intellectual reduction of phenomena from one field to an explanation lying solely in another. (To me, reduction is a useful scientific tool and not synonymous with science itself.) Rather, these projects focused on the combined, or emergent, interactions and connections among fields and disciplines. And, because few people pursue research in such an interdisciplinary—or even undisciplined—fashion, these interdisciplinary research projects have led to new intellectual advances, opportunities, and understandings.

Although I have no prior professional experience in queer feminist theory or in human developmental biology, I have worked repeatedly to find productive, nonreductive, intellectual connections among fields without becoming lost in them. Accordingly, this book is not a definitive statement or conclusion. I am not planting a flag to claim this intellectual territory as my own. Rather, I hope to build upon previous critiques of the biological conception of sex and the sexual binary in new ways, to expand the conversation among scientific and humanistic disciplines, and to contribute to a new direction in feminist scientific and humanistic inquiry. I hope to encourage a new generation of interdisciplinary researchers to pursue this area of research in newly productive ways.

This book is an appeal for an *intersectional* approach to the science, materiality, and culture of human sex. Continuing from deep intellectual roots in Black feminism, intersectional analysis was pioneered by Kimberlé Crenshaw as a critique of how "single-axis" frameworks of race, gender, class, and other social identities further social oppression. Cultural and social oppression cannot be understood productively if one is using historic definitions, categories, or frameworks that were established so narrowly as to ignore, or even prevent the perception of, the interactions *among* them—including sex, gender, race, ethnicity, class, caste, religion, education, natal language, migration status, and so on.[19]

How can science be pursued intersectionally? In a call for a broad conception of intersectional studies and methods, Sumi Cho, Kimberlé Crenshaw, and Leslie McCall write that "what makes an analysis intersectional . . . is its adoption of an intersectional way of thinking about the problem of sameness and difference and its relation to power."[20] They also recognize a "centrifugal" process by which the concept of intersectionality "travels from its groundings in Black feminism to critical legal and race studies; to other disciplines and interdisciplines in the humanities, social sciences, and natural sciences." In this spirit, this book is broadly, or "centrifugally," intersectional in its focus on how scientific concepts of sex, the genome, and the body support the power of the scientific and biomedical communities in the direct and indirect oppression of women and minorities—including lesbian, gay, bisexual, asexual, transgender, nonbinary, intersex, genderqueer, and gender nonconforming people—and extends their broader cultural control over the topic of sex and gender.

My—admittedly ambitious—aim is to undo the scientific justifications for categories and conceptions of a sex and gender binary that have contributed scientific support to sexual oppression. By abandoning single-axis frameworks of sex and gender, adopting the intersectional concept of gender/sex, and analyzing the material sameness and difference of human sexual bodies independent of binary categories, I hope to contribute to both scientific understanding and expanded opportunities for human freedom and thriving.

By deploying queer feminist intellectual tools in science, I also hope to show that such post-disciplinary thought experiments are not only possible but scientifically productive. As Donna Haraway observes in "Situated Knowledges," the scientific search for "translation, convertibility, mobility of meanings, and universality" becomes reductive "only when one language (guess whose?) must be enforced as the standard for all translations and conversations."[21] In this spirit, I am proposing a nonreductive, interconnected, sharable nature/culture framework for thinking about gender and sex that is rooted not in the traditional language of biological science, but in the conceptual and analytical language of queer feminist theory. As I will argue, this effort is not a scientific accommodation to contemporary culture or politics; rather, it constitutes a genuine intellectual and empirical *contribution* to biological science on its own terms.

This book is an experiment in pursuing scientific questions that I have previously investigated in very different ornithological contexts—the evolutionary origin and radiation of innovations; the development and hierarchical complexity of feathers; plumage coloration; vocal anatomy; song; song learning; and display behavior. But here I will be employing intellectual tools from feminist and queer theory to focus on the human body. In this way, I think the work of science can take place in many contexts and on many fronts outside of the laboratory, museum, clinic, or research station. Together, I want to us rethink what it means to engage in scientific inquiry simultaneously within and beyond the traditional edges of science.

I must admit, however, that I never expected to arrive at many of the conclusions that I have come to during the writing of this book. When I started the project, I had no reason, for example, to question the individual sexual binary, or any intellectual commitment to reconceiving the relationship between genes and the body. I arrived at new views on these fundamental issues through engaging with the literature—both scientific and cultural—and trying to think more clearly about the material body, and how it grows, functions, and evolves. As I explored my core idea of applying queer gender theory to the material body, I kept challenging my framework with more and different kinds of data. In response to every challenge, I found the idea to be ever more clarifying, powerful, and productive than I had understood at the start. I am as surprised as anyone at where I ended up.

I can sympathize with readers who may feel uncomfortable with the implications of the conclusions of this book, because I have made myself uncomfortable as well. However, at each stage, I came to think that conceptual changes will be necessary to understand and to grapple with all of the available evidence—both biological and cultural—about human sex, gender, the body, and its evolution. Whether it makes any of us uncomfortable is less important than whether it is good science that will contribute to a fully functioning science/culture of the future, to be pursued, refined, and fully realized by new generations of scientists, humanists, scholars, and people in general. In sharing this perspective, I hope I can bring you to see sex, gender, molecular biology, and the diversity of human bodies in new ways as well. As an "Ornithologist for Intersectionality," I also hope this book can support other, ongoing dialogues on the long list of difficult and urgent issues at the interfaces of biological science and culture, including racism, reproductive rights and autonomy, economic inequality, health disparity, sustainable food production, climate-change mitigation, biodiversity conservation with cultural respect and economic equity, and more.[22]

This is a book about biology for feminists, and a book about queer feminist theory for biologists. This is also book about the profoundly perplexing questions of gender and sex for young people looking to understand their own place in the material, biological, social, and scientific worlds, and for their parents, family, and friends seeking to understand the diversity of our individual becomings, and to be allied with them. And I hope it is a book that will inspire productive conversations among us all.

CHAPTER ONE

Performance All the Way Down

We are living through a time of enormous cultural change involving broad reconsideration of ideas about individual sex and gender, their boundaries, their meanings, and their mutabilities. There is a growing realization of the diversity of lived gender identities and sexual experiences. In many cultures, an ever-larger number of people are declaring transgender, nonbinary, gay, lesbian, bisexual, and other nonnormative identities and orientations.

These cultural changes have not gone unopposed. Having lost the legal battles in the United States to prevent marriage equality and protections against sexual discrimination in the workplace, political and religious conservatives have mounted a new wave of efforts to legally enforce strictly binary definitions of sex and gender in the United States. Under the Trump administration, the United States Department of Health and Human Services adopted a new federal definition of individual sex as "unchangeable and determined on a biologic basis." New federal rules established that "sex means a person's status as male or female based on immutable biological traits identifiable by or before birth," and that "the sex listed on a person's birth certificate, as originally issued, shall constitute definitive proof of a person's sex unless rebutted by reliable genetic evidence."[1] This legal change eliminated federal recognition of the over 1.4 million transgender Americans, which could have dramatic impact on their access to health care, legal protections, and civil protections in schools, jails, shelters, and other public institutions. This legal definition of sex has since been rescinded by the Biden administration, but the political challenges continue.

In 2020, Idaho became the first US state to permanently define an individual's sex as the sex on their birth certificate, and to prevent transgender girls from participating in scholastic sports.[2] Since 2021, a tsunami of state legislation defining sex as a binary fact established at birth, prohibiting transgender girls from participating in sports, and restricting or prohibiting medical treatments for transgender minors have been proposed or adopted in dozens of American states. In February 2022, the Texas governor instructed the Texas Department of Family and Protective Services to say that gender-affirming medical treatments, including puberty-blocking drugs and hormone therapies, constitute child abuse under Texas law.[3] These political efforts to constrain the rights of transgender youth, transgender adults, and their families are moving so fast that it is impossible to accurately summarize them here.[4]

In February 2021, in response to a trans pride flag hung across the hallway outside the office of another congresswoman, Marjorie Taylor Greene, the Republican representative of Georgia's Fourteenth Congressional District, placed a large sign outside her office door in the United States Capital building stating:

> There are only TWO Genders:
> MALE & FEMALE
> "Trust the Science!"

Putting aside Greene's refusal to trust the science on global climate change, evolution, the prevention of gun violence, vaccination, epidemiology, and a host of other vital issues, we have to ask ourselves, "To what science is Greene referring?" What *is* science communicating to the public about sex and gender that gives Greene the impression that science unequivocally supports her views?

On its face, Greene's sign is incoherent. Sex has generally been defined as the biological components of reproductive anatomy and physiology in contrast to *gender*, which has been regarded as encompassing the culturally and socially mediated components of sex and sexual roles. Whatever you think gender is, however, human history and current events demonstrate that it is subject to enormous cultural influences that are generally understood as outside of the traditional purview of science. Thus, human gender is not an exclusive, proper subject of science. However, Greene's admonition to "Trust the Science!" highlights that most scientific and biomedical research on sex over the past seventy years has assumed and reinforced a binary concept of biological sex as an essential scientific "fact" about human individuals, and even individual human cells. The scientific view that male or female sex is "determined" by the possession of specific chromosomes, genes, or hormone-expression levels has permeated our entire culture (including the Trump administration's regulations). As we will see, however, biology fails to deliver a clear, nonarbitrary definition of individual sex. Now, more than ever, it is important to scrutinize and reconsider the science of sex.

Many twentieth-century feminists embraced the male-female sexual binary as a source of positive social identification and political organization that could support the fight for equal rights and resistance against oppression. Some gender-critical feminists still do. However, many feminists came to consider that women's liberation was not furthered by using the category and identity of "woman," which has been constituted in a binary way that ensures the oppression that they are trying to overcome.[5] Furthermore, recognition of the intersection of sex with race, class, colonialism, caste, nationality, and so on contributed to the search for a broader political and ethical base for feminism. Thus, since the 1990s, many feminists have come to reject this traditional binary perspective.

On a parallel track, concepts of the category of gender have been evolving as well. The once progressive, feminist distinction between sex and gender has been criticized as perpetuating a false nature/culture dichotomy that does not account for the feedback loop between culture and science—the way cultural ideas about gender have inevitably influenced and infused the supposedly objective science of sex. As a result, the strict distinction between sex and gender is being replaced by a single, unified concept that treats sex and gender as a continuous, complex, and indivisible phenomenon. Following feminist biologists Sari van Anders and Anne Fausto-Sterling, I will use the term "gender/sex" to refer to the combined biological and cultural elements of human sex and social behavior.[6]

Among the advantages of a unified concept of gender/sex is that it suggests the strict distinction between sex and gender should be retired like other false dichotomies—including nature and culture, mind and body, and the like. The use of gender/sex focuses our awareness on how ideas about the cultural versus biological elements of sex and sexuality have necessarily influenced each other. Over the last twenty years, a new generation of feminist critiques of science have documented the historical and ongoing impact of cultural concepts of sex on scientific research into sexual development, sexual variation, and the search for differences between the sexes.

At this point, some readers may wonder, "Why should we question what sex is? Isn't it obvious? Hasn't biological science long ago established a definitive answer to this question?" The fact that the nature of sex seems so obvious to many of us has, I think, actually impeded serious scientific scrutiny and intellectual progress. Cultural confidence in the reality of an individual sexual binary has papered over the persistent biological challenges to defining individual sex. Over and over again, science teaches us that many things that appear obvious and simple to us—like the apparent flatness of the Earth, or the evident independence of space and time—are actually erroneous. More than ever, it is important to question, scrutinize, and think clearly about what sex is.

Material Feminisms

This book is largely "undisciplined"—that is, it is framed and written by ignoring the historic boundaries among traditional academic disciplines. But it has been made possible by decades of prior and ongoing research in material feminism—an area of philosophy and cultural critique that seeks to connect our understanding of gender, sex, and sexual politics to the scientific study of biology, the material body, and psychology. Like many academic arenas, the history, boundaries, and goals of material feminism are active subjects of debate in the field. This is not an attempt at a thorough review. Indeed, many of the researchers mentioned here have themselves presented such detailed and nuanced surveys.[7] However, in this section I want to connect my perspective to previous work in this area, and ultimately point out some ways in which my perspective will differ from previous work. (I present a more detailed review of recent work in material feminism in appendix 1.)

Historically, material feminisms are a response to what has been called the "linguistic turn" in feminist theory and analysis. Through the influence of literary theory and Continental philosophy in feminist scholarship during the late twentieth century, the problems of sex, gender, oppression, and political action came to be viewed as problems of language and culture. Materiality and the body were at times conceptualized as separate from language and culture; mute; and acted upon, or inscribed by, culture. In 2003 physicist and feminist philosopher Karen Barad summarized the broad frustration over this linguistic turn, writing, "Language matters. Discourse matters. Culture matters. There is an important sense in which the only thing that does not seem to matter anymore is matter."[8] In subsequent years, there have been renewed and growing calls to engage with the scientific literature and data, with the goal of finding new and incisive insights into the problems of science/culture entanglement.

These proposals have met with some resistance. For example, in "Gut Feminism," Elizabeth A. Wilson describes what she characterizes as the "anti-biologism" and the "distaste for biological detail" in second-wave feminism. Wilson laments that "too often, it is only when anatomy or physiology or biochemistry are removed from the analytic scene . . . that it has been possible to generate a recognizably feminist account of the body."[9] Following Wilson and other material feminists, this book strives toward a materialist-feminist account of the body that is fully engaged with the scientific details of genetics, developmental biology, and evolutionary biology.

The richness and diversity of material feminisms demonstrate that the field is thriving, but I seek to extend this discussion in new ways. I think that a new perspective on the growing blizzard of details from molecular biology, genetics, developmental biology, and evolutionary biology about the material body can make that these data more comprehensible and accessible to those outside these fields. Furthermore, conceptually reframing biology can make biology visible in new ways. For example, feminist mathematician Eugenia Cheng describes the view of the city of Chicago from atop the Hancock Center. Most vistas give a chaotic picture of the urban sprawl below. But viewed from a different angle along Michigan Avenue, the underlying grid of the city's streets becomes visible.[10] Of course, not all clarifying perspectives yield orthogonal grids, and the view I propose here will not be a simplification.

For decades feminists have critiqued the empirical problems of the gender/sex binary and biological essentialism. These critiques have instigated some important incremental changes in science, but have not, I think, gone deeply enough. The impact of feminist critiques on the practice and teaching of biology has been limited by the professional goals of philosophers, historians, and sociologists of science. Many of these scholars may be hesitant to roll up their sleeves and actually *do* science—try make science *work* better. Perhaps it is considered unprofessional to step out of the critical perspective and into an active/invested role. However, I feel no reluctance to *doing* science, and I think that trying to make science better is part of my job as a scientist. To me, the value of studying the history and philosophy of science is the prospect of improving science itself. In "Can There Be a Feminist Science?," Helen Longino writes that "criticism of the deep assumptions that guide scientific reasoning about data is a proper part of science."[11] Accordingly, even though this book ventures outside of many of the traditional boundaries of science, I see it as a "proper part" of scientific research on the material body, its development, and evolution.

In "Gut Feminism," Elizabeth Wilson proposes that "in alliance with the biological sciences, feminism could build conceptual schemata about the body that are astute both politically and biologically."[12] But critical and productive collaborative bridges can only be built if the foundations for them are laid down from *both* sides of the current science/culture divide, fostering new conversations, collaborations, and exchanges between feminists and biologists, lab scientists and humanists, students and their professors, and genderqueer people and their families and friends. Thus, scientists have a critical role in material feminism.

A Performative Continuum

At the heart of issues of gender/sex are persistent questions about how we each become ourselves—materially, psychologically, behaviorally, socially, and sexually. There is a long intellectual tradition from Freudian and developmental psychology of investigating the growth and development of our psychological selves. Likewise, the discipline of developmental biology has provided a scientific account of the growth of our material bodies. In recent decades, queer and feminist literature on gender/sex has also explored facets of the broader cultural question of the development of individual gender. Here, I will apply some of the intellectual tools developed in queer feminist inquiry to human becoming—namely the concept of gender performativity—to investigate the molecular-developmental biology and evolution of the sexual body.

Following the pioneering and influential work of feminist philosophers Judith Butler, Eve Sedgwick, Karen Barad, and others, many feminists have come to view gender as *performative*, meaning that each new individual is an iterative realization—a doing, or enactment, of cultural norms of gender presentation and behavior. According to this view, gender is realized through a performance of one's individual capacities and desires, facilitated and constrained by the broader social and cultural context in which we live. Gender performativity does not imply that individuals are free to simply make up, or construct, whatever gender they want. Although humans are highly variable, each one of us brings our own specific, individual body, psychology, and personal self to the process of our self-becoming. Rather, gender performativity refers to the way in which the realization of our individual sexual and gendered selves is facilitated and constrained by the social and cultural worlds in which we live.[13] (The history and details of the performative theory of gender will be laid out in more detail in chapter 3.)

Coming from a radical forefront of late twentieth-century feminist thought, the performative view of human gender is frequently understood as conflicting with, or simply irrelevant to, contemporary scientific accounts of sex, which are deeply rooted in the concept of a biological sexual binary determined by chromosomes, genes, or hormones, and arising from hardwired genetic and developmental mechanisms. As historical studies by Anne Fausto-Sterling, Sarah Richardson, and others have documented, twentieth-century biological theories of sex "determination" focused on a succession of phenomena as the decisive, material causes of binary sex, beginning with gonads, proceeding to hormones, then on to chromosomes, and now genes.[14] According to this prevailing biological view, the categories of male and female are essential features of individual organisms that can be defined by anatomical features determined by specific combinations of genes and chromosomes—XX for females, and XY for males.[15] In parallel with reductive trends in biology and biomedicine, these approaches share a commitment to the idea that there is indeed an individual, essential fact of the sex of the genome that is determined at the formation of the zygote—or fertilized egg—that can be identified definitively by scientific investigation.

Given the enormous differences between the performative gender theory and contemporary molecular-developmental genetics, one might think that connecting these scientific and cultural perspectives to sex and gender would face intractable obstacles and conflicts. However, we will see that scrutinizing the details of the genetics and development of sex dissolves these apparent obstacles.

In this spirit, we will examine the intellectual interface and interconnections between gender performativity and the genetic, developmental, and evolutionary biology of the human sexual body. I will argue specifically that the queer feminist concept of gender performativity is a profound intellectual discovery that provides powerful insights into the development and evolution of complex organismal bodies. I use the term *intellectual discovery* as a high compliment, so it is disconcerting for me to learn that some readers may have a visceral, negative reaction to my attempt at praise. To many people, the strength of the concept of gender performativity comes from its congruence with individual, lived experiences, and not from abstract intellectual observations or scientific claims. I will argue, however, that the personal and intellectual analyses of the queer experience that are explored in queer feminist studies have yielded *fundamental* insights into biological individuality and becoming that biological science has not yet perceived or recognized. In other words, I think that keen observation can be both a site of subjective insights *and* a scientific skill, and that it should be no surprise that these perspectives can work productively together. The real question is why biological science has been so far unable to recognize, or accurately articulate, the deeply performative nature of organisms and their bodies.

Feminist philosopher Judith Butler posed the powerful challenge of how to move closer to an understanding of "the materiality of sex that is not burdened by the sex of materiality."[16] In other words, how can we view the sexual body scientifically in a way that is not rooted in (or at least attempts to be conscious of) traditional, culturally influenced biological assumptions of sex as an essential, binary, and heteronormative fact about individual bodies and lives? In what follows, we will establish both scientific support and a common cultural/scientific vocabulary for a unified concept of gender/sex.

In biological vocabulary, I am proposing a performative intellectual framework for understanding the relationship between the *genotype*—the heritable genetic material shared by all the cells of an organism—and the *phenotype*—the sum of all observable features of an individual organism including its body, physiology, and behavior. Simply put, I propose that the phenotype is best understood—materially, biologically, and culturally—as the *performance, or enactment, of the self*. Furthermore, this *self* is not an essence, a definable state, or an end goal, but an ongoing individual becoming.

The closer one looks into developmental biology—the science of how genetic regulation and expression contribute to the material formation of complex organismal bodies—the more we come to understand that we are each enactments of ourselves, biologically, evolutionarily, psychologically, socially, and culturally. The ways in which a single-celled, fertilized zygote becomes a complex adult organism with differentiated tissues and organs, homeostatic physiological regulation, integrated neurological control, consciousness, psychological functions and mechanisms, gender, and sexual behavior are scientifically best described as a performative *continuum*. In short, we are *performance all the way down*.

To articulate and defend this proposition, I will need to present a wide variety of material from queer feminist theory, molecular-developmental biology, and evolutionary biology. I hope that all readers will be able to learn from the material that is new to them, and gain new perspective from more familiar material by seeing it juxtaposed in a new way.

Queer science studies scholar Angela Willey has observed that appeals to feminists to engage with "biological detail" and "biological data" often eliminate the conceptual space for feminist analysis and critique of science.[17] In contrast, the recognition of a performative continuum from cultural gender to molecular genetics creates a new queer intellectual space for feminist analysis and research *within* genetics, developmental biology, and evolutionary biology.

Feminist literary critic Eve Sedgwick writes, "As a general principle, I don't like the idea of 'applying' theoretical models to particular situations or texts—it's always more interesting when the pressure of application goes in both directions."[18] Although I cannot claim to be thinking, or expressing myself, as elegantly as Sedgwick, her comment reflects an important aspiration I have for this project. Throughout this book, performative theory and biology will each exert "pressure" on the other, forcing most of us to rethink some of our most basic, perhaps even cherished, intellectual commitments along the way.

In order to support the assertion that our material and sexual bodies are *performances all the way down*, I will explore the genetic and molecular mechanisms of sexual development, and their evolution, in substantial detail. However, I have included enough introductory information on feminist and biological topics that any engaged reader will find what they need to understand the entire book within these pages.

What Is the Role of Metaphor in Biology?

Throughout the book, I will point out what I think are unique intellectual contributions to science that can be made using a performative framework to investigate the genotype-phenotype relationship. (In chapter 8, for example, I present a number of novel performative scientific hypotheses about the mechanisms of sexual development.) I will argue that the properties of a performative process are not exclusive to human cultural phenomena, but are real, fundamental properties of organisms, genetics, and biological systems, and that adopting a performative framework will advance our scientific understanding of biology.

Here, I want to describe why I think performativity provides a useful language for scientific research. Biology and the other sciences abound with productive metaphors. Indeed, along with mathematical abstraction, instrumentation, and observation, *metaphor* is among the most fundamental and invaluable scientific tools. It is through metaphorical language that mathematics, instruments, and observations come to function in scientific explanation. As historian of science Lily Kay asserts, "Metaphor connects theory with nature."[19]

Thus, language and metaphor have a critical role in productive science. Discounting any concept—including the performativity of the phenotype—as "*mere* metaphor" is not a valid scientific critique. My scientific use of the term *metaphor* does not imply support for science as a social construction, a step back from materialism, or a lack of commitment to realism (at least, the active realism of Hasok Chang or the agential realism of Karen Barad).[20] Rather, accepting and using scientific metaphors is merely the recognition that language is an indispensable part of science.

Productive metaphors are already ubiquitous in biology, starting most notably with the concept of the gene. In the early twentieth century, the gene was defined abstractly as the unknown material cause of heritable similarities between parents and offspring. After DNA was shown to be the genetic material, the gene was reconceived as a sequence of DNA coding a specific protein—a piece of genomic real estate, if you will. But increasing confusion about the definition and limits of the "gene" has arisen with the discovery of the many complex and arcane ways in which gene expression—the production of proteins from DNA sequences—can be promoted, suppressed, or epigenetically modified, as well as the possibility of alternative RNA splicings that can produce different protein products from a single gene sequence. Whenever you try to materially specify exactly what is, and is not, a gene, your definition turns out to limit the intellectual function of the concept, and to inhibit productive scientific research. Fascinatingly, the vagueness of the gene concept has expanded accordingly in response to new empirical discoveries and challenges. This more metaphorical gene concept has become widely accepted over the same decades in which the concept has itself become ever more vital in biology and biomedicine.

Of course, genes or other scientific metaphors are only worth thinking about if it is scientifically productive to do so. By this standard, I argue that performativity is an essential concept in biology because it *captures* and *advances* our understanding of a real, fundamental issue in biology—the relationship between genotype and phenotype.

The performativity of the phenotype is a replacement for previous, highly influential metaphors for gene action in the world. But, as we will see, what is inherited through the genome from one generation to the next is *not* an instruction manual, a blueprint, a developmental program, an algorithm, or an essence of individual identity. Rather, the genome is a highly structured, material lexicon—an historically contingent informational resource—to be *used* by the nascent individual in the enactment of itself. (I discuss various contemporary models of the genotype-phenotype relationship, including gene by environment [GxE] interaction and reaction norms, in appendix 3.)

An important aspect of how science progresses—that is, learns more and more *from* and *about* reality over time—is the generation and upgrading of new models of reality. However, scientific language can also function in a broader way. As historian of science James Bono writes, metaphorical language can connect complex scientific ideas with other "social, political, religious, and cultural" ideas.[21] Metaphors can function both as a critical component of scientific explanation, and as a way of creating productive connections to human culture at large. By adopting a new model of the phenotype, I also hope to create scientific connections between biological ideas and cultural ideas outside of biology.

The philosophers of science Michael Arbib and Mary Hesse state concisely that "scientific revolutions are *metaphoric revolutions*, and theoretic explanation should be seen as a metaphoric description of the domain of phenomena."[22] In this spirit, I am calling for a metaphoric revolution in biology with profound implications for the relationship between biology and human culture.

What Is at Stake?

Scientists are not used to asking themselves a question that is fundamental to the humanities and social sciences—"*What's at stake?*" Perhaps we have been trained to take the explicit empirical goals of our scientific inquiry, however modest (like describing the song of a species of bird) or grand (like curing cancer), as obvious, sufficient justifications for our work. Most scientists have been trained to think that science should stand apart, aloof, above the culture, and removed from politics and everyday life. By seeking to expand science beyond its usual boundaries, however, I want to be explicit about what is at stake, and articulate *why* I believe it is important to reconsider the traditional binary view of individual "sex determination," and adopt a performative view of the material body.

Judith Butler has memorably described the problematic cycle by which cultural *assumptions* get unconsciously "taken up" into scientific concepts of sex and the body, similar to the "Assumption of the Virgin Mary," through which her material body was "taken up into a more elevated sphere."[23] These culturally infused scientific "assumptions" are then represented as pure, fundamental, foundational, and unquestionable scientific "facts" that exist prior to, and independent of, culture and politics. Then, deploying what Donna Haraway calls the "god trick" of addressing nature from an imaginary space outside of it, scientists represent their views as objective facts that are beyond cultural critique. But, surprise, science is done by humans! All of us are embedded in societies that provide a rich cultural broth that we take up into our conceptions of the world.

What is at stake in this work are the answers to a series of urgent questions: Can scientific descriptions of the materiality of our human sexual bodies be dissociated from historic, heterosexist, patriarchal cultural norms? Can the science of the sexual body be intellectually separated from the social, political, and physical abuses that have flowed from the unquestioned, scientific "fact" of individual binary sex? Can questioning the individual sexual binary actually improve the science of biology? Many biologists may prefer to leave these questions aside, and remain aloof from, or ignorant of, gender theory. But they can no longer pretend that their work is not imbued with cultural assumptions about gender/sex merely because they are not interested in thinking about it.

Scientists frequently celebrate the "power" of their instruments, methods, and analytical tools, but they are often blind to the ways in which they have used this intellectual and cultural power to constitute, even to conjure up, the subjects of their research—whether they are genes, biomes, binary sexes, or, in our eugenic past and increasingly genomic present, human races. How do we work toward a science that pursues and fulfills our genuine human curiosity about the workings of the material and biological world, including ourselves, in a way that does not further the history and ongoing potential for abuse of this power? How can scientists reckon with the historic legacies and ongoing cultural impacts of our work? How can science deploy its reductionist and experimental tools in ways that do not harm human individuality and diversity?

By engaging with one facet of this enormous challenge, I will argue that many apparent conflicts between queer feminist science studies and the genetics of human sexual development are not conflicts with the empirical *details* of the science, but conflicts with the *culture* of science. In *Sex Itself*, Sarah Richardson appropriately criticized biology's "geneticization" of sex, which has accelerated rapidly since the description of the structure of DNA.[24] However, I will argue here that a performative understanding of gene action provides a productive, new, alternative view of what being "genetic" actually means. Likewise, I hope to encourage scientists to consider that they may have a lot to learn about science itself from the creative work of people that are not scientists.

I am not, however, merely aspiring toward a mutual understanding between feminism and biology. Rather, I propose that the queer feminist concept of performativity makes vital and necessary contributions to the science of biology. Likewise, the recognition of a performative continuum—extending from the cultural and psychological self down to the genetic and physiological regulatory mechanisms of the body and its cells—can ground our conception of gender/sex in the materiality and evolutionary history of the organismal body. A workable intellectual concept of gender/sex will require a common framework for understanding what it means to become, to develop, and to grow as a material, biological, psychological, *and* cultural being.

I don't pretend that this will be easy. The individual sexual binary is among the most foundational intellectual commitments in biology. Questioning it, and rejecting it in favor of thinking about our sexual bodies in a new way will pose a profound intellectual challenge to many biologists. Likewise, some feminists may still view the intellectual and empirical tools of biology and genetics as hopelessly tainted by past histories of abuse. Meanwhile, in defense of a greater engagement with biology, many material feminists have recommended reducing the intellectual influence of the same linguistic/cultural tools of analysis that I am actively trying to recruit and apply to genetics and biology. So, there are many intellectual obstacles in the path ahead. But I hope to be able to document how science actually *needs* the intellectual contributions of queer, feminist, trans, and nonbinary people and points of view to scientifically understand our material bodies—all of our bodies.

The Stakes for Evolutionary Biology

The performative theory of the phenotype presented here also poses some fundamental challenges to broadly prevailing ideas in evolutionary biology, some of which have had a profound impact on contemporary culture. Specifically, the performative phenotype challenges the twentieth-century isolation of developmental biology from evolutionary biology and disputes the strict conception of gene-level selection.

In the mid-twentieth century, evolutionary biology was reorganized around the "New Synthesis"—an intellectual framework that deployed explicitly ahistorical population genetic models and species concepts to connect genetic evolution to the study of species and organismal diversity. This population genetic framework involved the redefinition of genetic inheritance—the evolutionarily crucial process by which organisms grow up to resemble the parents they inherit their genes from—as the slope of the regression of offspring and parental trait values. This new mathematical definition of heritability conceptually eliminated *any* role for organismal development in evolutionary process. At first, the elision of organismal development was simply a pragmatic convenience. Eventually, however, hubris set in, and organismal development was permanently set aside as irrelevant or distracting.[25] This intellectual simplification contributed to a rupture between developmental and evolutionary biology.

The New Synthesis also pioneered mathematical models of genetic evolution within populations that were developed in conscious emulation of the physical laws of thermodynamics. This desire to reframe evolutionary biology in the intellectual form of the material sciences—what I refer to as Physics Envy—led to the adoption of explicitly *ahistorical* concepts of evolutionary process, and the diminishment of the role of history itself in evolutionary biology. In sum, the New Synthesis led to the virtual elimination of development and phylogeny—both central themes to Darwin's historical worldview—from the intellectual core of modern evolutionary biology.

Furthermore, within New Synthesis models of genetic evolution, natural selection was seen as the sole signal-producing mechanism among the random effects of mutation, migration, and genetic drift. This contributed to the advance of adaptationism—the idea that natural selection is a strong, deterministic force that will dominate all other mechanisms of evolution.

Rather than constituting a true intellectual synthesis—a combination of ideas and perspectives from multiple disciplines to create a new, greater understanding—the "New Synthesis" brought about a profound intellectual flattening of twentieth-century evolutionary biology. Yet, the Synthesis was so politically and sociologically successful that it had an enduring impact on the intellectual values of evolutionary biology— what sorts of theories and research questions are legitimate, fundable, and count as advances in understanding. In previous writing, I have referred to the reductive, adaptationist simplicities of New Synthesis biology as "flatitudes"—faux-profundities that gain their intellectual appeal by flattening the actual complexity of the world.[26] In evolution, the reductive flattening of biological complexity became an intellectual value in itself. Respect was accorded to those ideas that flattened biology's emergent complexities into simplifying, lawlike principles acting at lower levels of organization—cooperation explained exclusively by kin selection, aesthetic mate choice explained as preferences for honest handicaps, and so on. The result has been a plethora of smaller, wannabe, faux-syntheses that were lauded as scientific progress, but left us with an impoverished capacity to explain the biological world. The problem is not that these evolutionary models did not make valid contributions. The problem came in proposing that they were all-encompassing explanations of biological phenomena, and that research that does not flatten complexity into ahistorical, population genetic conclusions cannot be progress.

In response, both phylogeny and developmental biology have had to struggle to regain recognition as central concerns within evolutionary biology. Yet, their intellectual restoration is far from complete.[27] In this book, we will see how a performative model of the phenotype creatively addresses these twin weaknesses of New Synthesis evolutionary biology, by incorporating both developmental *and* phylogenetic evolution of phenotypes as core issues of the discipline, and as emerging from our concern for the genotype-phenotype relationship.

A prime example of the intellectually impoverishing flatitudes of adaptationist New Synthesis evolutionary biology is Richard Dawkins's concept of the "selfish gene." According to this idea, genes are the only true replicators in biology, and organisms are simply trivial vehicles for the reproduction of those self-interested genes. From this perspective, adaptive evolution consists entirely of natural selection acting through competition among alleles (or gene variants) of individual genes. Although there is abundant evidence that gene-level selection does occur, the problem comes from the dogmatic insistence that gene-level selection provides a unifying synthetic explanation of *all* evolutionary biology and biodiversity.

In what follows, we will see how gene-level selectionism fails—even by its advocates' own admission—to explain the material production and evolution of complex multicellular phenotypes, and therefore provides an intellectually impoverished framework for understanding the evolution of biodiversity and ourselves. (See also appendix 6.)

Because of the long history of redefining intellectual progress as conceptual flattening, many evolutionary biologists may regard the recognition of the performative complexity of the phenotype and its evolution as a step backwards. (Earlier generations—likely even some of the same voices—previously held the same opinions about phylogenetics and developmental-evolutionary biology.) However, to flourish evolutionary biology needs to abandon Physics Envy, and set its goals on acknowledging the breadth and complexity of the empirical and theoretical challenge that biology really presents. A performative view of the phenotype can make this possible. (Discussed further in chapters 4 and 7–8 and appendixes 3–7.)

Mind the Gap

Many people's experiences and views of gender/sex are changing in Western cultures and around the world. More people are challenging traditional gender/sex norms by acknowledging and living a greater variety of gender/sex identities. Although trans, nonbinary, and genderqueer people have always lived among us, recent decades have seen a growing number of adolescents and young adults openly identifying as trans, nonbinary, agender, asexual, or other queer identities. These ongoing cultural changes are also being legally and culturally contested in many ways, contributing to a mixed picture of expanding freedom and possibilities with new and enduring threats to human rights, well-being, and individual flourishing. Of course, all this is accompanied by the scandalous vulnerability of queer and transgender individuals to social violence, murder, homelessness, and suicide.

Meanwhile, in recent decades biology has also undergone revolutionary advances in the fields of genomics, molecular genetics, developmental biology, and evolutionary biology. With some notable exceptions, however, biologists have largely remained professionally isolated, detached, and aloof from cultural changes in conceptions of gender/sex. This conservative inertia, complacency, or overt resistance to change is probably reinforced by the skewed age and demographic structure of scientific influence and power; older, white, mostly heterosexual, male professors and researchers have a predominant influence on what counts as legitimate and fundable topics for scientific research, publishable results, promotion-worthy accomplishments, and permissible topics of scientific discussion. Many biologists may think that cultural issues in gender/sex are irrelevant to their work, not in their interests, or even unprofessional. Students and younger professionals may also think that "keeping your head down" and adopting a similar attitude is a good way to get ahead in the highly competitive and highly conforming culture of science. Women, queer, minority, and untenured scientists may feel particularly vulnerable to professional judgment or retaliation if they exhibit a research or teaching interest in gender/sex or queer feminist biology. Whatever the explanation, the result is an alarming and growing gap between the expanding cultural reality of gender/sex diversity and traditional, essentialist scientific conceptions of sex.

This growing gap between people's lived experiences of gender/sex and what science continues to say about sex is scientifically and culturally dangerous because people can become increasingly alienated from, and suspicious of, science that fails to reflect, or even acknowledge, their lived experiences. Furthermore, cultural misconceptions and misuses of science can thrive when scientists remain aloof and disengaged from the cultural consequences of simplified, binary, essentialist sexual thinking. (Remember the call from right-wing American politicians to "Trust the Science!") I argue that "minding the gap" between the science and culture of gender/sex can contribute fundamentally to both.

To fans of the continued isolation of science in an idealized Ivory Tower, looking down in objective detachment on the concerns of culture, politics, and lived experience, I can only ask, "How's that working for you?" Science today may never have been less respected, more open to question, held in lower general esteem, or more subject to misinformed attack than at any time since its origins in the Enlightenment a few hundred years ago. Wherever science comes into critical contact with culture and politics today—in evolution; climate change; biodiversity conservation; pollution; eco-toxicology; or, most recently, vaccines, epidemiology, and public health—there is an organized and well-funded movement to undermine science, question its independence and authority, and portray it as another publicly funded special interest group, or "deep state" conspiracy. It seems clear that the socially detached, "voice from nowhere," Ivory Tower concept of science is failing to defend the sciences from cultural attack.

It is perplexing to me that so many scientists are hesitant to discuss and take responsibility for their place and actions in the culture. Science has not always adopted the currently popular position of cultural remove. After the broad realization of the genocidal atrocities of World War II, a large committee of anthropologists, evolutionary biologists, and geneticists from dozens of nations, organized under the auspices of the United Nations, produced a report on the biological concept of race that rejected the scientific eugenic theories of racial difference. In the 1950s, Linus Pauling led an international movement of biologists, chemists, physicians, physicists, and engineers to ban atmospheric testing of nuclear weapons to limit the harmful effects of radioactive fallout on humans and the natural world. Ironically, the same scientists that now argue for a cautious distance from engagement with cultural issues are universally proud of these historic instances of scientists using their knowledge to intervene decisively in the cultural and political world.

These examples show that we've done it before, and we can do it again. This book about the biology and culture of gender/sex is my call to my fellow scientists to advance a culturally engaged defense of science as a way of knowing that remains vibrant and relevant to the lives of people. To do this, more scientists will need to make risky moves outside of their comfort zones.

Why Queer Biology?

I could have written a book about the biological performativity of feathers, limbs, or flowers. Would it not have been simpler and clearer to abandon this awkward engagement with feminist theory, its jargon, and history, and just *do* biology? Although some of my biological colleagues have recommended this approach, I think not, and here is why.

Being queer in this world is not easy. And queer theory has been developed by people doing the hard work necessary to understand what it means to stand out from, defy, and conflict with the gender/sex norms that the rest of the world has taken for granted, and even justified as foundational, scientific facts. This work has led to novel, fundamental, and powerful insights into what it means for an individual to *become* in the world—insights that have been obscured or invisible to most people, perhaps especially to professionals in the biological sciences.[28]

To me, the scientific contributions of performative biology depend solely on its value as science—the ability to supersede alternative models, conceptions, or accounts of the phenotype, its development, and its realized and evolved diversity. But creating a queer scientific space in the intellectual heart of biology—centered broadly on the genotype-phenotype relation—is also a new way for science to confront the social injustice, violence, and precarity experienced by queer people everywhere. Queering biology is an appropriate way for science to reckon with the historical and ongoing contributions of biological science to the legitimacy of the cultural forces that tolerate, condone, and encourage these obstacles and harms to queer lives.

Where Are We Going?

In the chapters ahead, I will lay out the theoretical basis for my argument, apply those ideas to human sexual development, and then explore the broader implications of this framework. In chapter 2, I introduce, define, and clarify some critical concepts and vocabulary that will be necessary for the rest of our discussions. In chapter 3, I outline the intellectual origins and elements of performative theory in the humanities, drawing primarily from the work of Judith Butler and Karen Barad. I focus on aspects of Butler's and Barad's work on performativity that will connect directly to later discussions of biology, the material body, and gender/sex. In chapter 4, I outline a performative theory of genetics, developmental biology, and physiology. This chapter will provide an introduction to genetics and developmental biology for those with little background in it, but it will also provide new perspectives on these topics for biologists. In chapters 5 and 6, I discuss the sexual development of human embryos. Chapter 5 focuses on the roles of various genetic and hormonal signaling pathways in early human sexual development, and chapter 6 investigates the many possible variations on that process. Chapter 7 explores the implications of performativity for the evolution of the diverse mechanisms of sexual development, exploring some of the distinct evolutionary consequences of phenotypic performativity. Chapter 8 sketches out a view of the future for research programs in performative biology, including examples of distinctly performative hypotheses to direct future research. Finally, chapter 9 returns to consider the implications for this performative view of biology for our understanding of culture, gender/sex, and ourselves. I will present a history, or genealogy, of biological and cultural performativity, and argue that we are also *performance all the way up*.

To keep the focus as much as possible on the sexual and material body, I have gathered discussions of several important topics in a series of appendixes. These include additional discussion of material feminism, acquired immunity, current alternative models of the genotype-phenotype relationship, biological modularity, genetic assimilation, the insufficiency of gene-level selection, and internal selection. Unlike traditional scientific appendixes, which are usually packed with math, methodological details, or other less digestible materials, these appendixes are written in the same style as the rest of the book, and should be accessible to all interested, general readers.

CHAPTER TWO

Critical Concepts

To discuss the diversity, development, and evolution of sexual bodies, we will need to use various terms and concepts whose meanings have been historically problematic, actively contested, or directly involved in ongoing intellectual problems that we will be trying to investigate and clarify. In some ways, I think that the definitions of such terms could be properly thought of as *results* of the analyses and discussions that follow. *If* my arguments are lucid and well supported, *then* we can make some conclusions about how to speak clearly moving forward. Thus, these definitions might better belong among the conclusions of this book.

However, there are too many opportunities for miscommunication and misunderstanding along the way before we get there, so I will start with definitions and explanations of some of these critical concepts and terms. All of these positions will become clearer, I hope, as they are applied and further explained later. Although I cannot avoid all the problems of our history-laden sexual vocabulary, I hope to clarify what I intend by my uses of these terms.[1]

What Are Male and Female?

Like many organisms, adult human bodies vary in ways that are related to reproductive functions. These reproductive variations have been traditionally used to refer to individual organisms and their bodies as either male or female. To many people, including many scientists, these terms imply the existence of natural, binary categories that apply to all individuals. Males reproduce by producing small gametes, or sperm, and females reproduce by producing larger gametes, or eggs. But this definition is not really accurate because we also use the categories male and female to refer to many individuals that cannot, or do not, sexually reproduce, such as children, the aged, or the infertile. The apparent naturalness of a binary classification of the sexes is usually justified by assertions about the fundamental importance of sexual reproduction to biology, leading to the interpretation that individuals that do not or cannot produce fertile gametes are definably male or female because they have other, essential properties of these binary sex categories.

Thus, a common scientific justification for the reality of individual sex is that male and female can be defined by some critical or essential characters they possess—a specific anatomy, combination of chromosomes, genes, gene-expression levels, or hormone levels, and so on. Yet, as we will see, scientific observations, experiments, and evolutionary comparisons have demonstrated clearly that all such criteria fail to provide a clear, binary definition of male and female individuals. Although biologists often fall back on what are proposed as decisive differences in reproductive function of certain bodies, they rarely consider restricting the discussion of male and female solely to those reproductive functions. As we will see, to some scientists, even individual human cells have a definable sex.[2] Thus, the concept of individual binary sex functions more broadly in science and culture than in reproduction alone. In most ways, biology chugs along without reckoning with its failure to effectively define male and female, and without rethinking what research results based on its assumptions means for real people, or considering what the consequences are for the lives and families of queer, intersex, and sexually variable people who are marginalized by narrow concepts of sex.

Despite the problematic history, however, I will continue to use the terms *sex*, *male*, and *female* in this book. How can this be justified? In what way can an undefinable category exist? To understand how sex categories can be undefinable but real, we will need to take a brief philosophical interlude.

Historical Ontology

Asking "What are the sexes?" is a question of ontology—the branch of philosophy concerning the states or varieties of existence. In particular, questions about the material reproductive capacity of organisms are questions about *natural kinds*—those categories, or groups, of things that exist independent of human observation, description, naming, or conception of them.

When we think of natural kinds, we typically think about *classes*, which are natural kinds that can be defined by their shared properties—like helium or diamond. Helium is an element consisting of atoms with two protons and two neutrons in the nucleus, having an atomic number of 2. (By convention, an isotope of helium with one neutron in the nucleus is also referred to as the element helium, but that convention could be different if chemists preferred.) Likewise, diamond is a mineral composed of carbon atoms bound together in a tetrahedral crystal lattice. Anything at any time that comes to meet this definition—whether it is created by the profound pressure deep in the Earth's crust over a billion years ago, or synthesized in a laboratory in the last few months—is a diamond. Thus, natural kinds that can be defined by their shared features are examples of classes.

In contrast, everything in biology has a distinct origin. Everything in biology has a potential to thrive and to reproduce. Everything in biology has an eventual death, or a pending extinction. Everything in biology is also subject to temporal or evolutionary change. But everything in biology also belongs to natural kinds. So, what natural kinds are these? Well, they include individual organisms, species, higher taxonomic (monophyletic) groups, anatomical parts (such as limbs and livers), specific proteins, and specific genes. In short, every material thing in biology. All of them have an origin, are reproduced, diversify over time, and inevitably die or go extinct.

However, unlike helium atoms, diamonds, or the members of other classes, everything in biology has the inherent potential to evade—through evolution or aging—*any* definition of it that we could propose. In fact, every distinct thing in biology owes its distinctiveness to its own history of having changed or evolved from something else that was itself distinguishably different. New proteins and new species come from earlier proteins and species that were themselves different. Try as we might, our efforts to create definitions of biological entities based on essential features—to place biological entities into definable ontological classes like the elements—will fail. As a result, biology requires recognizing a different ontology—a different type of natural kind.

In contrast to classes, *individuals*, or historical entities, are natural kinds that exist despite the fact that they cannot be defined by their properties.

In contrast to classes, ontological individuals have a unique origin or birth; a potential duration, or thriving; often a possibility of reproduction; and ultimately an end, death, or extinction. Individuals are spatiotemporally restricted—that is, they originate and persist only in certain places and times. Biological things obtain their ontological status as natural kinds through their history of descent with modification. Individuals are natural kinds with history. Of course, the members of classes—like individual helium atoms—have individual histories as well, but they are usually trivial and unimportant for any scientific investigation we might make of them. In contrast, ontological individuals exist through their historical contiguity (genealogical and evolutionary descent), and many (most?) of their scientifically interesting features arise because of their specific histories. By history here, I do *not* mean human history, the history of human thought, or the history of human interactions with these entities, but rather the material existence, persistence, and iterative reproduction of these entities over a span of time.[3]

Examples of ontological individuals, or historical entities, includes individual human beings like you and me. Species like *Homo sapiens* and *Drosophila melanogaster* are also ontological individuals, that encompass many individual organisms within them. Higher organismal, or taxonomic, groups like Mammals, Flowering Plants, or Fungi are individuals as well, whose existence is manifest by their single, historic common origin (what biologist call monophyly), and their history of evolutionary descent and diversification. Biological parts like limbs, feathers, and flowers are historical individuals as well, each sharing a unique origin and a spatiotemporal restriction. Likewise, biological molecules like hemoglobin, collagen, and insulin are historical individuals, as are the genes that encode them. Both anatomical homologs and genes can reproduce by duplication and evolutionary divergence. Thus, the multiple globin proteins that combine to form the distinct adult and fetal hemoglobins in human red blood cells evolved by duplications of an ancestral globin gene in an ancient jawed vertebrate ancestor, later in a placental mammal ancestor, and then subsequent evolutionary divergence.[4] Historical ontologies apply to all of the material things in biology.[5]

Biological things cannot be defined by any essential properties because any such properties may themselves change over time through aging or evolution. For example, one might want to recognize tetrapod vertebrate animals as an ontological class defined by their shared possession of four bony limbs. But that definition would fail because numerous tetrapod lineages have evaded that definition by evolving into snakes, legless lizards, or caecilians, which all lack limbs. Nevertheless, the group tetrapods, including all these legless members, *is* a natural kind because of its singular history. Similarly, we might want to define the family of animal heat shock proteins according to their first recognized function—stabilizing other proteins in response to heat stress. However, this diverse family of proteins has actually evolved numerous functions in response to other physiological stresses, like cold and UV light, and they have even evolved to function optically in the lens of the vertebrate eye. Thus, trying to define kinds of proteins based on their function also fails. Even the once "universal genetic code"—the relation between messenger RNA codon base composition and the amino acids they specify—is evolvable, and therefore no longer universal.[6] Despite their evolved diversity and variations, all these biological entities retain their status as natural kinds through their historical contiguity—the iterative cycles of biological reproduction and phenotypic development that result in genealogical and evolutionary descent.

If we cannot define them, how do we recognize categories that are ontological individuals? We can only point to them, describe them, or diagnose their historical status by their distinctive (often unique, shared, and derived) historical features. However, we must keep in mind that none of these features is essential; they are all merely evidence of shared history. Indeed, it is quite possible for evolution to proceed so rapidly or for so long that diagnostic evidence of shared history has been erased. For example, after a certain amount of evolutionary change, DNA sequences of two historically related genes, or gene copies, may be indistinguishable from random similarities shared by historically unrelated genes or gene copies. But the erasure of this recoverable ancient history does not undermine their individuality.

Historical individuality does not invoke any notion that these natural kinds must have strict edges, entirely independent existences, rigidly defined boundaries, or tightly integrated parts. Ontological individuality is not about existential autonomy. For example, my own individuality is not violated or threatened by my body's dependence upon a diverse gut microbiome, the endosymbiotic mitochondria of bacterial origin in every one of my cells, or the parasitic chickenpox virus DNA sequences that were inserted into the nuclear genomes of some of my nerve cells when I was a child. Likewise, Dick Cheney's individuality is not threatened by his having had another human being's heart transplanted into his chest (though that may not stop one from being enduringly creeped out by this fact). Ontological individuality is also completely compatible with symbiosis—lichen species are perfectly good, historically persisting species. Indeed, ontological individuality captures precisely that which *is* contiguous, ongoing, and iteratively realized despite the permeability of boundaries and contents, the mutability of properties, or the necessity and complexity of ecological interactions—the historical persistence and contiguity of individuality itself.

Many nonbiological things are also ontological individuals, including the planets Venus and Mars, and the Andromeda Galaxy. Although black holes are a definable class of astronomical phenomena, the supermassive black hole Sagittarius A* at the center of the Milky Way galaxy is an ontological individual with a unique origin, spatiotemporal restriction, and inevitable extinction. Likewise, geological formations such as the Appalachian Mountains, Lake Baikal, and the San Andreas Fault are ontological individuals—with their own distinct origins, histories, and inevitable ends. Thus, like biology, astronomy and geology are historical sciences in which the participating objects have unique, scientifically relevant histories. Like biology, these sciences also include practitioners who deemphasize the individual historicity of the objects they study in search of universal laws, as well as natural historians—astronomical and geological bird-watchers—who are fundamentally interested in scientifically understanding the histories of individual instances. In this way, recognizing the historical ontology of the things under study can have a profound impact on the nature of scientific inquiry, and affect what qualifies as a valid scientific question and an appropriate scientific answer.

Historical ontology extends as well to human cultures, social organizations, and institutions. Human dialects, languages, and language families are ontological individuals that can only be diagnosed or described, rather than defined. Cultural organizations like the League of Women Voters or the Boston Museum of Fine Arts are social examples of ontological individuals. These social entities have the same ontological properties as biological, geological, or astronomical individuals. The existence of English is not put into question by the adoption of words from other languages like *segue* (French), *adobe* (Arabic to Spanish), or *anime* (English to Japanese and back). Likewise, the Boston Red Sox traded Babe Ruth to the New York Yankees without threatening the continued ontological status of either team.[7]

In summary, historical individuals are natural kinds that allow us to investigate ontologically real, yet undefinable, entities in the world—their continuity, diversity, and historical contingency—without seeking lawlike, reductive generalizations, or presuming or precluding what those histories or futures may be. The desire to intellectually emulate the laws of thermodynamics expressed by architects of early twentieth-century population genetics and molecular biology has focused too extensively on the search for lawlike, general properties, erroneously assuming that they can treat biological entities as classes. However, the historical sciences including biology, geology, planetary science, and astronomy also have long parallel intellectual traditions of investigating historical individuality, so the concept I advocate also has deep roots in these disciplines.

Sex *Is* a History

What does all this have to do with sex? Although I think that biologists and people in general have accurately perceived sex as a natural kind, I think we have made a categorical error in identifying what type of natural kind the sexes are. Most people, including many biologists, think of male and female as ontological classes that can be defined by essential properties, such as the production of sperm or eggs, the presence of the anatomical organs to produce them, the presence of other reproductive anatomies, the presence of specific chromosome combinations or genes, the expression of specific genes, the expression levels of specific hormones, or other characteristics. However, as we will see in detail in later chapters, the sex of individual organisms cannot be defined by any of these criteria. Even the production of differentiated gametes—small sperm and large eggs, a condition that biologist call anisogamy—is free to evolve in some organisms like green algae.[8]

We will see that male and female are not essential, binary facts about individual genomes, zygotes, or lives, but historical sets of (intra-acting) reproductive possibilities that are iteratively realized in bodies within each generation. (I hope future chapters will make this statement easier to understand.) In short, sex *is* a history.

The sexes are *not* definable as classes. Rather, male and female are ontological individuals or *historical entities*. The individuality of the sexes is a consequence of their descent through a historically contiguous chain of sexual reproduction, inheritance, iterative realization, and evolution. The sexes are coexisting, interrelating biological continuities that have evolved, and coevolved, through our long evolutionary history as sexually reproducing, vertebrate animals. Accordingly, *male and female are repeated, or iterated, clusters of embodied reproductive homologies*—that is, anatomical and physiological similarities that are shared due to common ancestry. (We return to a more ample explanation of homology in chapters 4 and 7.)

Likewise, homologous reproductive body parts—like the clitoris, labia, ovaries, penis, scrotum, and testes—are also ontological individuals that cannot be defined by any essential properties or features. Rather, like feathers, flowers, eyes, and limbs, each of these anatomical features is a historical individual with a unique origin, a flourishing of diversity, and a pending extinction. (If similar structures don't share a single, unique common origin—like the legs of insects and vertebrates—then they are not homologs but convergent structures, called analogs.)

An important consequence of the historical ontology of sex is that individual organismal membership in a sexually reproducing population, species (like *Homo sapiens*), or higher taxon (Mammalia) does not necessitate or dictate that an individual *have* a specific sex. Like all anatomical parts, each individual instance of reproductive anatomy and behavior is achieved developmentally by that individual—the subject of this book. Eyes and hands are heritable anatomical homologs, respectively, but one's eyes and hands do not create the eyes and hands of one's children. Each organism develops its phenotype itself, starting from a fertilized egg. So being a member of a sexually reproducing species does not mean that every and all individuals develop sex, any more than being a human means that all humans must have eyes and hands. Likewise, being a tetrapod vertebrate does not mean that an organism must have limbs. Furthermore, species—like horses and donkeys—exist and persist despite the fact that hybrid individuals—like mules—cannot be unambiguously recognized as either one of them. Historical individuals lack any essential, defining properties. Individuality persists, but it does not proscribe.

However, sex *is* biologically real and remains fundamental to the biology of humans and a myriad of other sexually reproducing species. All humans are the product of a human egg fused with a human sperm and gestated in a human womb, and only certain bodies can afford these reproductive possibilities. This constraint on the origin of human lives creates a sexual reproductive binary bottleneck in the human life cycle.[9] As a consequence of this persistent sexual constraint on the origin of individual lives, humans have coevolved bodies with variable and differentiated anatomies and physiologies to afford the possibility of sexual reproduction. As I hope to demonstrate in subsequent chapters, the mechanisms of the development of human bodies create a widely overlapping range of possibilities that cannot be clearly, individually distinguished or diagnosed into discreet binary categories. Furthermore, the biological mechanisms of individual sexual development do not allow for the input of any antecedent "fact," predetermined essence, or parental intention of individual sex. (By *essence*, I mean an inherent and indispensable quality that exists prior to, or independent of, individual embodiment or becoming.) Although they are long enduring and historically real, the ontological existence of male and female sexes does not mean that individual organisms *are*, or *must be*, one sex or another. No matter how central or fundamental sexual reproduction has been to our evolutionary history and to the origin of individual human lives, that evolutionary history cannot create essential facts about individual organisms, embryos, cells, or genomes.

As we will see in later chapters, the sexual reproductive nexus at the origin of every human life creates both the historical continuity of the sexes—that is, the evolutionary persistence of bodies characterized by clusters of anatomical homologies that afford sexual reproduction in specific and structured ways—*and* the coevolved functional interrelationship *between* the sexes—the fact that these sexual homologies facilitate sexual reproduction through anatomical, physiological, and functional co-correspondences with one another. The former is the source of the historical ontology of sex. The latter is a source of functional constraint on the evolution of sexual bodies, and has facilitated the amazing, evolved diversity of sexual morphologies *among* organisms. In other words, the fundamental necessity of sexual reproduction to create new biological lives places functional limits on the reproductive possibilities in the present, but the coevolved correspondence between the sexes—what we will later refer to by Karen Barad's term *intra-action*—facilitates a great diversity of sexual reproductive possibilities in the long term.[10]

Among these possibilities is the existence of many sexes or the elimination of one sex. Many fungi sexually reproduce with a large number of mating types—organisms that can sexually reproduce with any other individual of the species with a different mating type from their own. In such species, mating type diversity can be hard to measure. On the opposite extreme, some formerly sexual animals—including some lizards, fishes, and many invertebrates—have evolved to reproduce without males, a reproductive mode called parthenogenesis (discussed further in chapter 7). However, in mammals, it is scientifically justified to state that there are only two sexes (even though that does not mean that every individual, cell, or genome is one sex or the other).

Some previous material feminist definitions of sex come close to this concept, but they miss an important intellectual opportunity made available by this historical ontology. For example, evolutionary biologist Malin Ah-King and marine ecologist Eva Hayward conclude that "sex may be better understood as a dynamic emergence with environment, habitat, and ecosystem," a statement that is highly congruent with the performative view. In this and other works, Ah-King and colleagues also propose that sex, reproductive behavior, and parental investment should be viewed as a "reaction norm"—an experimental genetic method that exposes phenotypic variation elicited by environmental variation from a common genotype.[11] (I discuss this proposal further in appendix 3.)

In *Sex Itself*, historian and philosopher of science Sarah Richardson writes, "Sex is a relational property of individuals within a (sexual) population or species," and she concludes that "sex is a *dynamic dyadic kind*." While I agree that sex is a relational *possibility* and often dyadic, thinking of sex as "a property of individuals" implies the existence of essential or deterministic sexual features of organisms that I do not think is supported by the biological data.[12] However, Richardson's concept of sex as a dynamic dyadic kind is an accurate description of the more specifically historical definition of sex that I develop here.[13]

In the rest of this book, I will argue that, like all other aspects of the organismal phenotype, the capacity to reproduce in a specific, structured way is an individual organismal enactment, performance, or achievement that creates relational *reproductive possibilities*. The impossibility of identifying all human individuals as either male or female is incompatible with the idea of a sex as an essential binary characteristic of individuals, but it is entirely consistent with the concept of male and female as real, historical, natural kinds whose persistence is not undermined by their coevolved interdependence or their fuzzy, overlapping, queer absence of distinct boundaries.[14]

So, why do we still need the labels male and female if they are undefinable, historical categories? Why not abandon these words entirely? In my view, we still need names for the sexes to both discuss the biology of the diverse reproductive possibilities that evolutionary history has produced, maintained, and diversified, and to untangle the history of uses and misuses that science and culture have put them to. I do not think that naming male and female is a kind of scientific possession, colonization, or control over their meanings or their future possibilities. Rather, unlike helium, diamond, and other ontological classes, the names we give to historical individuals are truly *proper nouns*—like Jane, Denali, *Homo sapiens*, or the New York City Ballet. Recognizing an historical individual by its proper name is an act of respect, not a possession or a colonization. Indeed, I hope to use the proper names Female and Male—capitalized from here on to recognize their historical individuality—in ways that can open up our understanding of what they mean and can be. Once again, referring to these historically iterative clusters of reproductive homologies as Female and Male does not imply that all human individuals must be one, the other, or identify with either. The historical fact of sex does not imply a universal fact of individual binary gender/sex.

Unfortunately, I am pretty confident that the distinction I am trying to make here—between (1) the traditional view of the sexes as definable classes that imply essential facts about all individuals, cells, genomes, and bodies, and (2) a view of the sexes as enduring, reiterative, historical clusters of reproductive homologies that cannot be defined independently of their realizations—will be repeatedly and persistently misconstrued and confused. Although I will try throughout the book to make this distinction clear, my own experience has taught me that it can be very hard to change how people think. For example, in previous scientific research on the evolutionary origin of feathers, I contested the long-held notion that feathers evolved from elongate scales through natural selection for the evolution of flight. Without getting into the details, my alternative, developmental theory proposed that the feather placode—the developmental precursor of the embryonic feather—evolved from the duplication of a scale placode, and that all subsequent events in feather evolution were novel to feathers. Although the differences between the two ideas are profound (as least as far as feathers are concerned!), the subtlety between them has frequently been lost, even by professionals in the field with an advanced interest in the topic. Despite repeated clarification, many still consider that feathers have simply "evolved from scales."

But I am hopeful that many people will realize that understanding the perplexities of gender/sex requires thinking subtly and differently about issues that we have been culturally educated to think are simple, natural, and obvious.

Sex Difference versus Sexual Difference

There are many good reasons to be scientifically and biomedically attentive to the variations in sexual anatomy among human and other animal bodies. For example, in order to have uterine or prostate cancer, one needs to have a uterus or a prostate gland, which have distinct, primary functions in sexual reproduction. Of course, we could avoid many damaging generalizations about proposed sex differences by treating the risks of uterine or prostate cancers as dependent on having a uterus or a prostate gland, rather than a risk of being Female or Male. Distinguishing among the variations in sexual anatomy of individuals is different from distinguishing among classes of individuals based on sex.

Biology and psychology have a long history of investigating, searching for, insisting upon, and indeed conjuring up *sex* differences—the differences in various traits *between* the sexes. This huge, contested, and important topic has been discussed at length from many perspectives, but it is not the subject of this book. Even so, a few comments are appropriate here.

Since I think, and will try to demonstrate, that there are no essential properties that can define, or identify, all human individuals as either Male or Female, I am particularly skeptical of most sex difference research. In addition to having a long-running history of sexist biases and often being deployed to reinforce those biases, much sex difference research is simply not that interesting. Imagine how much you would understand about humans by looking at sex differences in height or body weight. The answer is not much. You cannot identify a person's sex by their height or weight. Because the variations in height and weight within the "sexes" are so much greater than the differences between them (not unlike race), studies of such differences are usually scientifically missing the forest for the trees. Furthermore, the underwhelming sex differences in height and body mass are *much* greater than most of the features that are the subjects of sex difference research. In other words, not all statistically significant findings are actually worth investigating, especially when they require imposing unwarranted assumptions about the categories they propose to investigate.

How can we tell when a proposed sex difference could be important? When presented with evidence of sex differences, we can simply ask, "How big is the difference between the sex categories compared to the total range of variation?" This quick smell test asks whether the trait, or feature, has a truly bimodal distribution—in other words, does a frequency distribution of the trait in all individuals show two, distinct, separable peaks for each sex. (For the statistically minded, the difference between the means of two normal distributions have to be greater than the sum of their standard deviations to produce a bimodal joint distribution.[15]) This is a very high bar. So, do human body mass or height differences pass the smell test? The answer is no. The sex differences in human height and weight are nowhere near different enough to produce two distinct peaks in a frequency distribution including all adults. If the trait distribution is not bimodal, then it is quite likely the sex difference researcher is asking either the wrong question or a boring question, or is seeking to create differences for some other (perhaps unconscious) cultural or social purpose. This is not an argument in favor of ignorance, but a caution against meaningless, or even harmful, scientism. (I return to discuss this idea again in regard to sex and race in chapter 9.)

For example, recent declarations that neuroscientists have documented a "fundamental sex difference in human brain architecture" based on fMRI data do not pass the smell test. Like height and weight, you cannot identify the sex of an individual from an fMRI of their brain, or from the "structural connectome" modeled on that data. Studies that brag about the "power" of their analyses based on over one thousand individuals are actually papering over their weaknesses. The large sample sizes required in such studies simply demonstrate how difficult it is to find sex differences in human brains in the first place. The samples size must be large in order to find statistical differences so small. Although the findings of increased intra- and interhemispheric connectivity in brains of Males and Females, respectively, are reported with extremely high p values, the results are presented completely without statistical effect sizes—that is, the magnitude of the difference relative to the total variation in the sample. Thus, statements about "fundamental sex differences in the structural architecture of the human brain" in absence of effect sizes are virtually impossible to evaluate.[16] The question becomes whether research on brain "connectomes"—the description of the connectivity among neurons and regions of the nervous system—has the goal of understanding brain connectomes, or the goal of using connectomes to justify the assumption of differences between the sexes?

I am not proposing that the results of such studies are fudged, although there is a long history of that in sex difference research (see, e.g., the historical details of such flawed research described in Jordan-Young and Karkazis's book *Testosterone: An Unauthorized Biography*).[17] The correlations between variations in the data and the category of sex may be clear, but that does not mean these correlations are causal. Given how difficult it is to actually establish these sex differences, it is almost certain that some other scientific hypothesis can do a better job of explaining individual variation in human connectomes, but simply hasn't been thought of and deployed yet.

Another poorly explored feature of sex difference research is the intellectually lazy mapping between reductive measures of biological variability—like hormone levels, gene expression, or brain fMRIs—and presumed differences in organismal function. In fact, variation among individuals may be *compensatory*: physiological, metabolic, hormonal, or developmental *differences* among individuals may exist *in order to compensate for* individual variations in genes or environment, and to *minimize* variation in phenotypic function. In other words, individuals may vary internally in order to be *more the same*. Physiological or hormonal sex difference can be a mechanism to achieve greater phenotypic *uniformity*, not sex difference. For example, a recent study demonstrated that the human Y chromosome copy of the EIF1A gene, which has a critical function in managing gene expression, is *over* expressed compared to the X chromosome copy. This overexpression has evolved to compensate—indeed, *overcompensate*—for a noncoding variation in the sequence of the Y copy of the gene, especially in heart muscle.[18] In this case, XY Male gene expression *differs* from XX Female gene expression in order for different individuals to be functionally more *similar* to one another. We do not know how many proposed biological sex differences are actually compensatory because researchers are not required to demonstrate bimodal differences in function before claiming to have found meaningful sex differences. Future studies should be required to do so.

Although I find most sex difference research to be uninteresting or misfocused, I do find the broader and distinct concept of *sexual difference* to be quite useful and productive. Sexual difference refers to variation in sexual anatomy, physiology, and behavior *among all* individuals of a species or population independent and regardless of sexual classification or category. Sexual difference encompasses total sexual variation, not simply difference *between* imposed sex classes or categories. The concept of "sexual difference" became evident to me in my consideration of the biology of sex. Interestingly, it was also developed independently decades ago by Continental feminist philosophers—including Luce Irigaray and Elizabeth Grosz—as a framework for describing women and sexual politics not derived from comparisons to men. Grosz writes that sexual "difference is not seen as a difference *from* a pre-given norm, but as a *pure difference*, difference in itself, difference with no identity."[19] As feminist literary critic Barbara Johnson writes, "The differences between entities (prose and poetry, man and woman, literature and theory . . .) are . . . based on a repression of differences within entities."[20] This is as true in the sciences as it is in the humanities.

Sexual difference is difference among individuals and individual bodies. The philosophical concept of sexual difference does not privilege any a priori category, norm, or binary, and implies autonomy from such categories. However, this Continental feminist framework also captures the biological idea of investigating individual variation in sexual anatomy and physiology in comparison to the full scope of human variation—not framed as variation *between* or *within* sexes, but as pure, individual differences themselves.

Biological investigation of sexual difference can be very productive. For example, in later chapters, we will see that the mammalian clitoris and penis are sexually homologous organs—that is, the same anatomical structure realized in a variety of ways in different bodies. As Anne Fausto-Sterling has documented, the observed variation among humans in the size of this homologous structure and urethral position is continuous, yet decisions about whether any given anatomy is considered a penis, a clitoris, or a pathological example of one or the other is entirely determined by social expectations based on assumed binary sex categories.[21] A productive alternative to pathologizing unexpected, intermediate morphologies is to regard variation in the clitoris/penis as varieties of a common structure without references to binary categories.

Sex and Race

One of the most important and persistent issues in thinking about sex and proposed sex differences is the intersection of gender/sex and race. In recent decades, a large body of research has documented how scientific and cultural concepts of the individual binary sex were developed concurrently with scientific and cultural concepts of race, the expansion of Anglo-European colonialism, and market capitalism. In "Pelvic Politics," for example, Sally Markowitz documents the influence of Victorian racism and human eugenic theory on the development of Anglo-European sexology at the turn of the twentieth century, specifically in the influential work of Havelock Ellis.[22] Within the racist intellectual scheme of the time, the concept of sex difference—that is, difference between Male and Female—was modeled explicitly upon the racist ideal represented by elite, educated, wealthy, white Anglo-European Maleness and Femaleness. In this way, specific features of elite white Anglo-European gender/sex roles were idealized as both biologically natural, *and* as evolutionarily advanced adaptations. Accordingly, variations in gender/sex roles and relations among races, cultures, classes, and castes were not regarded as evidence refuting the biological naturalness of Anglo-European cultural norms of masculinity and femininity, but rather as evidence of the inherent evolutionary and adaptive inferiority of other races, cultures, and classes.

For example, in *Psychopathia Sexualis*, the pioneering German sexologist Richard von Krafft-Ebing wrote explicitly, "The higher the anthropological development of the race, the stronger these contrasts between man and woman, and vice versa."[23] In Markowitz's words, Ellis "struggled to maintain a cross-racial sex classification—that is, one that divides the human world exhaustively into men and women—that did not challenge what he believed to be the hierarchical differences between the races."[24] Ellis resolved these tensions through a convoluted, hierarchical analysis of the relations and differences between men and women as a whole, between men and women of different races, and between men and women of a particular race. Ellis concluded that broad hips were the ideal feature of Anglo-European Female sexual beauty, and that "broad hips, which involve a large pelvis, are necessarily a characteristic of the highest human races, because the largest heads must be endowed also with the largest pelvis to enable their large heads to enter the world."[25] Here, Ellis conflated racist concepts of differential intelligence, phrenological associations of intelligence with head size, sex difference, and Victorian erotic aesthetics to reify Anglo-European Victorian sexual ideals as universally superior biological adaptations—all in the absence of data. Far beyond the simple "intersection" of sex and racial difference, Markowitz documents their *co-constitution*—the historic embedding of eugenic racism in the origin of the science of sex difference, and our modern concepts of masculinity and femininity. Ironically, Havelock Ellis is still commonly regarded as a progressive social reformer for his generally "sex positive" attitude.

Philosopher Ladelle McWhorter has further analyzed the historical development of scientific concepts of sexual perversion during the same historic period. McWhorter shows that investigating, classifying, and policing sexual variation was identified in the late nineteenth century as a strategy to defend the genetic purity, cultural supremacy, and economic productivity of the white race. Accordingly, variations in gender/sex from a married, monogamous, heterosexual ideal were conceived as constitutional weaknesses and moral corruptions that involved a reversion to the evolutionarily more primitive, bisexual state of lower races. This early scientific study of human sexuality spawned an entire cast of sexually perverse scientific caricatures—the masturbator, the feeble-minded whore, the Black rapist, the invert predator, the cross-dressing seducer, the wily manipulative lesbian, and so on—that were explicitly motivated by efforts to defend the white race from succumbing to corrupting genetic and moral influences. Masturbation (which encompassed both self-stimulation and any mutual stimulation other than vaginal intercourse) was considered a moral threat and constitutional weakness that would lead to nervous excitability, physical debility, and sexual corruption. Succumbing to any source of sexual corruption would further reduce an individual's capacity to resist ever more perverse sexual temptations. The result was a decades-long, cross-continental, eugenically motivated crusade to prevent the sexual corruption of the white race by policing masturbation by generations of white youth.[26]

The intellectual association of sexual perversion with hereditary taint and moral decline is documented in many of Krafft-Ebing's case studies, which typically begin with a recitation of the psychological, physical, and moral ailments of the relatives of each patient. The hereditary, evolutionary, and eugenic implications of these "data" were considered obvious. As McWhorter writes, "Along with genocide, sexuality was the main medium through which populations, races, and the Race could be shaped."[27]

Thus, racist and eugenic cultural and scientific theories had a decisive role in the construction of twentieth-century concepts of normative sex difference. Our contemporary concepts of the existence of masculinity and femininity as biological facts and the naturalness of binary gender/sex are themselves shaped by this late nineteenth- and early twentieth-century eugenic science. But the refutation of eugenic science and the broad contemporary abhorrence of racism has simply driven the racist origins of our contemporary concepts of masculinity, femininity, and sex difference underground and out of our conscious awareness.

Even though racism and racial purity are not motivating factors for contemporary biological research on sex difference, scientists have a special obligation to be sure that their implicit assumptions and choices of research methods do not further propagate this intellectual history. I suggest that reframing the investigation and discussion of sexual variation as the investigation of sexual difference—that is, individual variation in the absence of a priori binary categories of individual sex—can contribute productively to addressing this intellectual history. (We will return to this topic in chapter 9.)

Sexual Development and Differentiation

The concept of sexual difference as pure, individual variation in sexual anatomy and physiology connects directly to our concepts about the biological processes that give rise to our individual sexual bodies. Biology has a specific vocabulary to refer to different aspects of organismal growth. The word *growth* itself is both too general and not specific enough. *Growth* can refer to increase in body size, increase in cell number, changes in anatomical shape or cellular composition, or all three. Yet many specific capacities of bodily development are not captured by the word. *Morphogenesis* refers to the organization of the cells of the body into functional structures via coordinated cell proliferation, cell migration, and cell death. *Differentiation* refers to the processes by which unspecialized cells become specialized into the many diverse cell types and tissues of the body. *Pattern formation* refers to the processes by which embryonic cells form ordered spatial arrangements of differentiated cell types, tissues, and organs.

Following a long tradition in developmental biology, I will also use the terms *differentiation* and *morphogenesis* to refer to the process of development of cell, tissues, and organs with reproductive functions. Human bodies begin as single-celled zygotes—the fusion of an egg and a sperm—which divide and proliferate into many, many cells that become more and more specialized, more hierarchically organized (i.e., anatomically and spatially nested), more different from one another, and more and more different from their original homogeneous states. The development of a complex body involves the origin and maintenance of differences among many types of cells organized hierarchically into different tissues and organs with diverse functions, anatomical relations, and physiological interactions. Biological development is the realization of specific hierarchically organized differences among cells, tissues, and organs within the body. This developmental process is referred to as *differentiation* because the result is the structured creation of hierarchical differences within the body. Thus, biologically, *differentiation* refers to the origin of the differences *within* the individual body over space and time, *not* to differences of one body from another body.

Differentiation involves the realization, or materialization, of a specific, individual self from among the infinite, yet bounded, range of human possibilities. As we will see, development involves the specification of the emerging identity of cells, tissues, organs, and so on through the *exclusion* of other possibilities. A cell cannot simultaneously become a motor neuron *and* a liver cell, a cardiac muscle *and* a corneal cell. Likewise, one body cannot realize all possibilities at once. So, developmental differentiation involves self-materialization, specification, and the exclusion of possibilities.

Lastly, although the term *development* is traditionally used by biologists to refer mostly to the early stages in life—usually embryogenesis, childhood, and sometimes adolescence—*development* is also the appropriate term for aging over the entire life span. Thus, development continues throughout all stages of life through senescence and death.

The Sexual Phenotype

Our topic is the relationship between the genotype—the genetic material of the individual—and the phenotype—the fully realized organism. Most discussions focus on those components of the human phenotype related to sexual reproduction. I will refer to the biological components of the body that have evolved through selection for sexual reproduction as the *sexual phenotype*.

Applying Richard Dawkins's concept, the "extended phenotype" of an organism encompasses the material self *and* all of its consequences in the world. For example, the extended phenotype of beavers includes the dam they construct, *and* the complex cascade of ecological consequences caused by the flooding of the forests upstream of the dam. Thus, the extended human phenotype includes individual psychological and social behavior, including explicitly cultural phenomena. The continuity between genetic, material, environmental, and cultural inputs and influences on the development of the human phenotype means that the biological concept of the sexual phenotype properly extends outward, and out of the boundaries of the discipline of biology, to connect to the cultural concept of gender/sex. Indeed, biologists who think the feminist notion of gender/sex is awkward, radical, or too overtly political can simply consider gender/sex to be the extended sexual phenotype of an individual. This feminist concept is precisely congruent with Dawkins's extended phenotype (although it doesn't function as a purely adaptive extension of selfish genes in the genome, as Dawkins hypothesizes). When necessary, I will refer to this broadly encompassing concept as the gender/sex phenotype. Of course, it is also important to note that the individual gender/sex phenotype includes the nonbinary possibility of not identifying with a distinct gender, or the asexual possibility of lacking sexual feeling or desire.[28]

In practice, the sexual phenotype can have fuzzy boundaries. Scientific ambiguity arises because bodily traits can have both reproductive and nonreproductive functions. For example, the mammalian penis with an enclosed urethra has both reproductive and excretory functions. In a complicated way, bodily structures that function in reproduction can also be taken up by cultures to have nonreproductive functions in gender/sex norms. For example, breasts with developed mammary glands can function in feeding infant offspring, but breasts have also come to function in many cultures in nonreproductive aspects of gender/sex communication and norms. Furthermore, many features of the body that have not evolved for any sexual functions—like scalp hair, eyelashes, and fingernails—have also come to function in gender/sex communication. Lastly, I will entirely exclude from the sexual phenotype variation in features like height and body mass that have nothing to do with reproduction. Throughout, I will restrict my use of sexual phenotype to apply only to those iteratively embodied features that have evolved for reproductive functions, but I will not be able to clarify or reemphasize this at each turn.

Sex Determination and Sex Reversal

I will not be using two terms that are nearly ubiquitous in the scientific literature on sexual development—"sex determination" and "sex reversal"—and here I will explain why.

In science, determination can refer to an observation, calculation, or measurement—like the determination of current atmospheric CO_2 concentration on Earth. But in developmental biology, the initiation of the development of gonad sexual phenotype is often referred to as "sex determination."

The problem with the concept of sex determination is that it assumes that sex is an essential, prior, binary, genetic fact about each individual human genome or cell that can be "determined" materially during the development of the physical body. The idea that a developmental process is a mechanism of "determination"—as opposed simply to a process of the individual development of reproductive possibilities—assumes that the process of development materially reflects something that is a prior binary essence of the genome, embryo, or individual. In this view, sex determination is the material representation of something that is already manifestly true about the genome, cell, or zygote. Accordingly, variations in the process of "sex determination" are viewed as inherently pathological—mis-determinations—and the investigation of bodily variations in sexual development involve scientific "determination" of the actual, factual sex of every variable individual.

However, as we will see in later chapters, the material body is not a representation of an essential fact present in the genes or chromosomes, but a material *doing*, a performative becoming, an individualized enactment of the organism. The concept of sex determination as both a development process and a goal of empirical investigation of individual bodies tells us more about the social need to fulfill the anxious expectations of the sexual binary than about the actual process of bodily development and its individual and evolutionary variations.

An initial process of sex determination is usually conceived of as a decisive event following which all other events in sexual development are merely predetermined consequences, like a row of toppling dominoes. The extreme version of this logic, applied to humans and other placental mammals, is that Male sex is determined by a "master-switch" gene found only on the Y chromosome, and that, in absence of this gene, Female sexual development is simply a default.

I will critique and reject this view of sexual development from multiple directions. However, at this point, I want to emphasize that the idea of sex determination as a decisive moment, switch, or event fails because all the later developmental events downstream of, or subsequent to, that moment are themselves contingent upon *other* genes, pathways, and anatomical agencies that the earlier genetic "switch" does not, and cannot, control or determine. As we will see, sex cannot be determined because the realization or enactment of the sexual phenotype is not a moment, but an ongoing process of becoming.

The essentialist commitment of the concept of "sex determination" is confirmed by the frequent use in scientific literature of the phrase "sex reversal" to refer to the effects of those genetic and environmental influences on sexual development that conflict with a researcher's concept of an individual's "real," or essential, binary sex which *can* be asserted by investigation of chromosomes or genes. A Google Scholar search for the exact phrase *sex reversal* yields over 51,400 publications.[29] Ironically, a concept of "sex reversal" is only required because of the adoption of the erroneous idea of "sex determination." It relies on the assumption that specific combinations of chromosomes, individual genomes, or zygotes can be defined as inherently Male or Female, and that variations from the developmental outcomes presumed by such genetic definitions constitute a material *reversal* of an individual sexual essence. As we will see in detail in chapter 6, examples of supposed "sex reversal" involve deviations from the expected, normative paths of sexual development whose very existence demonstrates the fallacy of trying to define a priori the sex of genes, chromosomes, or zygotes. There is no individual sex independent of, or prior to, the process of its individualized enactment or realization.

Discourse

My argument will draw heavily on the concept of discourse developed in sociology, literary theory, and Continental philosophy. Discourse refers broadly to communication among individuals, but the idea captures more than simply the transfer or exchange of information. Drawing on post-structuralist analysis of the role of language in literature, the arts, and society, the term *discourse* captures the way in which modes and structures of communication create and constrain what it is possible to communicate; what kinds of communications have value, meaning, or consequence; and who is authorized to be a legitimate speaker or listener. This broad notion of discourse is central to Butler's definition of performativity as the "reiterative power of discourse to produce the phenomena that it regulates and constrains."[30]

In subsequent chapters, I will extend this idea of discourse substantially to describe the highly structured, *molecular*, and even electric communications among cells, tissues, and organs of the complex, developing, and physiologically functional multicellular organism. I will argue that discourse is more than an apt turn of phrase; it is an intellectually powerful concept that can help us to better understand the details of molecular, developmental, and evolutionary biology.

Agency

Agency is a capacity for action. Traditionally, agency has been conceived of as freedom of action contingent upon an individual's mental or psychological state. These mental states are interpreted as causes of individual actions. From multiple directions, however, many scholars have argued for the expansion of the concept of agency beyond exclusively anthropocentric definitions to encompass nonhuman organisms and even aspects of the material world.[31]

For example, philosopher Xabier Barandiaran and colleagues propose a broader sense of "living agency" defined by the combination of individuality (distinction from the environment), interactive asymmetry (actions distinct from the environment), and normativity (acting according to individual goals).[32] They argue that living agency can been attributed to almost all living organisms, including bacteria.

Similarly, anthropologist Bruno Latour analyzes previous textual descriptions of Prince Kutuzov—a character from Leo Tolstoy's *War and Peace*—the Mississippi River, and the human corticotropin-releasing factor receptor. Latour concludes that the agency ascribed to each of these entities is "a property of the world itself, and not only a property of language about the world."[33] It is not accidental that Prince Kutuzov, the Mississippi River, and the human corticotropin-releasing factor receptor are all ontological individuals (see above).

Throughout this book, I will apply and expand Barandiaran's and Latour's concept of agency to individual organs and cells within the multicellular body, to hormones and other signaling molecules and their receptors, and to individual genes and their regulatory binding sites. All of these biomolecules and molecular features share historic/ontological individuality, and distinct capacities for action that are *contingent* upon their histories of selection. Following Latour, I will use the scientific language of biological action to describe the agencies of organisms, organs, cells, biological signaling molecules, genes, and gene regulatory networks. These biological entities have agency by virtue of their evolutionary histories of selection to hone and diversify their functions. Furthermore, as we will see, many of these cellular and molecular agencies are shaped by coevolution for specific functional *intra-actions* with one another—a concept to be introduced in detail soon in chapter 3. Thus, like the traditional framing of human agency as contingent on prior and current mental states, these biological agents pursue specific actions that are contingent on both their evolutionary histories *and* their current physiological/biological environments. Agency in biology is also a contingent capacity for action.[34]

Queer and Queering

During the post-Stonewall fight by gays, lesbians, and trans people for cultural acceptance, legal rights, and personal liberty, the formerly pejorative term *queer* was advocated and adopted by many as a reference to alternative sexualities that was itself unburdened by essentialist concepts and categories that participated in their own oppression. Perhaps best dated by the publication of Eve Kosofsky Sedgwick's *Between Men: English Literature and Male Homosocial Desire* in 1985 (though any specific origin would be controversial), various strands of *queer theory* arose in literary criticism and Continental feminist philosophy.[35] Queer theory investigates the ways in which cultural categories and conceptions of sexuality and identity have been constructed in ways that privilege heterosexuality and patriarchal social and sexual control. Over subsequent decades, this emphasis on liberation and politics contributed directly to the development of diverse, interconnected literatures connecting queer theory to the study of Marxism, racism, post-colonialism, migration, caste, nationalism, environmentalism, and more. For example, Roderick Ferguson's "Queer of Color Critique" explores the intersections of race, political economy, gender, and sexuality in literary theory, philosophy, and aesthetics.[36] Similarly, queer ecology, in the words of Catriona Sandilands, seeks to "disrupt prevailing heterosexist discursive and institutional articulations of sexuality and nature, and also to reimagine evolutionary processes, ecological interactions, and environmental politics in light of queer theory."[37]

Within all of this work, the term *queer* serves an important and distinct intellectual purpose. According to Annamarie Jagose's definition, *queer* refers to any perceived or realized mismatch between an individual's gender/sex and the biologically or culturally expected norms of sex, gender, and desire.[38] David Halperin explains further that *queer* "acquires its meaning in opposition to the norm. Queer is by definition *whatever* is at odds with the normal, the legitimate, the dominant. *There is nothing in particular to which it necessarily refers.*" To Halperin, then, queer is not a "positivity" but a "*positionality vis-à-vis* the normative."[39]

This book uses this anti-essentialist and positional sense of the word *queer* to critique normative assumptions in the genetics, developmental biology, and evolutionary biology of sex. At various points in subsequent chapters, I will also refer to "queer bodies" of individuals in exactly this sense, based on various differences in sexual development from normative or canonical, anatomical, or physiological expectations. I do this to avoid reflexively pathologizing individual sexual differences, and to highlight that our responses to them are often dominated by their "positionality" relative to normative expectations and not by the potential of these individuals to thrive and live healthy lives. However, I want to acknowledge that everyone, including developmentally different, intersex, and trans people, has the right to claim their own personal and embodied identity as queer or not queer.

Halperin's definition of queer provides a good basis for understanding the term *queering*, which implies the active change to, disruption of, or dislocation of a concept, discussion, culture, or population by the introduction, existence, or recognition of queer variation from the normative. Thus, queering is the normatively disruptive consequence of admitting the queer into consciousness or discussion. We will be trying to do exactly that in the chapters ahead.[40]

CHAPTER THREE

Gender Performativity

In Ludwig van Beethoven's opera *Fidelio*, the honorable Florestan has been unjustly imprisoned by his cruel political enemy Don Pizarro. To secure his freedom, Florestan's wife, Leonore, has assumed the identity of a man, named Fidelio, and gotten a job as an assistant to Florestan's kindly jailer (go figure?), Rocco. Fidelio is resolved to be a dutiful and motivated employee, and quickly earns Rocco's trust and respect. Just as quickly, the handsome, kind, up-and-coming Fidelio attracts the warm affections of Rocco's daughter, Marzelline. As a consequence, Marzelline must fend off the pesky entreaties of her hapless ex-beau, Jaquino, whom she dropped like a rock when Fidelio came to town. In a gorgeous first-act quartet, Marzelline, Fidelio, Rocco, and Jaquino each express their private hopes and emotions—Marzelline ecstatic at the prospect of marital bliss with this upstanding young man; Fidelio anxiously hopeful at her progress in liberating Florestan while regretting her deception of Marzelline; Rocco proud that his daughter has found such a fine catch, and fantasizing that they will soon be happily married; and Jaquino jealous and forlorn about his unexpected turn of fortune.

When Rocco blesses the future couple, Fidelio presses him to trust her further, and allow her to help him tend to the secret prisoner in the dungeon. When Rocco warns that great courage is necessary to face such terrible things, Fidelio declares, "I have the courage! Love can bear great suffering for a great reward!" Rocco and Marzelline hear Fidelio's expression of bravery as a sign of his commitment to be a dutiful husband to Marzelline, but Fidelio's true commitment is to liberate her husband

and restore justice. In a heartrending trio, Rocco commends the couple to join hands and "tie the knot" with tears of sweet joy. Fidelio confides in an aside, "I joined hands in sweet commitment, and it cost bitter tears," referring ironically to her tragic marriage to Florestan, and not to a happy future with Marzelline. (Who is the secret prisoner? No spoilers here! To learn what happens, you will have to see the opera for yourself.)

Fidelio succeeds as an opera because Beethoven is able to powerfully express the complex, textured, often conflicting emotions and motivations of these characters in the music itself. The musical drama portrays both the inner realities and the outer social presentations of each character, creating a rich nexus of emotional connections and tensions with each twist and turn in the story and the harmony. Drawing on the long, historic tradition of gender-bending "trouser roles" in opera and the contemporary French play *Léonore* by Jean-Nicolas Bouilly, Beethoven and his librettists created the unforgettable character of Fidelio—a brave, determined, powerful, and capable cross-dressing woman who fights injustice to free her partner.

Fidelio exemplifies that social drama involves complex, sometimes flexible, context-dependent presentations of the self. With its gender-passing theme, *Fidelio* emphasizes that gender in Western culture has been seen as a fundamental and ubiquitous feature of the everyday self and that being gendered functions in nearly all social interactions. More than two hundred years ago, Beethoven was exploring in *Fidelio* aspects of what we can now describe as the *performative* qualities of human gender.

What Is Performativity?

The word *performative* has a catchy and intellectual mouthfeel. Recently, as a consequence, the term has moved rapidly from academics into popular culture. Unfortunately, along the way, the term has taken on new meanings that are divorced from the actual history of the concept. In popular parlance, *performative* has come to describe self-conscious, inauthentic, or insincere social expression, posing, or virtue signaling. However, in this book, I propose that the entire organism—including its material body, physiology, behavior, and psychology—can best be understood as "performative." Because there is nothing more fundamental, authentic, or real to an individual organism than its self-performance, the popular connotation of *performative* is quite precisely incorrect. Given this popular misunderstanding about the word, it is important to carefully outline here the intellectual origins of the concept of performativity, in which performative enactment is an individual achievement, not a superficial or demonstrative faux-wokeness.

The concept of gender performativity draws on several traditions in philosophy, the humanities, literary criticism, and the social sciences. Performativity starts with (but is not limited to) the idea of an artistic performance—such as by an actor in a role from the *text* of a play, or a singer in a role from the *score* of an opera. The text or score is given prior to the performance, but the role itself only becomes realized in the world through a specific, individual performance. An actor's individual realization of a role is drawn with variable fidelity from the text, and shaped by the actor's body, voice, language, talents, and capabilities. The actor's realization of the part is also influenced by the material, cultural, and social environment of the performance including the other cast members, the stage, set, lighting, director, budget, audience, stage crew, price of admittance, and even whether the cast and crew are members of a labor union. No account of a specific performance of a play would be complete without incorporating considerations of both the prior text *and* the means and conditions through which it is realized in the world.

This dramatic concept of performativity has arisen in several different contexts in twentieth-century social sciences and humanities. Margaret Mead introduced the idea of gender roles as culturally mediated performances in *Sex and Temperament*,[1] her pioneering 1935 portrait of variation in gender/sex roles in three different human cultures. Possibly influenced by Mead, the influential sociologist Erving Goffman later proposed the idea of social personality as performance in *The Presentation of Self in Everyday Life* (1956).[2] Goffman analyzed interpersonal social interactions with the tools of dramaturgy—the analysis of structure, composition, and representation in the theater.

A second source of today's concept of gender performativity is of a quite different origin. In the 1950s, ordinary language philosopher J. L. Austin distinguished referential statements that describe details about the world—"*My mother's hat is on the table.*"—from performative speech, in which the utterance is itself a kind of action. Descriptive speech consists of "constative utterances," or statements that are (truthful or not) representations of facts in the world. In contrast, performative speech would include the wedding vow "*I do*," the commitment "*I promise . . .*," or the action "*I bet you sixpence that it will rain tomorrow.*"[3] Such statements do not refer to, describe, or represent the world. They actually *do* something *in* the social world. As Eve Sedgwick writes, "Austinian performativity is about how language constructs or affects reality rather than merely describing it."[4] Recognizing the potential power of language to shape the world had a radical impact on twentieth-century philosophy and literary and feminist theory.

In the late 1980s and 1990s, the concept of gender performativity was introduced into contemporary feminist thought by Judith Butler, feminist philosopher at University of California at Berkeley, in *Gender Trouble*, *Bodies That Matter*, and other works.[5] Butler observed that the once progressive idea of gender as "culturally constructed" was insufficient to account for the role of our material bodies and personal psychologies on our realized genders. In contrast, Butler conceived of gender as a performance of the self that is initiated by the individual, but facilitated and constrained by individual experiences of cultural norms, language, social rewards, punishments, and legal sanctions. To Butler, gender is "an identity instituted through the *stylized repetition of acts*."[6] Butler further states, "Consider gender . . . as a corporeal style, an 'act,' as it were, which is both intentional and performative, where 'performative' itself carries the double-meaning of 'dramatic' and 'non-referential.'"

In gender theory, then, performativity refers to the social and psychological system by which cultural norms and expectations are reiterated in new individuals as they are born, mature, and become part of society. Spanning Mead, Goffman, Austin, and Butler, performativity describes a complex way of becoming that recognizes an individual agency *and* the social, cultural, environmental influences and constraints on the realization of that agency. As Butler asserts, quoting Simone de Beauvoir, "To be a woman is to have *become* a woman."[7] Gender is a process, or a doing, by an individual in the social world.

Gender performativity has had an enormous influence on queer feminist thought and the humanities, but most critical to this project is the work of physicist and feminist philosopher Karen Barad of the University of California Santa Cruz, who developed and proposed a concept of *material performativity*. Barad describes matter itself as performing its material agency. This "posthuman" performativity extends beyond the social and psychological realms to encompass the entire material world and the human, scientific enterprise of investigating it. Influenced deeply by the philosophy of the pioneering quantum physicist Niels Bohr, Barad uses the concept of performativity to model the scientific investigation of the material world. To Barad, science is a "material-discursive practice"—a way of knowing about the world through interacting with it—that creates the phenomena that scientists observe and describe.

For example, Barad discusses how light can exhibit properties of either a wave or a particle depending upon the scientific apparatus used to investigate it. In this view, the human-made apparatus collaborates with the material world to create the phenomenon of a light wave *or* a photon. To characterize this scientific engagement with the world, Barad coins the term *intra-action* to refer to those special interactions among components that actually create a phenomenon. An intra-action is an active, reciprocal relation among entities or parts that have no genuine actions or functions other than their *shared, mutual* actions. In other words, an intra-action is an interaction that creates, or invents, the action itself. Thus, depending upon the specific design and deployment of the scientific apparatus, light and the apparatus intra-act to create the scientific phenomenon of a light wave *or* a light particle. Barad called her posthuman, materialist expansion of performativity "agential realism."[8]

Of course, since Butler's and Barad's proposals in the 1990s and 2000s, there have been many new developments in material feminist and queer scholarship that I will not review here. Rather, I will focus on the concept of performativity developed by Butler and Barad because I think it has special, productive relevance to our understanding of our physical, material, and cultural bodies. Although these works may now be considered "classics" in queer feminist studies, the concept of performativity has become somewhat less central to ongoing research in the field. In a way, then, I will be calling for a return to, and a revitalization of, the concept of performativity in queer feminist thought because of the vital contribution it can make to our understanding of the materiality of sex and the body.

Elements of Performativity

To develop a performative account of the organismal phenotype in later chapters, I want to explore theoretical aspects of the concept that will be relevant to the application of performativity to biology.[9]

An important element of contemporary gender performativity bridges the theory of language and literary criticism. As Butler writes, *performativity* "carries the double-meaning of 'dramatic' and 'non-referential.'"

To achieve this, Butler expands on Austin's view of performative speech to analyze gender development as a *discourse* between the individual and other agents in their social environment. A discourse involves more than simply the perception of available information or cues; it includes an active social exchange through communication. According to this account, from the moment of birth, humans receive verbal and nonverbal messages about their bodies, behavior, gestures, postures, and manners—including demonstrative examples, advice, instruction, praise, encouragement, condemnation, censure, or punishment—from their parents, siblings, peers, authority figures, books, stories, and, more recently, from an expanding blizzard of social media. Thus, gender/sex development involves a discourse—a structured social exchange of communications. Exchanges between the developing individual and their social environment are also communicated through nonverbal facial and bodily expressions, language, and symbols that enact cultural norms, traditions, rules, and laws governing gender/sex realizations and behavior.

In other words, the concept of performativity captures both connotations of "performance" as the playing out of an historically derived role in a social context of other interacting/observing individuals, and as a "becoming" through "doing"—an enactment through discursive action of an individual, constitutive agency. One aspect focuses on the social, theatrical context, and the other on the assertion of individual agency through discourse.

Conceiving of gender as realized through performative discourse has focused feminist analysis on the way in which language itself—from the grammar of pronouns to the texts of bedtime stories, medical texts, Instagram posts, and criminal codes—acts to constrain and facilitate gender realization and sexual behavior. For example, traditional, English-language titles of polite address—Mr./Miss/Mrs.—require gendering people and, for women, further categorizing them in terms of their availability for monogamous heterosexual marriage.

Barad's extension of performativity to the material world broadens the concept of discourse to include what she calls material-discursive practices, or "iterative intra-actions through which matter is differentially engaged and articulated."[10] In her performative metaphysics, Barad rejects the strict Cartesian distinction between "words and things," and presents a view of the material world that is guided by Bohr's insights into the scientific study of physics. Like a scientific apparatus, matter constitutes an apparatus that intra-acts with other matter in the world to create phenomena and exclude others. These phenomena include objects, the boundaries between them, waves, and the physical forces of the material world. This ongoing process is the world's performative becoming. To Barad, scientific measurement—like an experiment using an apparatus to manipulate light—shows us that "meaning making" involves physical *intra-action* with the world. Thus, discourse is always rooted in, or entangled with, material intra-actions, or material-discursive practices. This view establishes the possibility of entirely nonhuman discourses in the world.

In *Bodies That Matter*, Butler analyzes the "gendering" of human embryos through the use of (then new) sonogram technology, illuminating the process by which a fetus is "girled." This naming is both "the setting of a boundary and also the repeated inculcation of a norm."[11] To Butler, the embryonic body is given, and becomes gendered by language and culture. Barad presents a fascinating extension of this analysis focusing on the physics and engineering of the piezoelectric crystal, which is the sound wave–emitting element of the sonogram apparatus that also changes its shape in *response* to the polarity of the *echoes* of sound bouncing back from within the body being imaged. In Barad's analysis, the piezoelectric crystal is simultaneously an instrument for intra-active intervention into the material world, and a *receiver* of the material-discursive responses of sound waves *from* the material-biological world. These intra-actions create the phenomenon of an image of the embryonic body, which then feeds into cultural meaning-making of fetal gender/sex.[12] Thus, Barad's performativity extends the concept of discourse beyond human social behavior into the material world. In later chapters, we will explore the biological performativity of embryonic development itself—the biomaterialization of the embryonic body with which this ultrasound technology intra-acts.

Another crucial aspect of performativity is the *iterative* property that Butler (drawing on literary theorist Jacques Derrida) refers to as *citationality*. Like academic citations of previous literary, artistic, or intellectual expressions, citationality captures the idea that individual gender realizations exist in reference to, or as *citations of*, other previous or contemporaneous enactments of gender. Citationality describes the relation between individual instances of gender in different individuals at different places and times. So, when a rural kid is empowered to expand their gender boundaries after watching episodes of *RuPaul's Drag Race*, these subsequent gender realizations may be considered citations of prior performances in the culture.

Like tracing stylistic influences in literature, art, and art criticism, unraveling citational relations in gender requires historical analysis to reveal sources of influence, inspiration, and innovation. Such historical approaches draw on philosopher Michel Foucault's concept of *genealogy*, which refers to the historical analysis of the cultural uses of language, especially the role of authority and power in controlling the meaning of language concerning sexuality, morality, and truth.[13] An iconic queer intellect of twentieth-century Continental philosophy, Foucault is commonly associated with a strongly anti-science critique, but in *The History of Sexuality* and other work, Foucault reveals a deep interest in the material body. As we will see, several aspects of Foucault's philosophy of sex are strikingly relevant to the scientific understanding of sex and the body. Both citationality and genealogy will be central to developing a concept of biological performativity.

If gender is performative, then gender is not a material representation or expression of an individual inner essence of the self. In other words, gender is not a "constative utterance," or a material/cultural statement of a prior truth. Consequently, the concept of performativity is also anti-essentialist. Essentialism is the philosophical belief, or commitment, that things have essences—inherent, indispensable, and definable characteristics or qualities that make them what they are. In this context, essentialism is the belief that binary sex and gender are fundamental truths about individual organisms that can be defined and determined by the study of individual genetics, chromosomes, hormones, or transcriptomes.[14] Rather, gender/sex is a self-realization, or enactment of the sexual self. Profoundly, this means that the still common feeling of the "rightness" or "naturalness" of binary heterosexual gender categories is itself a cultural/psychological phenomenon that functions to obscure the performativity of gender/sex, and reinforce the legitimacy of the scientific sexual binary and heterosexual patriarchy.[15] This conclusion will have precise parallels in biology. However, this observation does not imply that the individual feeling of the rightness of one's own gender identity is not justified. The profoundly performative basis of the material body is entirely consistent with the emergent psychological experience of, and commitment to, an individual gender/sex identity.

As a progressive political concept, one of the distinct features of the performative account of gender is that it simultaneously describes how gender/sex norms have historically constrained and oppressed women and sexual minorities, *and* provides an account of how variations in gender can transform those cultural norms. Through iterative "citation" of prior instances of gender variation in successive generations, gender performativity facilitates the expansion, diversification, and queering of gender boundaries as well as the cultural creation of new possible sexual selves and new ways of being. In this way, performativity establishes an account of both oppression *and* pathways toward social and cultural innovation, expanded autonomy, freedom, and civil rights for sexual minorities and genderqueer people.

Butler states that "performativity implies agency."[16] Indeed, the focus on the individual *agency* is another reason why performativity provides a more fruitful account of gender than previous concepts, like gender construction or gender expression.[17] The traditional focus of performativity has been on the agency of individual persons. Recently, however, Judith Butler has elaborated a performative theory of mass assembly.[18] In this work, Butler focuses explicitly on how mass action can give political voices to otherwise invisible people through their joint material presence.

Performative agency and mass (i.e., hierarchical) assembly are also invoked by Barad's innovative and indispensable concept of material intra-action. When a process is realized through intra-action among entities, the outcome is performative. Thus, to Barad, the aggregate, bulk properties of nonliving matter arise from the hierarchical intra-actions of subatomic particles, atoms, molecules, and objects of the material world. Subatomic quarks, leptons, and bosons have no particular functions or consequences in the world except through their intra-actions with each other to create protons, neutrons, atoms, and other phenomena of the physical universe. Thus, both Butler and Barad establish *hierarchical* concepts of performative agency that apply simultaneously to phenomena at multiple, nested levels of organization—agencies of, within, and among. This hierarchy of agencies will be fundamental to understanding biological performativity.

In gender performativity, the agency of individuals is constrained by the controlling force of social norms, which Butler analyzes as the regulatory *power* of a patriarchal society. Following Michel Foucault, however, this power is not seen as a top-down form of central, unitary, or authoritarian control, but an emergent effect of the innumerable individuals in a network of social relations. To Foucault, power is a broadly distributed, diffuse, bottom-up, self-organizing, and self-regulating force. Likewise, resistance to power is simply another strategy that is not separate from power, but arises in a similarly distributed way at various foci within a social network. Foucault's view of power is intellectually revolutionary in that power becomes creative as well as constraining, innovative as well as regulatory, and frequently discursive in addition to physical. Ironically, to Foucault, the "logic" of power is always clear but has no "author." Power is always exercised with a series of aims and objectives but is not the choice of an individual subject. In parallel, we will explore the regulatory and creative powers of the processes of organismal development, natural selection, and sexual selection.[19]

Critical to incorporating the sciences, Karen Barad expands the concept of performativity to be explicitly *posthuman*—that is, intellectually conceived or framed without human beings as the sole or central organizing focus.[20] Accordingly, performative agency is not exclusively about human agency and social discourse, but is an outcome of intra-actions of the material world—subatomic particles, atoms, photons. Agency is not restricted to human individuals and their social networks, but is omnipresent and immanent in the material world's propensity for intra-action, and its ongoing becoming. In this view, the scientific investigation of materiality involves an intra-active discourse with material agents that are mediated by scientific apparatuses. These human and material agents intra-act through an apparatus to create a specific observable, measurable scientific phenomenon, just as the wave or particle phenomena of light are created by the apparatus used. Thus, Barad expands the concept of performativity to encompass both the iterative social/cultural enactment of individual gender, *and* the continuous, active becoming of the material world.

I will return to each of these properties of performativity—discourse, agency, citationality, constraint and innovation, regulatory and creative power, and posthumanity—repeatedly in later discussions of performative biology.

Performativity and Trans Experience

Butler's performative theory of gender has been enormously influential, but also controversial. Without tracing the various applications and criticisms that it has inspired, I do want to mention important debates on gender performativity in transgender and transsexuality studies.

In his influential book in trans studies, *Second Skins*, Jay Prosser seeks to root transsexual identity and experience to the material body, and to counteract what he perceives as the erosion of that connection by performative gender theory.[21] Prosser characterizes Butler's avowed interest in the materiality of sex as too shallow and argues that the concept of gender performativity obscures the role of the material body that is necessary to understand trans experience. Prosser also opposes "the appropriation of transgender by queer" studies, given that many trans people do not identify as queer. Prosser emphasizes that, "there are transsexuals who seek pointedly to be nonperformative . . . , quite simply, to *be*."[22]

As Gayle Salamon and other trans theorists point out,[23] however, Prosser mischaracterizes Butler's view of the contributions of the material body. Butler's performative project was explicitly inspired by a rejection of gender constructivism to account for the constitutive variability *among* individuals in their gender/sex identification. Furthermore, by portraying performativity as an exclusively cultural theory in opposition to material accounts, Prosser's view forestalls the work by Barad and me to use performative theory to bring the material body into clearer view. In a direct challenge to Prosser's view that performativity places the "material body even further out of our conceptual reach," this book uses performative theory to focus on and clarify the molecular, cellular, genetic, and developmental details of our embodied sexual becomings.[24]

In contrast to Prosser's anxiety about trans people being co-opted by queer politics, transsexual activist, author, and molecular-developmental biologist Julia Serano is deeply concerned about the exclusion of transsexuals from queer feminist culture and comprehension, and aspires to full inclusion of transsexuals in queer and feminist movements. Serano proposes that biology supports a "holistic" model of the development of individual gender that is neither constructivist nor essentialist. Rooted in her understanding of developmental biology and genetics, Serano's holistic account of the development of gender/sex is highly congruent with the performative account that I will present here, but it is less detailed.[25] I will return to discuss the implications of a performative theory of the sexual phenotype for trans experience in chapter 9.

Between Butler and Barad

In recent decades, a broad movement of material feminists have advocated for a better account of the cultural consequences of our material/sexual bodies (see appendix 1). However, many of the goals of a fully engaged material feminism—including a comfortably queer conceptual and investigative space in the intellectual heart of biological science—have yet to be fully realized. Even though the concept of gender/sex implies an intellectual continuity across cultural and biological phenomena and our methods of studying them, there is still a need for a detailed proposal about what that continuity might look like. In subsequent chapters, my goal is to outline in detail an intellectual continuity in the explicitly biological space between Judith Butler and Karen Barad—applying, extending, and modifying their performative theories to engage productively with biological and cultural perspectives on the materiality of the body.[26]

However, I do not want to ignore the intellectual differences between Butler and Barad. Barad and other material feminists framed many of their views as critiques of the discursive, Continental style of feminist theory represented by Butler and others. Thus, Barad's retort against feminism's "linguistic turn"—"the only thing that does not seem to matter anymore is matter"—was aimed as squarely at Butler's work as any others. However, I hope that my analysis of genetics, developmental biology, and evolutionary biology can trace new connections between both their views, while pushing back in both directions.

In the opening of *Bodies That Matter*, Butler asks, "Is there a way to link the question of the materiality of the body to the performativity of gender?" Butler further challenges feminists to "return to the notion of matter... but as a process," and more specifically as "a process of materialization that stabilizes over time to produce the effect of boundary, fixity, and surface."[27] Here, Butler invites an explicitly performative account of the development and materiality of bodies but does not yet provide one. As Barad asserts, Butler is absolutely correct to look for an understanding of material body in the physical world, but "does not follow this impulse through to an actual account of materialization that is general enough."[28] To Barad, this failure occurs because "Butler's [literary] theory... reinscribes matter as a passive product of discursive practices rather than as an active agent participating in the very process of materialization." By adopting Barad's concept of material-discursive practices in the analysis of genetics and developmental biology, my goal is to provide an account that explicitly connects performativity to the core biological processes that create our material bodies.

A critical element of the performativity of organisms and biology is the recognition that discourse—intra-active communication among agents—is not limited to humans, culture, or scientific investigations; rather, it is deeply engrained in the biological mechanisms of genetic regulation, cellular physiology, and cell-to-cell signaling. This recognition extends the intellectual tools of Butler's performative theory deep into what Lewis Thomas referred to as *The Lives of the Cell*.[29] Likewise, Barad's agential realism strongly roots discourse in the material world. But my analyses of the material performativity of the body will extend beyond Barad's, almost entirely nonbiological examples.

Barad's expansion of performativity to encompass the scientific investigation of the material world creates a solid foundation for a scientific, posthuman concept of performativity, but her singular focus on physics leaves a large explanatory gap in the crucial space between physical, inorganic materiality and the cultural dimensions of gender/sex. This gap includes the biological processes of the development, physiology, and evolution of organismal bodies. Barad's performativity clearly recognizes nonhuman material agencies, but her oft-repeated reference to "human, nonhuman, and cyborgian forms of agency" leaves explicitly nonhuman, biological agencies yet unspecified and unanalyzed.

Barad clearly understands "apparatuses of bodily production" to include entirely nonhuman phenomena—for example, her fascinating analysis of the arrays of optical lenses embedded in the exoskeletons of brittle stars (Ophiuroidea).[30] Barad explicitly states that "the importance of the body as a performance rather than a thing cannot be overemphasized."[31] Yet, Barad does not analyze any explicitly biological processes by which those bodies are performed, creating the opportunity for this work.

Performative analysis of biology provides the next step toward a *fully* posthuman performativity—a theory in which nonhuman, physical, and biological agencies, and cellular discourses are understood as part of to the world's richness and our intellectual responsibility to explain. This posthumanity is rooted in the material, connected to biology, and fully encompassing the biodiversity of life. But it requires a substantial extension of Barad's framework, which focuses primarily on how humans learn things about the world—that is, the performativity of scientific inquiry. Although I will rely on Barad's concepts of scientific apparatuses and the performativity of scientific investigation at various points, my goal here is to theorize the performativity of biological agencies and organismal bodies themselves, independently of the specific methods used to study them.

In the chapters that follow, I propose to address Butler's opening question in *Bodies That Matter*: Is there a way to link the materiality of the body to the performativity of gender? I will explore the performativity of biological phenomena in the broad intellectual space between Butler and Barad in order to establish a clearer understanding of the material production of human sexual bodies. I will apply aspects of Karen Barad's agential realism to molecular, organismal, and evolutionary biology.

CHAPTER FOUR

The Enactment of the Biological Self

The abstract distinction between the genotype—the inherited genome of an organism—and the phenotype—the entire material, functional organism—was conceived in the early twentieth century to facilitate experimental research in genetics. Specifically, the goal of genetic research was to investigate *how* variation in genotype can explain variation in phenotype. To operationalize this genotype-phenotype abstraction, geneticists conducted controlled experiments on inbred strains of model organisms, like fruit flies and mice, lacking any genetic variation and raised in special, uniform environments. In the century following, as population geneticists Peter Taylor and Richard Lewontin have observed, biology has not generally recognized "the need to reintegrate what has been abstracted away." In other words, biology has not reoriented the genotype-phenotype concept from the study of genetically invariant organisms raised in strictly controlled experimental conditions to the study of natural populations and species, including ourselves. Yet, the genotype-phenotype relationship has remained a central question within genetics, developmental biology, evolutionary biology, and the philosophy of biology.[1] (For more discussion of current models of the genotype-phenotype relationship, see appendix 3.)

The most concise way to express the thesis of this book is that the phenotype is a performative enactment of the individual organism. Each individual organism begins with an inherited genotype and some (modest or extensive) material contribution from its parents. The resulting organism is the outcome of a performative process realized in interaction with the organismal environment. In this chapter, I present a broad description of this view of the performativity of the phenotype.

Just as gender can be understood simultaneously as dramatic performance *and* a nonrepresentational expressive action, biological performativity describes a complex mode of individual becoming as a form of doing, of self-enactment. Performativity involves the realization of a constitutive individual phenotype through discursive action. This view also recognizes how interactions—or, rather, Baradian *intra-actions*—among multiple biological agencies facilitate and constrain one another. To understand how this could work, I will introduce and explore the molecular genetic mechanisms for the growth and differentiation of the complex organismal bodies. Of course, to support the assertion that we are "performance all the way down," the biology will get rather specific and detailed. But this attention will pay off in later discussions of the development and materiality of sex.

Because it is so obvious that it often goes without saying, I want to emphasize here that complexly differentiated and functionally integrated body parts *matter* to organisms. Your liver and heart *matter* to you, to your bodily functioning, and to your continued persistence and thriving. Likewise, roots *matter* to plants. The sea water circulatory system of a starfish *matters* as critically to its survival, thriving, and fecundity as our own blood vascular system matters to us. Thus, development is the means by which organisms make matter *matter* to themselves, and the investigation of developmental and evolutionary biology fits squarely within the traditional feminist inquiry of how bodies come to matter. Developmental biology, physiology, functional morphology, evolutionary biology, and phylogeny are sciences about the mechanisms and history of mattering in both senses, which makes them all central to the feminist project. As Evelyn Fox Keller asserts, developmental biology is a "feminist cause."[2]

In the subsequent discussions, I will show how gene regulation and development constitute nonhuman, molecular discourses among a hierarchy of biological agencies, that networks of biological discourses regulate and constrain the development of the phenotype and themselves, and that these processes reoccur iteratively through the developmental materialization of each individual organism. Thus, I document how these biological processes meet Judith Butler's definition of performativity as the "reiterative power of discourse to produce the phenomena that it regulates and constrains."[3]

Genes and Development

Developmental biology is the discipline investigating the genetic and physical mechanisms by which complex, multicellular organisms grow. Developmental biology emerged out of the earlier, more descriptive, and comparative discipline of embryology, which investigated the anatomical details of organismal growth over time. Today, developmental biology focuses mostly on genetic and molecular mechanisms that regulate cell division, cell fate (i.e., self-renewal, differentiation, or death), morphogenesis (i.e., the creation of diversified anatomical parts), and pattern formation, which all give rise to the cell types; tissues; organs; and complex, modular but integrated parts of multicellular plants, fungi, and animals.

Developmental biology addresses an intellectual challenge of staggering complexity. Each one of us begins life as a single-celled zygote.[4] How do the trillions of genetically identical cells in our mature bodies achieve their current diversity of forms and organization within our mature, physiologically stable, and functionally integrated adult selves? Although each of the cells in a human body shares the *same* genome, they achieve many distinct shapes, sizes, molecular compositions, and functional states.[5] The dead cells in the outer epidermal layer of your skin contain abundant α-keratin proteins polymerized (i.e., bound together) into long chains to create the tough external surface of your skin. Muscle cells, in contrast, contain interlaced arrays of myosin and actin protein filaments that slide past one another to achieve the muscle contractions that pump our blood and allow our body parts to move. Such descriptions could go on and on.

The myriad variations among the different cell types of the body result from different uses, or enactments, of their shared genome. The relationship between the sequence of the genome and the diversity of mature cell types in the body is very complex. Like the script for a play, the genome is an antecedent informational resource to be realized in the material world in a particular context through an enactment that involves means, actions, and agencies outside of the text or script itself. Only a portion of the entire genomic text specifies the details of the structural components of the cell—the molecular materials that constitute the bricks and mortar of the body. The vast majority of the genome is comprised of other genes and gene-flanking regions that function in the regulatory control of gene expression, and long spacing sequences that are something like linear packing peanuts between the genes. Together, these genomic elements function like stage directions in the script; they facilitate, but cannot determine, the realization of any performance.

In short, genes in the genome either specify the structure of stuff, or the molecular tools to manage and regulate the production of stuff. But what are these gene regulatory tools? They are best understood as resources for the molecular discourses of development, physiology, and homeostasis (i.e., the maintenance of physiological set points in the body). By itself, the genome cannot do or cause *anything*. The genome is not an actor in the world. As we will see, to become embodied, the genome of an individual must be materially converted into enacting expressions—or, linguistically speaking, material Austinian performatives (see chapter 3). The closer one examines the details of gene regulation, the more discursive and performative it appears.

To understand how bodies develop through gene regulation—how organisms make matter *matter*—we will need to review some details of the structure and function of the genome. For those who would like more background or a refresher, the genome is a long polymer, or molecular chain, of deoxyribonucleic acids, or DNA, which form a double-stranded, ladderlike, helically twisted molecule (figure 1). Each rung in the ladder is made of a pair of nucleic acids: either adenine and thymine, or cytosine and guanine (abbreviated as A, T, C, and G). If all the DNA coiled up in the nucleus of a single human cell were stretched out to form a single ladderlike thread, it would be nearly two meters long. Linearized, the nuclear DNA within the approximately thirty trillion human cells in an adult human body would extend more than sixty billion kilometers, or farther than the distance from the Sun to the dwarf planet Pluto.[6] That's a whole lot

1. Structure of DNA

Top: DNA consists of parallel, ladderlike strands of nucleic acids—adenine (A), cytosine (C), guanine (G), and thymine (T)—with deoxyribose sugar backbones. The rungs are formed by hydrogen bonds between adenine-thymine and cytosine-guanine pairs. The ladder twists into a spiraling double helix. *Middle:* The DNA wraps around clusters of histone proteins to form round nucleosomes. Chemical modification of the histones can open regions of the DNA, making it accessible for transcription, or keep it tightly coiled, and inaccessible for transcription. *Bottom:* The coils of coils are packed together to form a chromosome. Illustration by Rebecca Gelernter.

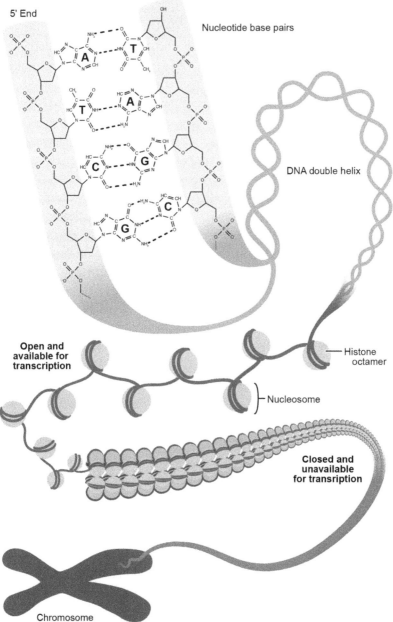

of DNA in your body! Of course, what really matters is not how long the genomic DNA is, but how our complex bodies are realized using a massive number of duplicate copies of the same DNA base pairs.

In the sequence of its nucleic acids, the genome specifies information that is necessary (but not sufficient!) for synthesizing the proteins of the body. Each of the sixty-four possible sequences of three nucleic acids, called codons, specifies one of twenty different possible amino acids, or the beginning or end of the protein sequence. Proteins are themselves long-chain polymers of amino acids that fold into specific shapes, allowing them to achieve their diverse functions as biomolecules.

At a higher spatial level, genomes are also highly structured. The specific DNA sequences that encode the amino acid sequences of proteins are generally referred to as "genes." In organisms with nucleated cells, the coding sequences of individual genes are often broken up by intervening, noncoding regions, called introns. The genes are also interspersed with vast stretches of noncoding DNA, which include critical promotor and enhancer sequences that function in the regulation of DNA expression, and long spacers that allow the genome to be folded up tightly, and selectively unfolded, within the nucleus (more on these topics later).[7]

Before the discovery of DNA or any genetic regulatory mechanisms, "genes" were defined abstractly as the material causes of heritable similarities between parents and offspring. After the first experimental transformation of bacteria by transfer of DNA, the gene was reconceived as a finite sequence of coding DNA—an expanse of genomic real estate—but the strong association of genes as biological causes was not broken. Experimental methods in genetics were devised to further the concept of the traditional population genetic view of genes as "hardwired," deterministic causes of variation in organismal phenotypes. Along the way, we learned a lot about how genes are regulated in various life forms—from bacteria and yeast to mice and human tissue cultures—and we now know that it takes a lot more than a gene's coding sequence to create and regulate gene expression or protein production. However, most practicing biologists have not spent a lot of time revisiting whether "genes" are actually the material causes of the phenotype.

Really looking at the development of multicellular organisms, however, you realize that genetic causes are not hardwired. In fact, there aren't any wires at all, and nobody is in charge. Bodies are not "constructed" like buildings to match the specifications of a blueprint. Furthermore, buildings also do not grow up by "doing" the business of buildings, the way biological bodies do. Bodies are not external presentations or expressions of those organisms' essential natures. Computationally inclined biologists have taken to describing the genome as an algorithm or a developmental program. Although some of the structures of computer programs—with "If-Else" statements and "Do-While" loops that encode decision-making rules—share some logical features with some genetic regulation networks, the open-endedness, flexibility, redundancy, contingency, and robustness to mutation and environmental perturbation of organismal development are clearly unlike any normal algorithm or computer program.[8] The performative phenotype should replace the blueprint and algorithm concepts of the genotype.

As the multicellular human body grows from a single cell to trillions of highly differentiated cells that are highly organized into tissues and organs, no cell or organ can *make* or force another cell to do anything. Even the immune systems killer T cells only attack other cells of the body in response to surface signals initiated by *those* cells.[9] We may think that cellular signals *cause* other cells to perform actions. For example, we say that insulin produced by pancreatic beta cells "makes" other cells in the body take up glucose. However, such hormonal and physiological actions are all contingent on those other cells deploying the appropriate molecular machinery to receive, process, and respond to these signals—in this case transmembrane insulin receptors, etcetera. But cells do not topple deterministically like rows of dominoes. Each cell has its own agency. Each cell is its own locus of distributed and self-regulatory power, provided by its own internal mechanisms for regulating the expression of its own copy of the organism's shared genome. So, how does anything get done in a complex body?

What Is Molecular Discourse?

To grow a complex body, the cells of an organism engage in intercellular molecular conversations—a kind of self-organizing intercellular *suasion*—that results in cellular developmental decisions that lead to a variety of cell identities; differentiated structures; and the boundaries among organized tissues, organs, and body parts. In other words, gene regulation in developmental biology can be best described as a molecular *discourse*—a series of structured, communicative, material intra-actions (*sensu* Barad, see more below) among and within cells—that results in the self-organizing, hierarchical becoming of the body. In Donna Haraway's memorable phrase, developmental gene regulation constitutes an "apparatus of bodily production"—a multitude of cellular decisions about what to do, what cell types to become, what boundaries to recognize, how to function, and how to interact with other cells.[10]

Why refer to these intercellular molecular discourses as *suasion*? Suasion, I think, refers appropriately to communicative social influences in general without the assertion of, or perhaps even having a stake in, the *specific* outcome of that influence. In self-organizing into a complex body, tissues, organs, and individual cells pursue their goals of participating in the development of body parts—a pancreas, patella, prostate, etcetera—but they are not dedicated to convincing *specific* cells to do so. It only matters that enough cells do so. (How many is enough? Negative feedback loops will help determine that number by transforming the molecular discourse of recruitment when the appropriate "quorum" of cells has been reached.) By contrast, persuasion implies communication with the goal of exerting social influence toward a *specific* outcome. Persuasion is directed, forceful, and specific, but suasion is suggestive, seductive, and general.

Developmental biology involves the study of how the molecular signals among and within cells give rise to cell identity, hierarchical anatomical structure, and the boundaries and spatial organization among tissues, organs, and modular parts of the body. Although the genome *does* encode information, we can now understand that this genomic information is *not* an instruction manual, a descriptive blueprint, an algorithm, or a developmental program. Rather, like a material lexicon, the genome is a pre-constituted, historically derived set of informational and regulatory resources for the discursive activity of realizing the organism. Like language, the elements of the genetic discourse function in very specific, particular, and highly structured combinations to meet a diversity of ends. Like language, this molecular communication involves flexible and redundant components whose downstream "meanings" and consequences are highly dependent on developmental and physiological context. Finally, like Austinian performative speech acts, gene expression involves converting abstract genomic information into a *material action* in the world, including the creation of enzymes that catalyze specific biochemical reactions and signaling molecules and receptors that function in cell-to-cell communication. Thus, gene expression constitutes a physical, biochemical *doing* by the cell.[11]

Some biologists may bristle at the suggestion of genetic regulation as a molecular discourse, but only a little reflection will show that the science of genetics is already loaded (perhaps larded?) with the vocabulary of language. The hundreds of differentiated cell types in the human body are the result of what biologist's call differential gene *expression*—whether specific genes are "turned on" and making a functional protein product or not. Gene expression involves the *transcription* of gene sequences from DNA into corresponding *messenger* RNA, often involving *editing* out noncoding regions, sometimes alternative *splicing* of these RNA transcripts, and ultimately leading to the *translation* of messenger RNAs into the material chain of amino acid sequences that form a protein. These historic terms all reference the discursive properties of gene expression and regulation.

During the twentieth century, however, linguistic analogies of genome function took, I think, a problematic turn. It became prevalent to describe the genetic "code," "text," "language," or "message" as a "blueprint" of the organism that can be "read" like a "Book of Life."[12] These are all *representational* analogies in which a genomic text describes the essential, intended details of the organismal body—that is, thigh bone connected to the hip bone, etcetera. However, the information in the genome is *not* a representation. Rather, the genome is an informational resource deployed in the *performative* enactment of the individual organism. Gene expression is a material-molecular doing in exactly Austin's linguistic sense. Performative expression is the only way in which genomic information is made real, is materialized in the world.

The "central dogma" in biology that traces the flow of genetic information from DNA to RNA to protein classically overlooks the fact that even after a long chain of amino acids has been linked together, the resulting protein must be folded into its appropriate, complex three-dimensional structure in order to function as an enzyme, a structural protein, or regulatory molecule. This folding process can be envisioned as something like doing origami under water with long, sticky strands of molecular linguini. Although a protein's own amino acids foster the appropriate folding by weakly attracting and repelling each other at particular points in appropriate ways along the molecule (another example of coevolved Baradian intra-actions), most proteins are so large and complicated that there are many possible alternative ways they can stick to themselves, and misfold into nonfunctional or pathological molecular structures. For example, prion diseases, like Creutzfeldt-Jakob disease, are the results of pathological aggregations of misfolded proteins. Consequently, the folding of new protein molecules often requires a special class of *chaperone* proteins that foster the achievement of functional shapes while inhibiting pathological misfolding. Thus, each protein must not only be informatically specified and accurately strung together, but must also be appropriately *performed* in three dimensions through intra-actions with other previous expressed chaperone molecules in order to become a functional part of the phenotype. This is another detailed example of a discourse that produces "the phenomena that it regulates and constrains."[13]

These intercellular signals are the basis for the molecular discourses that are used to enact the material body, and they will play a big role in the discussions in this book. So a general introduction to them will be helpful. The fundamental elements of the intercellular discourse in physiology and development are signaling molecules made, and released, by one cell that bind to specific, corresponding molecular receptors on the surfaces, or inside, of other cells, potentially influencing their gene-expression states. Hormones, or *endocrine* signals, are molecules made by excretory cells within special organs and tissues of the body called glands, released into the blood, and circulated throughout the body (figure 2). A second, parallel system of local, or *paracrine*, signals are molecules released by individual cells that diffuse nearby and bind to receptors of cells in their neighborhood (figure 3). To receive an endocrine or paracrine signal, a cell must express the gene for the corresponding, coevolved receptor that will selectively bind to that specific signaling molecule. The signaling molecules and their receptors have coevolved complementary shapes and binding affinities that attract and engage one another specifically. Their only biological functions are to bind to one another, and to activate or suppress subsequent, downstream genetic regulatory or physiological events, conforming precisely to the definition of a contingent molecular intra-action. Most of these receptors lie across the cell membrane with their receptive/binding surface outside the cell and their signal transmission portions extending into the cytoplasm of the cell. Other receptors, including most steroid hormone receptors, float freely in the cell cytoplasm.

2. Endocrine signaling

Hormone molecules are produced by cells in endocrine glands and released into the blood circulation (*top*). The hormone molecules then travel in the blood throughout the body. A hormone signal can only be received by those cells that are actively expressing the gene for the corresponding receptor that binds that specific hormone (*lower left*). Hormone receptors can be located across the cell membrane (shown) or in the cell cytoplasm (for example, some steroid receptors). Illustration by Rebecca Gelernter.

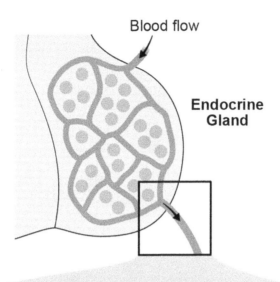
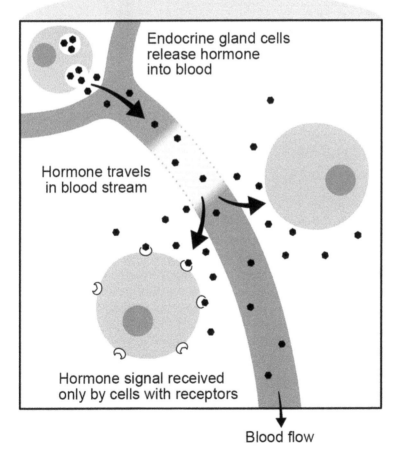

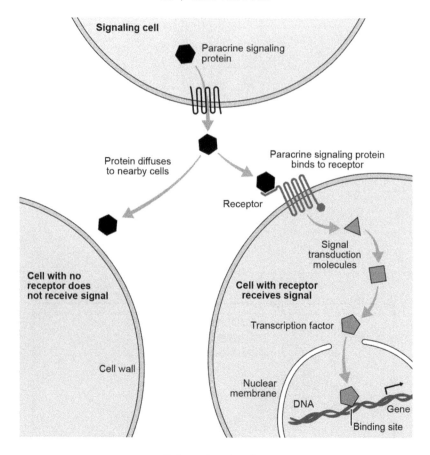

3. Paracrine signaling

Paracrine signaling molecules are released by one cell (*top*) and diffuse locally to nearby cells. Cells with appropriate receptors (*lower right*) will bind the signaling molecule, initiating a cascade of molecular interactions among intermediate signal transduction molecules in the cell cytoplasm, which may lead to the translocation of a transcription factor protein into the nucleus. In the nucleus, the transcription factor may bind to a specific, corresponding binding site on the DNA and act to enhance or suppress the expression of a particular gene. Cells that do not express the corresponding receptor gene (*lower left*) will not receive the paracrine signal. Illustration by Rebecca Gelernter.

Once an extracellular signaling molecule is bound to its receptor, the receptor can initiate a cascade of other molecular signals within the cell. These specific networks of intermediate signaling molecules can lead to activation or inhibition of the expression of specific genes within its genome (figure 3). These complex signaling cascades constitute a molecular decision-making process of the cell, which balances among the multiple influences from the various different signals received from other cells in that cell's environment, and ultimately contributes to differential gene expression, and the establishment of cell type identity of the cell.

Starting from the single-celled zygote, organismal development proceeds from a naive, indeterminant beginning to increasingly specified, specialized, and diverse outcomes. To an individual cell in a complex body, development consists of deciding who you are going to be when you grow up. In cell development, as in human life, you can either listen to your parent(s) (i.e., the progenitor cell that divided and gave rise to you), or you can listen to your friends (i.e., other cells). In some animals, like the roundworm *Caenorhabditis elegans*, developing cells largely "listen" to their parent cells.[14] This highly constrained system has lots of power in facing some material obstacles, but it also has many limitations. In more complex plant and animal bodies, a naive, young cell decides who to be when it grows up based upon molecular "conversations" with other cells through endocrine and paracrine signals.

Want to Go to the Movies on Friday?

To think about how intra-active communications among multiple cellular agencies can lead to the organization of differentiated cellular identities, bodily structures, tissues and organs, and anatomical boundaries, I want to compare endocrine and paracrine signaling systems to what used to happen in my own household in years past with three teenagers as the school week progresses and the weekend approaches. On a Wednesday, one of my teenage boys receives a text from a friend with a proposal to go to a new movie on Friday night. "Awesome!" is the response. Soon, a larger group of other friends are recruited by further texts to the same collaborative project, or developmental fate. By Thursday night, however, after the group has grown larger and other interactions occur during school, more texts go out expressing problems or alternatives. "That movie sucks!" That person "dissed me." Or, "My parents will be out of town, so I am having a party!" By Friday morning, the developing social group has now fissioned, multiple alternative possibilities have been envisioned, additional new individuals have been recruited, and each individual must decide to pursue a specific activity, which precludes other possibilities. In parallel, the group texts have also fragmented into separate, differentiated discussions with fewer or no overlapping individuals. At school on Friday afternoon, the principal gives an announcement over the intercom admonishing all students to summon their best behavior for the upcoming "no-homework" weekend, and sends an identical email to all students. By the time Friday evening arrives, all the parents in all the families are inevitably trying to figure out where the hell their children are.

The text messages among teens are like local, paracrine signals. The principal's announcements and emails act like widely broadcast hormones. The effectiveness of either in influencing teen behavior depends on a balance among all incoming messages, their relationships with other signalers, and whether anyone is even listening (i.e., expressing the gene for the appropriate receptor for that particular molecular signal). Each teen brings their own individual agency to these social interactions, but none is entirely free from suasion, and decisions are made in common with multiple others in likeminded associations or cliques.

Like the texts among teenagers planning their weekend activities, many intercellular paracrine signals can be characterized as excitatory or inhibitory messages. For example, the vertebrate paracrine signaling pathways Sonic Hedgehog and bone morphogenetic protein 2 (BMP2) often interact in concert to create stable morphological patterns in developing vertebrate tissues.[15] (Yes, the Sonic Hedgehog gene and protein were named after the Sega computer game character; another rich example of historic contingency that I discuss further in chapter 5.) Sonic Hedgehog frequently provides an excitatory message that stimulates cells to delay differentiation and maturation and, often, to divide and reproduce. Sonic means, "*Don't worry! Be happy! Reproduce!*" On the other hand, BMP2 typically sends an inhibitory message that favors maturation, commitment, and differentiation. BMP2 means, "*What are you doing with your life? Settle down! Get a job!*" A fascinating discovery from developmental biology—first predicted by mathematical theory developed by the queer English polymath, cryptographer, and computer science pioneer Alan Turing—is that stable anatomical pattern formation often requires the interactions among excitatory and inhibitory signals broadcast over different spatial scales.[16] Thus, stable biological structures with appropriate identities, boundaries, and morphologies arise out of undifferentiated uniformity through the content and spatial structuring of cellular discourses.

Like the fragmenting cliques of text-messaging teens, once a group of naive cells have co-recruited to share a newly established developmental identity, their paracrine signaling pathways also individuate. Individuation occurs by turning off genes for the receptors for paracrine signals deployed by other nearby tissues or structures, and establishing new, private, local signaling pathways that are not shared by other nearby cell types, tissues, or structures. Likewise, some students will be highly receptive to the principal's *inhibitory* messages, which are—as Turing predicted—broadcast *broadly* over the entire school community, while other teens don't care at all. In a similar way, development creates morphological identity through identity-and-boundary-defining individuation in intercellular molecular discourses.

These channels of intercellular communication are best called discourses because their networked structure both includes some and excludes other cells from participation. Furthermore, the "content" of these developmental discourses is not strictly factual information—representations of data about current physical or chemical conditions—but excitatory and inhibitory messages about the alternative, possible, to-be-enacted futures for different groups of cells and cell types. To achieve the diverse outcomes required for the many cell types, tissues, and organs in a complex body, these discourses must be buffered from noise and confusing molecular cross talk that drift or echo from other such discourses in the body, requiring increasing discursive differentiation and individuation. As specific identities are developed, other possibilities are excluded.

Thus, it is not an accident that the issues of identity, difference, boundary, and agency are central concerns to both the humanities and developmental biology. They arise within both disciplines through intra-agential discourses.

How Discourse Becomes Genetic Action within Cells

Between the binding of an extracellular signaling molecule by a receptor and a corresponding change in gene expression in the receiving cell, additional molecular players inside the cell, called *transcription factors*, relay the signal by moving from the cytoplasm into the nucleus, where they can bind to specific, coevolved promoter or enhancer sites on the DNA itself, and initiate or inhibit gene expression. These actions are facilitated by a host of additional intermediate signaling proteins, modifying enzymes, and cofactors in the cytoplasm and the nucleus.

A fundamental mission of developmental biology has been the discovery and description of the intracellular signaling pathways that transmit paracrine and endocrine signals into differential gene expression. These pathways involve a cascade of binding, unbinding, or modification events that are achieved by recruitment of multiple molecular agents, ultimately contributing to a possible change in transcription factor action.

The molecular details of these intracellular signaling pathways are often daunting and baroque in complexity, but these details all reinforce the discursive properties of gene regulation. Molecules that function in inter- and intracellular communication must correspond—biophysically match one another—in order to bind and enact their functions. Hormones and paracrine signaling proteins must precisely align in molecular structure and binding affinity to selectively bind to their receptors and not to others. Likewise, transcription factor proteins and the regulatory binding sites in the DNA must co-correspond in molecular structure for them to bind selectively to the promoter site and affect gene expression. Yet again, we must emphasize that these biophysical correspondences that facilitate binding or enzymatic alteration are contingent on their coevolutionary histories.

Like utterances and their comprehended meanings, these biophysical correspondences are arbitrary—that is, their specificity is unrelated to any other function beyond their correspondence, and does not encode any of the details of their downstream effects—but they are not accidental or random. Rather, they are historically coevolved—having no inherent meaning or objective output *except* in the specific context in which they have evolved to function together toward some stable phenotypic outcome. Similarly, the sounds and meaning of most words used in languages are deeply arbitrary and contingent.[17] The sounds of words only have meaning through the culturally coevolved correspondence between the particular phoneme combinations by a speaker and a listener's cognitive comprehension of those sounds as particular words in grammatically ordered sentences. Likewise, on their own, hormones, paracrine signals, intermediate signaling molecules, and transcription factors don't have any intrinsic developmental or physiological meaning either. They only acquire their biological meanings through their context-dependent intra-actions with other molecules that have coevolved to biophysically correspond with them in order to create the possibility for those actions. Indeed, the complex details of endocrine and paracrine signaling pathways are premiere examples of biological intra-actions—interactions among agents that have no inherent functions except through their coevolved, physical affinity to one another, and their joint functions within a broader genetic regulatory pathway.

The biological "meaning" of endocrine and paracrine signaling systems—that is, their impact on gene expression, cellular identity, differentiation, morphology, physiology, and behavior—are a product of the specific contexts in which they function. A major discovery of developmental biology is that these paracrine signaling modules are actually quite limited in number, and have been evolutionarily co-opted, reutilized, and redeployed again and again in the regulation of the development of *many* different parts of the same body. Like words, the same hormones and signaling molecules can be used in a variety of developmental discourses in different cell types or tissues at different times. They can function in different combinations to regulate different expressive actions.[18]

Thinking of gene regulation as a grammatically structured discourse can actually help us understand the baroque details of gene signaling pathways. For example, transcription factors often bind together, working as cofactors in regulating one gene, and have distinct regulatory functions from other such combinations, just as words can have different meanings in different combinations. As evolutionary biologist Günter Wagner points out, the protein-protein binding interactions between transcriptional cofactors function like the conjunction "and," which further requires the coordinated regulatory expression of two gene inputs to specify a single regulatory output.[19]

Furthermore, biological signaling molecules can be physically modified in tiny ways that dramatically influence their gene regulatory functions. For example, some transcription factors can be phosphorylated or acetylated—the addition of a phosphoryl ($-PO_3$) or acetyl ($-COCH_3$) group—at specific locations on the protein, leading to striking differences in their binding affinities and subsequent intra-active functions. Just as a small change in a noun can, depending on the language, make it plural, possessive, the subject of a verb, or a direct or indirect object of a verb, greatly influencing its grammatical function in a sentence, these tiny molecular modifications of signaling molecules work like grammatical declensions that affect the potential biophysical correspondences between that molecule and other coevolved receptors or binding sites in the gene regulatory system, creating new discursive possibilities, while excluding others that use the same syntactic elements.

The purpose of all this grammar-like molecular complexity of gene-expression regulation is not to materially represent an essential genomic plan for, or prior fact about, the organism, but to regulate the discourse through which the organism enacts itself. In this way, gene regulation is best understood as a performative action—a material *doing* in the world.

The Choreography of Gene Expression

Differential gene expression and the development of cell type identity is made possible by another highly choreographed physical phenomenon that takes place on the scale of the entire nuclear genome. The genomes of all organisms with nucleated cells (called eukaryotes) are subdivided into chromosomes. To pack the entire two linear meters of genome into a nucleus that is only five microns in diameter (that's five-millionths of a meter, or roughly two-ten-thousandths of an inch), the DNA must be tightly coiled around groups of globular proteins called histones, and coiled again into a highly structured clusters of coils, like a massively tangled telephone cord (figure 1). The tightly coiled state of the DNA prevents access to these regions of DNA, and makes their genes untranscribable. As cells mature, differentiate, and develop more specific identities, they must relax the DNA coils in certain areas across the genome to make it possible to express those genes that are necessary to achieve that cell's identity and functions. Certain coils of DNA on various chromosomes can be covered with an additional layer of dense protein, called heterochromatin, to further prevent them being transcribed and expressed. Because the same transcription factors can be used to regulate the expression of many different genes in different cell types at various stages of its development toward a mature state, the cell must also keep other sections of the genome tightly coiled to prevent the transcription factors from binding to the promoter or repressor sites for other genes that can bind it. So, the cell must regulate different context-dependent possibilities by physically and hierarchically exposing certain opportunities for binding to the DNA rather than others. The specific physical state of the genome's unwrapping is a necessary, choreographic condition for becoming a mature cell of a specific cell type.

The hundreds of cell types in the body and other physiological states of cells are each characterized by a distinct physical pattern of local and regional uncoiling, or relaxation, across the genome. This selective unfolding of parts of the genome is managed by chemically modifying particular sites in the scaffold of the DNA. Specifically, the phosphorylation, acetylation, or methylation—the addition of a methyl group ($-CH_3$)—of specific sites on the histones can affect whether a particular stretch of DNA remains tightly coiled, or opens and becomes available for gene expression. These complex unfolding mechanisms are managed by specific *pioneer transcription factors* that can bind to certain sites in the DNA when it is still completely coiled up and modify it, so that it locally unwinds. These kinds of transcription factors are called "pioneers" because they can open up entire regions of DNA, and establish the possibility of distinct, new sets of gene-expression opportunities.

The *genetic* performance of cell identity is facilitated by these complexly choreographed actions at the level of the chromosomes—a hierarchical unfolding of physical states facilitated by the expression of regulatory molecules. Specific genetic expression states *physically* preclude others because not all regions of the genome can be simultaneously accessible for gene expression. The cell cannot physically express all the appropriate genes for being, say, a skin cell, a neuron, a muscle cell, and blood cell at the same time. Cells mature toward specific cell identities by physically excluding other possibilities.

During development and throughout an organism's life, a few select cells must retain undifferentiated states, delaying commitment to a final, mature cell identity. Biologists call these cells *stem cells* because of their capacity to divide and produce descendent cells that will become new tissues in an embryo, or replace cells in existing tissues in the organs of mature bodies. Stem cells must retain a "naive" and uncommitted state, delaying any definitive performance of their own so that generations of their progeny may do so. The arrest by stem cells of any definitive self-performance requires the maintenance of a specific physical state of the genome. To retain this naive state, many stem cells must be physically sequestered in spatial isolation from the noisy, molecular hubbub of endocrine and paracrine signals in the "real" world of the body. Called a stem cell niche, these isolated sites function like a monastery where the stem cells can be cloistered, and maintained with special, limited molecular discourse just for them. For example, our blood stem cells are housed deep in the physically isolated crypts in the channels of marrow of our long bones. Hair stem cells, by contrast, are kept in a tiny pocket on the side of each hair follicle (where they wait eagerly to replace any hair that molts, or is plucked out). To replenish the cells of its many tissues, our bodies must maintain these special stem cell lines waiting in the wings to produce the future actors on the organismal main stage.

The Performative Phenotypic Landscape

More than seventy years ago, developmental biologist Conrad Waddington posed the famous analogy for organismal development as a ball rolling down a hilly landscape with multiple valleys (figure 4A). Although each ball takes a unique path down the hill in response to its individualized genetic makeup and specific environmental perturbations, all balls tend to end up near one of a few possible end points at the bottom of one valley or another. Waddington described this surface as an *epigenetic* landscape—a field of limited possibilities created by the regulatory interactions between the cell cytoplasm and the genes themselves. (In *Epigenetic Landscapes: Drawings as Metaphor*, Susan Squier presents an illuminating and insightful history and analysis of Waddington's iconic diagram.[20])

Waddington coined the term *canalization* to describe the tendency of organisms to develop similar morphologies—that is, that they end up within particular tracks or "valleys" in the developmental landscape *despite* genetic and environmental variation. Canalization describes the property that many genetic variations or environmental perturbations have limited impact on the developmental path to repeatable, robust, stable outcomes. The specific pathways down the valley may vary, but the endpoints at the bottom of the valley end up being quite similar. However, perturbations immediately before a bifurcation between alternative valleys, or developmental watersheds, can result in a very distinct phenotypic outcome—namely, entry into a whole different valley. Likewise, limited canalization—that is, flat, broad valley bottoms—will result in wider variability of alternate developmental outcomes and more variation in phenotype (figure 4B). Waddington's epigenetic landscape thus provided a powerful analogy for the *robustness* of the phenotype, meaning its resistance to developmental disturbance in the face of environmental or genetic perturbations; its evolvability, or its capacity for evolution of new and different outcomes; and its creativity, the idea that while branching points in development eliminate possibilities, they also open up new downstream alternatives.

At the time that Waddington conceived of the epigenetic landscape, the structure and function of DNA were still unknown. Consequently, Waddington envisioned differential gene expression as the result of biochemical processes in the cytoplasm, rather than genetic processes in the nucleus. Waddington's view of the cytoplasm as having a distinct agency from the genome in gene regulation was an early, inchoate reference to all the gene regulatory mechanisms we have been exploring: cascades of paracrine and

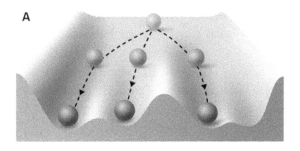
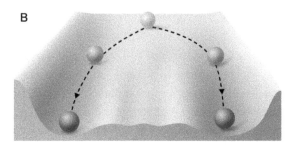

4. Developmental landscapes

A, Waddington's "epigenetic landscape" featured a complex, hilly topography on which development proceeds toward a few, stable developmental outcomes, like a ball rolling downhill. Canalization refers to the restricted impact of genetic or environmental variation on morphology. Redrawn from Waddington (1940). B, Weak canalization can result in a developmental landscape in which morphological outcomes are less well defined with multiple stable, variational possibilities in between them. C, A revised, performative developmental landscape with organisms climbing up toward peaks in complexity. This performative landscape recognizes growth, gene regulation, differentiation, and morphogenesis as active processes resulting from individual organismal agency. Illustration by Rebecca Gelernter.

endocrine signals; receptors; transcription factors and cofactors; the cytoplasmic mechanisms of acetylation, methylation, and phosphorylation enzymes; and so on, which were largely unknown at his time.[21] Thinking in strictly biochemical and enzymatic terms, Waddington envisioned the cytoplasm's role in development as a kinetic process involving a loss of potential energy leading to a stable ground state, which he expressed with the analogy of a ball rolling downhill under the force of gravity.

Here, I propose a reconceptualization of Waddington's epigenetic landscape as a *performative developmental landscape* that better captures the active agencies of the developmental process leading to the breadth of achievable but canalized possibilities. Viewed performatively, development is not an inert outcome of mass and gravity plus noise; rather, it is an active *doing* by an open system of hierarchically nested agencies to create an embodied self—an inherently more complex, more differentiated, more organized state—through the production or consumption of energy from the environment and the proliferation and differentiation of cells. Instead of merely rolling passively downhill, the individual phenotype builds itself actively by *ascending* the developmental landscape toward an achievable and sustainable phenotype peak (figure 4C). Developmental canalization arises from the constraints on the pathways available to ascend toward particular peaks. Developmental decisions over which path to take are made actively through the discursive developmental signaling and cell/tissue/organ individuation. Multimodal variation among phenotypes is a consequence of the availability of multiple possible stable phenotypic peaks resulting from different developmental trajectories. Alternative developmental "tracks" are not necessarily always strictly bifurcating, but may share multiple potential crossover points along the way.

As innovative and creative as it was, Waddington's concept of the epigenetic landscape employed a reductionist view of the organism as passively shaped by external forces acting upon it. A performative inversion of Waddington's epigenetic landscape better captures the dynamics of the materialization as an active agential becoming through molecular genetic regulation.[22] We will revisit Waddington's landscape again in thinking about the sexual phenotype in chapter 5.

Canonical versus Performative Pathways

In developmental biology, the descriptions of the roles of specific signaling pathways and molecules in development are typically communicated through narratives and diagrams that describe a clean, if complicated, causal chain of events—like toppling molecular dominoes. These standardized molecular-developmental stories are known as "canonical" pathways. Each discovery of another unexpected feature of these biological Rube Goldberg machines has been celebrated as an important new discovery, justifying a series of new experimental papers and the publication of newly revised review papers. However, problems arose when it was discovered that paracrine signals, transcription factors, and other intracellular intermediate molecules invariably have multiple possible functions resulting in bifurcating, alternative pathways. The "canonical" pathways are named to distinguish them from these many alternative modes of regulatory action by the same paracrine signals involving different intra-active cascades of downstream receptors, intermediate molecules, DNA-binding sites, and potential developmental actions.

In short, developmental biology has its own "canon." However, these hardwired, canonical accounts of gene action in developmental biology are always incomplete and inevitably wrong. And I don't mean that they simply do not include enough of the relevant details. Literally no written or graphical account could do this. Like trying to write a definitive description of how a short story works, or a definitive synopsis of a performance of *King Lear*, canonical accounts of gene regulation in developmental biology are useful but inevitably fail to capture the nature of the process they are describing. Each performance of *Lear* draws upon, emphasizes, and realizes a different aspect of the underlying text to create a performance that defies easy synopsis.

Canonical accounts of developmental mechanisms, on the other hand, mis-frame the goals of developmental biology around the wrong kind of answers—answers that are comfortably situated in the world of controlled, experimentally inferred causality but cannot appropriately account for the actual role of genetic and environmental variation in the development of non-model organisms in natural populations. Because each step in a gene regulatory cascade is contingent upon the coevolved biophysical intra-actions with other molecular agents, the canonical story can never capture reality. The cellular outcomes to express or suppress this or that gene are themselves products of multiple, often antagonistic, signals that the cell receives, making the isolation of any one "causal" pathway entirely artificial. Of course, geneticists and developmental biologists know this.

However, adopting a performative framework provides an intellectually productive alternative. This does not mean abandoning the experimental investigation and description of developmental and physiological mechanisms. Experiments *are* powerful tools for the investigation of developmental mechanisms, but these experimental results do not simply add up to an understanding, or a portrait, of the phenotype because each of these individual experiments requires eliminating (i.e., "controlling for") the myriad of other individual genetic and environmental variations that have critical roles in the realization of every phenotype. Thus, we need to think about experimental results in a manner that exposes, and contributes to, the reality of the performativity of gene action rather than exclude it from conceptual and empirical visibility.

How Does the Body Regulate Growth over Space?

Development cannot be explained solely in terms of differential gene expression. Organismal development requires actions, and the coordination of actions, over spatial scales and dimensions that cannot be described by genetics alone. In the next two sections, I will discuss two such additional dimensions—space and physical forces.

To be a functioning organism, it is not sufficient for the cells of the body to develop the distinct identities of muscle, skin, neuron, leaf, root, or flower petal. Beyond their identities and gene-expression states, the cells of the body must also be spatially organized to create functional, multicellular structures, organs, tissues with layers, and distinct yet functionally integrated parts of the organism. The development of coherent spatial organization of the body involves hierarchical assembly over *space*. This space is created by the proliferation of a volume of differentiating cells—by the materialization of the body itself.

Anatomical pattern formation takes place over spaces that are often characterized by gradients in paracrine signaling, which are called *developmental fields*. Because paracrine signals diffuse away from signaling cells in the intercellular space, their effects can organize axes and regions of the body during development. For example, the development of the vertebrate limb is characterized by multiple, orthogonal (i.e., at right angles) spatial axes—from base (proximal) to tip (apical), from top (dorsal) to bottom (ventral) surfaces, and from inner (medial) to outer (lateral) surfaces. Decades of molecular-developmental biology research have established that development and differentiation of the presumptive limb bud along these morphological axes is achieved through intra-actions between the cells and tissues of the limb bud and spatial gradients of paracrine signals produced by groups or aggregations of cells, referred to as signaling centers. The nascent limb bud begins growth as a small protuberance with a group of cells at its tip producing the paracrine signaling protein fibroblast growth factor (FGF). This signaling center, called the apical ectodermal ridge, encourages proliferation of the mesenchymal cells below, fostering the outward growth of the limb. Later in development, digit identity—pinky versus ring finger, etcetera—is established within the paddle-shaped tip of the limb bud in response to a gradient of Sonic Hedgehog (SHH) signals produced by a band of cells on the lateral, or outer, margin of the tip of the limb bud—a signaling center called the zone of polarizing activity (ZPA). Digit primordia cells that receive stronger Sonic signals establish pinky finger or pinky toe identities; those receiving the weakest Sonic signals establish thumb or big toe identities. Lastly, the third, dorsal-ventral axis—which runs between the palms and backs of your hands—is established by WNT7A signaling from cells on the dorsal surface, and BMP signaling by cells on the ventral surface.[23] These intra-acting signal gradients create spatially unique molecular coordinates that organize the development of the complex structures characteristic of the limb. Cells at every position within the limb receive a distinct combination of molecular signals. Of course, a reductionist model could describe this process as a code representing an adaptive developmental program. But this view obfuscates the fact that it is the cells of the limb bud that are spatially and anatomically *self-organizing* to produce a functioning limb. Unlike the mathematical axes on a graph, or the frame of reference and scale on a blueprint, the molecular-developmental axes of the limb bud are created by the cells of the limb bud itself. They are self-organizing their own spatio-material becoming into a functioning limb.

Signaling centers, like the ZPA, are examples of the ephemeral, or transient, agency of groups of self-coordinated cells to pursue local or regional effects in morphogenesis. Like a volunteer, pop-up information kiosk at a once-a-year street festival, the cells of these signaling centers use molecular signals to coordinate developmental events in space, but once their job is done, these cells lose their spatial coordination function, and resume maturation to become normal functioning cells of the body.

Developmental fields and signaling centers are biologically expedient means for the emergence of localized or regional, agencies that are components of the performative apparatus of bodily production conceived of by Haraway and Barad. This observation provides further evidence of the intellectual roots of material feminisms in developmental biology; Donna Haraway's first book explored the intellectual history of developmental fields and other spatial metaphors in pre-DNA embryology.[24]

Another spatially emergent agency in organismal development is the placode. A placode is a molecular signaling center that establishes that a specific developmental event will take place *here*. For example, integumentary appendages like hair, glands, scales, teeth, and feathers, as well as many neural structures begin development with the establishment of a placode that specifies the place within a developmental field of undifferentiated tissue that will develop into a distinct structure. Placodes function as the organizing center for subsequent morphological differentiation, and typically inhibit the development of other placodes in their immediate vicinity, resulting in distributed spatial patterning. This process gives rise to the orderly spatial distribution of teeth in our jaws, hairs on our scalp, and sweat glands over the surface of our skin.[25]

What Is the Role of Physical Forces in Development?

Physical forces created by the cells, tissues, and organs of a body can themselves participate in the performative development of the phenotype. Living cells and tissues constitute a form of material that physicists call *soft condensed matter*—a phase state of matter between a solid and a liquid in structure and physical properties. Soft matter is basically squishy stuff. Soft matter can produce a variety of surface and internal forces in response to various physical changes including increases in volume, drying, hydrating, or osmotic changes in chemical concentration. Think of the cracks formed in drying mud, or the buckling skin of your fingertips in a hot bath. During development, tissues can develop in ways that utilize these aggregate physical forces to produce certain morphologies. In other words, tissues and bodies have evolved to deploy the physics of squishy soft matter to build themselves.[26]

An example of the biological development that employs soft condensed matter physics comes from my own work on the intracellular optical nanostructures that produce the blue and green structural colors in the feather barbs of many bird plumages, including bluebirds (*Sialia*), jays (Corvidae), parrots (Psittaciformes), and cotingas (Cotingidae) (see the prologue). Like the bubbles in champagne or the swirls in a cooling bowl of miso soup, the spongy matrix of β-keratin and air in the medullary cells of the feather barbs are created by arrested *phase separation*, or unmixing, of the β-keratin protein polymer from the liquid cytoplasm barb medullary cells. Thus, the feather cells of blue birds have evolved to create the appropriate physical-chemical conditions for the nanoscale, soft matter, pattern-generating physical mechanisms of molecular unmixing to create their brilliantly colorful nanostructures. Because most of these colors are used in social and sexual communication within the species, these colorful optical nanostructures function through display to other conspecifics.[27]

In the last twenty years, developmental biologists have clearly established that mechanical forces can be crucial to the regulation of gene expression in multicellular animals and plants. Layers or aggregations of interconnected cells can exert mechanical forces on one another as they grow in size or change in shape. These mechanical forces can be sensed by cells, and influence their gene regulatory states. For example, in all animals descended from the most recent common ancestor of you and an earthworm, the development of mesoderm—which becomes our muscles, heart, blood cells, kidneys, gonads, etcetera—is induced by the *mechanical stress* caused by cell migration or invagination during the formation of the gastrula—the early embryonic stage characterized by a hollow ball of about two hundred to three hundred cells. Even though the mechanism of gastrulation have diversified dramatically over five hundred million years since its origin, the mechanosensitive mechanism of mesoderm specification has been retained.[28] Similarly, in vascular plants, the leaves, shoots, branches, and roots all develop from proliferative organs called meristems. As plants branch, the identity of new, lateral meristems to the sides of a main shoot apex are initiated by the distinct, conflicting stresses imposed on cells in the saddle-shaped regions formed at the crux between the main shoot and the side branches.[29] Only those cells that can sense these conflicting saddle-like mechanical stresses can become new plant branching points.

The capacity of cells to alter gene regulation in response to mechanical forces during development might seem, at first, to be a simple observation that organisms can respond dynamically to their developmental environments. However, on closer scrutiny, we realize again that these cells have evolved to respond to the mechanical forces exerted by the growing cells and tissues of the organism *itself*. These forces are not exerted *on* the organisms by their environments, but arise *from* the material consequences of the organisms' own becoming.

5. African Grey Parrot (*Psittacus erithacus*)

Unlike most birds, parrots (Psittaciformes) have zygodactyl feet with toes one and four opposing toes two and three. The reversal of the position of the fourth (outer) toe results from toe muscle contractions acting on the ankle cartilages during incubation in the egg. Juniors Bildarchiv GmbH / Alamy Stock Photo.

An example of the large-scale mechanical forces in development comes from the unusual configuration of the toes of parrots (Psittaciformes). Most bird feet have the first digit—the big toe—reversed to oppose digits two, three, and four, a condition called anisodactyly. However, parrots, cuckoos, woodpeckers, toucans, and their relatives have feet with both digits one and four reversed to oppose digits two and three, a condition called zygodactyly (figure 5). From the phylogeny and fossil record of birds, we know that anisodactyly is the original, ancestral foot type, and that zygodactyly is evolutionary derived from it. In a fascinating series of observations and experiments, evolutionary developmental biologist João Francisco Botelho and colleagues have shown that the development of the zygodactyl foot in parrots is mediated by *physical forces* of leg muscles *tugging upon* the developing cartilages of the fourth toe in the egg, gradually yanking it backward into an opposing position before its cells become completely ossified. Parrot embryos whose legs are treated experimentally with a muscle paralyzing chemical develop foot bones with the ancestral anisodactyl shape.[30] In this case, physical forces created by embryonic muscle and connective tissues actually shape the organism's foot bones. Thus, the performance of the phenotype may involve physical *work* produced by the muscles, nerves, and nervous system of the developing embryo itself.[31]

An exciting new area of research in developmental biology is the investigation of bioelectric fields. This developmental mechanism is possible because all living cells have an electrically charged state. The lipid bilayer membrane of cells requires that they each manage the concentrations of charged ions by actively pumping them across the membrane using protein ion channels in the membrane. The charged state, or membrane potential, of individual cells varies extensively both within and among cells. Although we are generally familiar with the uses of membrane potential in the functions of excitable tissues like muscles and neurons, an increasing breadth of data implicate electrical states of cells and tissues in developmental morphogenesis and pattern formation. Bioelectric states of a cell can be coordinated or conducted between cells through gap junctions or nanotubes—neighbor-to-neighbor, or long-distance connections—between the cytoplasms of different cells. When layers or aggregations of cells become electrochemically coupled, the local charge state can provide a morphogenetic signal to regulate developmental genes, creating electric developmental fields and signaling centers at the levels of tissues, organs, structures, or even, for smaller organisms, the whole body.[32]

The mechanisms by which coupled electrical states of cells and tissues regulate gene expression during development are less well understood, but there is abundant evidence that cells are sensitive to electrical fields, and that variation in them influences gene expression and development. Numerous developmental variations in vertebrate body structure—limbs, hair, fins, etcetera—are known to be associated with mutations causing gain or loss of function in genes that involve voltage-regulating ion channels and gap junctions. Experimental manipulation of potassium ion channel genes can disrupt the development of frog eyes, or even produce the development of "ectopic" eyes in other places on the larval body—such as near the end of the tail. Early development of the vertebrate limbs and fins is marked by regional shifts in electrical currents, and the creation of a specific limb-forming electrical field. In other words, another dimension of the spatial signaling in the limb bud described above are the coupled electrical potentials of different tissues over space.[33]

Regulation of bioelectrical cues in development is also associated with adaptative radiation. For example, the flying fishes (Exocoetidae) have evolved very long fins that allow them to glide out and over the surface of the ocean for tens of meters in order to escape fish predators. Genomic analyses and experiments indicate that the elongation of fins in flying fishes is associated with rapid evolution of several genes involved in ion channels and amino acid transport. Gain-of-function mutations in these same genes in zebrafish—a laboratory model organism in developmental biology—produces very similar changes in fin size and proportions to those of flying fish.[34] Although the precise molecular signaling pathways through which the potential of electrically coupled cells influences cell growth, differentiation, and morphogenesis are not yet known, the role of electrical fields created by charge-coupled arrays of cells in animal development is, at this point, well established.

Some biologists may respond that these physical properties and forces—gradients over area or volume or mechanical or electrical forces—are themselves merely consequence of gene expression, and therefore nothing new or distinct. But this view contributes nothing more to the explanation of the phenotype than pointing out that all organisms are composed of organic molecules contributes to genetics. Genes themselves cannot create spatial gradients, mechanical forces, or electrical potentials. The role of the physical material phenomena realized by cells, tissues, structures, and organs cannot be captured by the mere description of genes, or gene expression. Because they are irreducible to genetics, these spatial and physics phenomena require us to recognize the agencies of cells and aggregations of cells, their physical dimensions, and their mechanical and electrical properties in the generation of the complex phenotypes.

Performativity of Cellular Discourse

Inspired by the complexity of biological pattern formation and the capacity of bacteria to evolve antibiotic resistance, feminist anthropologist Vicki Kirby has asked whether our "understanding of language and discourse may extend to the workings of biological codes and their apparent intelligence."[35] Likewise, Karen Barad's concepts of material-discursive practices and apparatuses of bodily production, as we've discussed, explicitly extend the concept of discourse beyond the human. Yet, these conceptual forays have not yet been applied explicitly to gene regulation and developmental biology.

Paracrine signals, transcription factors, the host of intermediate pathway molecules, modulators, cofactors, and binding sites are *not* assertions of, references to, or constative descriptions of fact—representations of the physical conditions or physiological states—but enactments of a cell's intention to influence the gene-expression states of other cells. Expression of paracrine and hormonal receptor genes are manifestations of a cell's capacity to receive, and attend to, specific signals from other cells. The consequences of the signaling intention of one cell are contingent upon the developmental contexts and agencies of other cells. The signaling cell cannot control whether other cells are indeed "listening" to its performative utterances because expressing the appropriate receptor to receive that signal is ultimately up to the other cells themselves. This is why cells have individual agency. Genetic mechanisms of regulating multicellular development comprise a molecular discourse—a form of social communication and exchange among agents. Furthermore, the expression of any specific gene functioning in that discursive signaling pathway is not, by itself, an isolated cause of developmental differentiation.

This discursive view of organismal development provides a new perspective on the biology of complex bodies. Like Austin's performative speech acts, the expression or reception of paracrine signals cannot induce further actions or events in the material world without appropriately corresponding intra-actions from other agents. To extend Austin's example, one is not truly married if one's potential spouse does not hear your vow and also respond, "I do" at the appropriate place and time. It is erroneous, then, to conclude that a specific regulatory gene is "necessary and sufficient" to cause the development of a phenotypic trait—a common intellectual criteria in molecular-developmental biology—when the actions of any such genes rely upon complex cascades of other molecular agents, each one of which shares causal parity, or has an equivalent stake, in the cascade of events that initiates a change in gene expression.[36]

Some physiological communication events *can* be representational. *Salty*, *blue*, *sharp*, *hot*, *soft*, and *fetid* are all adjectives we use to describe certain representational messages initiated by sensory neurons and processed in our brains. Although we don't have a similarly colorful vocabulary for them, our brains and organs are also receiving similarly representational messages about blood pressure, glucose levels, water balance, carbon dioxide concentration, etcetera that are important in homeostasis. However, the hormonal, nervous, or other chemical communication systems that enact a homeostatic response to representational messages are themselves biological *doings*—enactments of physiologically compensatory responses—and therefore performative.

The role of chemical communication in development was first discovered in the 1920s through a series of experimental tissues transplants between frog and salamander embryos by Hilde Mangold and Hans Spemann. Mangold and Spemann discovered that a specific cluster of cells—later known as the Spemann Organizer—could induce the development of the nervous system. However, in later decades, the process of *induction*— that is, one tissue *inducing* a developmental identity change in another adjacent tissue—was later found to be contingent upon the capacity of the developing tissue to respond to the signals of induction, which Conrad Waddington called *competence*. We now understand that induction and the competence to respond to induction correspond, respectively, to the production and release of paracrine signals by one tissue, and the expression of genes for the receptors and other downstream signaling molecules by cells in another tissue. These and other milestones in the history of experimental developmental biology involved the gradual exposure of the discursive properties of organismal development.

Are Genes Causes?

A performative perspective on gene regulation has profound implications for our understanding of genetic causality in development. Mutations in some genes for structural components of the cell can, in some instances, be viewed as simple direct genetic causes when they encode actual material components of the phenotype—the bricks and mortar of cells. For example, loss-of-function mutations in single genes can constitute the causes of sickle-cell anemia, cystic fibrosis, hemophilia, and variations in color perception. It is interesting to note, however, that these rare "single-gene diseases" all involve physical interactions between biological proteins and inorganic molecules or physical energy, which are themselves *incapable* of coevolving and therefore incapable of actively partnering in biological intra-action. Hemoglobin binds O_2 and CO_2. Cystic fibrosis involves chloride transport channels. Visual pigments absorb photons of particular wavelengths. Thus, the explicitly inorganic/physical functions of these proteins constrain the role of discursive coevolution in these structures, and create the unusual opportunity for single-gene disease.[37]

In contrast, however, most genes do not function simply in the physical material of the body, but in the body's own internal physiological and genetic regulation. These genes function only within the complex, coevolved, intra-active molecular signaling networks. In development, whether, or how, a cell receiving a specific endocrine or paracrine signal responds to that signal will depend upon a detailed hierarchy of other conditions and features of that cell, including the simultaneous influences of *other* signaling pathways that may either enhance or inhibit each other. (Think of the mixed messages teenagers receive from various friends, parents, and authority figures.) Proteins that provide the physical materials of the cell and body—bricks and mortar—can also be intra-active if they collaborate with other biological molecules in their realization: for example, biochemical intra-actions among collagen proteins, fibronectins, and proteoglycans to form the extracellular matrix.

As Barad explains, "The notion of intra-actions constitutes a reworking of the traditional notion of causality."[38] Causality usually involves attribution of a difference in effect between distinct entities whose properties and functions are generalizable beyond the observed events in question. For example, a diamond can cause a mark on glass, and diamond or glass can cause a mark on wood. But neither glass nor wood can cause a mark on a diamond. We can make inferences about marks caused by the interactions of these materials because the relevant functional property of diamond, glass, and wood—hardness—is definable and measurable in each material *independently* of the others. But intra-actions—like biological paracrine and endocrine signals—are interactions among entities that have no distinct, independent functional properties *except through* their joint enactions. Paracrine signaling molecules, transcription factors, and intermediate pathway molecules cannot be said to have any function independent of their capacity to bind selectively to the appropriately corresponding receptors, cofactors, or promoter/enhancer sites, and consequently to influence cell physiology or gene-expression states. Thus, genes, molecules, and DNA-binding sites that are involved in paracrine and endocrine signaling have evolved specifically to intra-act with one another in these specific ways. They cannot be independent, separable causes of the gene regulatory or physiological effects that they jointly enact.[39]

This challenge extends to the heart of experimental methods in genetics and developmental biology, and to the use of genetic and environmental controls on the variation of other genes and molecular agents within the cell and the physical environment. Accounts of gene action are typically established through experimental research on inbred strains of model organisms—like the roundworm *C. elegans*, *Drosophila* fruit flies, zebrafish, or mice—that have been specifically created to eliminate all variation other than the experimental manipulation to be made by the investigator. Inbred strains eliminate genetic variation both *between* and *within* individuals (i.e., so that each of the two copies of every gene and regulatory region are identical within each individual). Developmental genetic experiments control out of existence all of the innumerable genetic variations that are found in virtually any wild population of organisms. Indeed, genetic experimental manipulations in homogeneous inbred strains create the a priori causal status to the genes being perturbed, giving a false impression of independent causality.[40] By creating inbred strains, biologists have invented artificial *ceteris paribus* worlds in which all other things are literally equal. The results of these powerful interventions have provided startlingly detailed views into many pathways of gene regulation.

The problem, however, is that the communication of genetic research results often proceeds as if confusing these controlled experimental methods for the world itself, which is, unlike these experimental systems, inherently and overwhelmingly variable. As Barad has shown in physics, material-discursive apparatuses do not allow for the distinction between the scientific intervention and the phenomenon observed. The details of the optical apparatus and experiment *create* the phenomenon of light being *either* a wave or a particle. Similarly, the inbreeding and environmental controls in genetic experiments in developmental biology constitute material-discursive apparatuses, and their deployment *creates* the phenomenon of strict gene causation. Prior to the specific experimental controls and interventions, the causal independence of the gene being manipulated does not exist. The "measurement" of a genetic effect requires controlling the inherently variable natural systems in precisely the ways that are required to create the opportunity for an observed effect and the ability to assign a cause.[41] These cause-and-effect relationships are not independent features of the world itself, but designed opportunities created by the experimental apparatus.

The recalibration of genetic causality that is required to bring developmental biology in line with the reality of the developmental processes of real organisms does not constitute an abandonment of causality itself, but a recognition that the cause of the organismal phenotype is the entire process of organismal becoming, its genetic actions, its hierarchy of multiple agencies, its environmental contexts and interactions, its random perturbations, its individuality, and (in some organisms) its cultural context—in short, the organism's enactment of itself. The long tradition of experimentally dissecting developmental process with controlled interventions has been highly productive (as we will see in subsequent chapters), but this progress cannot change the fundamental fact that these tools only work at describing the world by altering it, flattening its complexity, and eliminating its rich individualized differences. As a result, the sum of all these reductive interventions fails to explain the process of individual organismal development.

Agency in Developmental Biology

If "performativity implies agency," what agencies are being enacted in the development and physiology of the body?[42] The development of the body is a manifestation of the agency of the embryo pursuing its own embodiment. The body is a phenomenon that takes part in its own realization, its own becoming. But the organism is not the only active locus of agency in this process. Just like people in a performative assembly, as Butler has theorized—with their embodied, plural action in political speech through mass aggregation—individual organisms are simultaneously both performative agents and assemblies of agencies. As Barad writes, "The space of agency is not only substantially larger than Butler's performative account . . . but also, perhaps surprisingly, larger than what liberal humanists propose."[43] Agency, therefore, is not limited to conscious mental choices of humans.

Barad describes agency as "an enactment, not as something that someone or something has." To Barad, agency is intra-action "reconfiguring the material-discursive apparatuses of bodily production."[44] Drawn from an analysis of quantum mechanics and optics, Barad's view provides a precise template for the description of gene regulation, tissue differentiation, bodily morphogenesis, and physiology, which constitute a material-discursive apparatus of the body's own material becoming.

The hierarchy of biological agencies implies that many biomolecules, cells, tissues, and body parts have an evolved capacity to *intra-act*, and thereby *enact* events in the development of the organismal self. Although genes, cells, and anatomical parts do not have conscious intention, they have been shaped by the history of selection for specific organismal purposes. (For more detail, see appendix 7.) In the absence of a disease state like cancer, the purposes of these agents are shared with the goals of the whole organism itself. Biological bodies are the product of a hierarchical assembly of distributed, coordinated, and self-regulated agencies. These biological intra-actions are distinct from simple actions because they are contingent upon other coevolved agents. Thus, the actions of insulin or Sonic Hedgehog are not direct consequences of their mass or ionic charge (although changes in these properties may affect their functions). Rather, their actions are a consequence of their distinctive capacities to bind to specific, coevolved receptors, and affect other physiological or gene-expression events.

Performativity is the means by which the organism's evolved agency is enacted. Biological performativity is achieved by the rampantly hierarchical assemblage of agencies. The performative continuum extends *down* from the agency of the individual organism *into* the body to encompass the agency of genes, signaling proteins, intermediate molecules, DNA-binding sites, hormones, enzymes, microRNAs, cells, tissues, signaling centers, developmental fields, organs, and body parts that participate in the creation of a complex yet unified and well-regulated phenotype of the organism. This nested hierarchy also extends *above* the level of individual organismal bodies in certain species, such as eusocial ants, bees, and termites, colonial jellyfish, naked mole rats, and humans, in which social groups have agency, heritability, and the capacity to evolve biologically or even culturally.[45]

The recognition of the diverse agencies of genes, biomolecules, cells, anatomical parts, organs, and organisms all arise from their individuality—that is, from their status as undefinable, but real, historical entities.[46] Because each of these examples of biological agency is descended from prior individuals who survived, thrived, and reproduced, the forms, functions, and intra-actions of these genes, cells, body parts, etcetera have been shaped by their histories of natural and sexual selection. It is this history of selection and the consequent derived structures and functions that give these biological entities their agency—their contingent capacity for action.

These performative biological agencies are entirely compatible with Richard Dawkins's famous assertion of the agency of "the selfish gene."[47] Dawkins accurately conceives of the agency of genes as emerging from the history of natural and sexual selection acting upon them. But Dawkins's paradigm fails as a broader explanatory framework for biology because he conceives of the adaptive agency of genes as an all-powerful, reductive, and controlling force—the *singular* biological agency that determines all the subsequent structure, physiology, and behavior of organisms. It is this strong version of gene-level selectionism that is so problematic. Data on the developmental agency of signaling molecules, cells, tissues, and developmental fields themselves—almost entirely amassed in the decades since the publication of *The Selfish Gene* in 1979—argue overwhelmingly for a rejection of the simplistic, and ultimately naive, reductionism of Dawkins's gene-level adaptationism. There is no way to account for the evolution of differential gene expression among the trillions of genetically identical cells within the human body solely through competition *among* alleles at single genes. (For further discussion of the gene-level selection see appendixes 6–7.) Organisms are not simply vehicles for selfishly replicating genes. Furthermore, the agency of genes cannot be realized without being mediated by other agencies beyond their own.[48]

Our contemporary understanding of gene regulation is more compatible with a concept of the *performative gene*—an expressive action by the cell that utilizes its structured, historically derived informational resources in a material-discursive *doing* in the world. Dawkins's once radical conjecture of the "selfish" agency of genes has become entirely mainstream in evolutionary biology. The performative gene requires no further intellectual license than is called for by Dawkins's selfish gene concept. It simply rejects the reduction of all biological agencies and causes to the level of alleles at genes.[49] However, I would not simply replace selfish genes with performative genes. This buzzword swap would obscure the many indispensable agencies of alternative protein isoforms, organelles, cells, tissues, signaling centers, bioelectric fields, organs, etcetera that share causal parity with genes in organismal becoming. Rather, my goal is to abandon naive gene-level reductionism entirely.

The extension of biological agency that I advocate is not a neo-vitalism; it involves an entirely mechanistic understanding of biology. Just as the emergent agency of selfish genes cannot be explained entirely as simply a consequence of biochemistry alone (in which the phenomena of populations, heritability, fitness, and natural selection cannot be defined), the emergent agencies of molecules, cells, tissues, and organs in development and physiology cannot be explained as mere consequences of selfish genes. Richard Dawkins has not been accused of vitalism for conceiving of the agency of gene as "selfish." Likewise, it is not vitalism to state that the biological agencies necessary to explain the development of the phenotype are not shaped by selection acting solely at the level of alleles (genetic variations) in single genes, but by selection acting on the phenotype created by aggregations of millions, billions, or trillions of differentiated and individuated but genetically identical cells. (See further discussion in appendixes 6–7.)

Citationality and Homology

The details shared between gender performativity and biological development and evolution extend far beyond their common discursive structures. For example, the concept of citationality that is central to gender performativity closely corresponds to the broader biological concept of *homology*—the relation between similarities shared among organisms due to common ancestry. Coined by Victorian comparative biologist Richard Owen in 1843, homology has remained a central concept in biology. Examples of biological homologs include limbs, digits, teeth, hair, feathers, and flowers. Like gender presentations, each instance of a biological homolog comes into existence iteratively in a space and time distinct from other instances, yet each of these instances is also interrelated to the others by a shared history of genealogical and evolutionary contiguity of descent. The anatomical similarity between the bodies of parents and offspring do not arise by imitation, mimicry, duplication, reflection, or mentoring. Rather, each individual human body and body part must be enacted anew from a single-celled zygote through its own autonomously directed means of material becoming.

Because the body is an enactment by the organismal self, each homologous body part is likewise an iterative enactment of itself. Each flower is an iterative performance of flowerness. Each pair of limbs with hands is an independent realization of limbness and handness. And because hands only develop in individuals that belong to a specific, more restricted set of four-limbed, or tetrapod, vertebrates, complex homologs constitute a nested hierarchy of historical relations. Right now, I am typing with the digits on the hands of my forelimbs, which are simultaneous performances, *and* citations of the undifferentiated, ancestral tetrapod limbness, a more differentiated forelimbness, mammalian handness, primate digitness, human opposing-thumbness, and my own chirally inverted left and right hands.

Like you and me, the birds, and the planet Venus, all homologs are ontological individuals—historical entities that have an origin; a potential thriving; a potential for reproduction and diversification; and an inevitable end, death, or extinction. (However, as the planet Venus demonstrates, not all ontological individuals have homologs.) An inevitable consequence of the capacity of homologs to thrive and reproduce is the evolutionary diversification and adaptive radiation of homologs—including genes, developmental mechanisms, tissues, morphological structures, organs, and behaviors. From a single original form, a diversity of forms with a variety of functions have evolved. Thus, despite the tremendous variation in form encompassed by vertebrate limbs, my arms are homologous with giraffe forelegs, bird wings, bat wings, whale flippers, and the mysteriously tiny, two-digit arms of the otherwise gigantic *Tyrannosaurus rex*.

Gender citationality and biological homology are both names for the relation between the commonalities shared by phenotypes arising from iterative performative discourses—one cultural and the other biological. In biology, the homology relation arises from the historically contiguous properties of the evolved, and evolving, genetic-development discourse. In culture, citational relations reflect horizontal discursive influences of culturally proximate individuals. Although the developmental-evolutionary and cultural processes underlying the iterative continuity of historic entities have many differences, it is important to recognize the commonalities between these components of gender/sex, specifically, and the material and cultural phenotype as a whole. Thus, like cultural variations in gender presentation, variations in vertebrate limb morphology retain their citational relation to the first, original limbs and to all their other, descendent varieties.

Further examination of the molecular details of biological homology reveals additional opportunities for performative understanding of the phenotype. For example, evolutionary biologist Günter Wagner proposes a genetic model of morphological homology that posits the necessity of: (1) positional information signals, (2) a character identity network, and (3) character state realizer genes. Positional information signals provide spatial information about the relative position of cells in the body. The character identity network encompasses a series of co-expressed regulatory genes that: (i) are consistently associated with the production of that character (i.e., body part or cell type), (ii) sustain each other's expression, and (iii) suppress the enactment of alternative character or cell identities by disrupting other character identity networks, and regulating the activity of the downstream realizer genes. The realizer genes produce the morphological product, or the specific morphological character state. In this model, a character is a type of structure—like a hair or a limb—and the character state is a specific realization of that character—like the scalp, eyebrow, or armpit hairs on a single body, or the arms, bat wings, and whale flippers among different bodies. The character identity network is highly conserved evolutionarily, compared to the positional information and character state realizer genes.

Character state identity networks function through the molecular regulatory mechanisms that we have described above, providing important mechanistic insight into *how* homologous body parts persist over evolutionary time despite ongoing adaptive change (i.e., diversification in character states and their realizer genes). From a performative perspective, regulatory and historical stability arises from the arbitrary coevolved, intra-active correspondences between endocrine and paracrine signals and their receptors, among signaling intermediary molecules, transcription cofactors, and their DNA-binding sites. These networked relations constitute modular components of the organism's overall apparatus of the bodily production—the material-discursive practices that are elicited by appropriate positional and temporal information and the competence to respond to it. (For more discussion of modularity in biology, see appendix 4.) The performative perspective further emphasizes the agencies of all intra-actors in the network, including gene products and gene regulatory binding sites.

While consistent with a performative view of the phenotype, Wagner's model defines character identity networks as networks of *genes*, to the exclusion of the agency of cells, tissues, placodes, signaling centers, developmental fields, mechanical forces, electric fields, and so on. Wagner's character identity networks are thus conceptually limited by an exclusive focus on *genes* as the agents in the development and evolution of homologies. A performative perspective on the phenotype expands the hierarchy of intra-active agencies involved in the generation and evolution of homology to encompass these extra-genetic players. Accordingly, research on the evolution of any particular homolog can focus at any appropriate level of phenomena.

Recently, however, Wagner and philosophy of science collaborators James DiFrisco and Alan Love have made a parallel intellectual move to expand their explanatory model of homology from conserved character identity networks (ChINs) to conserved character identity mechanisms (ChIMs).[50] The latter include the former, but they also incorporate the active agencies of cells, tissues, signaling centers, and organs in the development of body parts. With further expansion to include electrical fields, mechanical forces, and whole organism behavior (like the leg muscle contraction involved in the development of the zygodactyl toes of parrots), a homology concept based on ChIMs is entirely congruent with a performative view of the phenotype. However, a performative model of the phenotype also brings along with it the additional intellectual tools and insights of intra-action; hierarchical agency; discursivity; distributed power relations; and a productive posthuman connection to culture, cultural evolution, and society.[51]

The existence of the sexual phenotype creates another type of homology. Sexual homology is the relationship between reproductive characters, or homologs, that are realized, or enacted, as variable character states for reproductive functions. As we will see in the next chapter, many anatomical components of the human sexual phenotype are sexually homologous character states, including the ovary and testis, the clitoris/labia minora and the penis, and the labia majora and scrotum. Because each of these pairs of sexual homologs are derived from a common embryological tissue, their actual realizations intergrade indistinguishably with one another.

Posthuman Power

Performative views of gender have been critically informed by Michel Foucault's influential cultural theory of power as an emergent, distributed, self-regulatory, *and* creative effect of stabilizing discourses within social networks. Foucault conceives power as a cultural means of social control over populations of people, their bodies, and especially their sexualities. In modern societies, this cultural power is enacted and enforced by social discourses that now draw on expert opinion from the disciplines of biology, psychology, mental health, and criminology. However, this power is specifically realized through self-regulatory control and individual conformity to implied cultural norms, rather than by centralized institutional force or explicit physical coercion. The effects of Foucauldian power are the "docile" subjects he views as the cultural mainstream.[52] Although Foucault is frequently portrayed as antagonistic to biological science, his work on sexuality actually involves a genuine interest and engagement with material bodies, and has productive implications for biological science.

Do the cultural and psychological power relations that play such an important role in understanding the performativity of gender have any counterpart in biological performativity? Indeed, the performativity of organismal development and physiology support the relevance of an extended, posthuman concept of power that is a product of the creative, distributed, regulatory biological discourses. Foucault's analysis of power is precisely congruent with our understanding of the emergence of integrated anatomical order and physiological homeostasis in the body through communicative intra-actions among a hierarchy of biological agencies. The understanding of the cell-to-cell paracrine and endocrine signaling as a molecular discourse constitutes an entirely material, distributed power that provides the capacity of a vast network of trillions of genetically identical cells to cooperatively realize a complex, hierarchically organized, integrated, well-regulated, and functional body.

Following Foucault, this material power is enacted through constraining and regulatory discourses, but these paracrine and endocrine signaling pathways are also *creative* instruments. Like cultural power, the logic of material power is a well-integrated, productive, and purposeful whole. The "authorless purpose" of Foucault's concept of cultural power exactly parallels the adaptive functional integration of organisms that arises from natural selection. Just as the capacity of centralized legal systems, prisons, and departments of "corrections" to control violence in society is limited, the power of intercellular suasion is limited because some cells can simply ignore such signals, or may actually go rogue and resist the organism's singular, prevailing purpose. In this way, Foucault's distributed concept of power explains why the nestedness of complex organismal bodies does not constitute, or imply, a hierarchy of biological power relations within the body.[53]

If the well-integrated cells of the functional organism are the "docile bodies" of the complex phenotype, then cancer can be understood as a form of localized Foucauldian *resistance* to the distributed self-regulating power of the organismal body. Following Foucault's analysis of the continuity between resistance and power, the potential for cancerous proliferation is distributed everywhere, omnipresent, and continuously regulated by the material power of the body. It arises as the result of the potential for selfish pursuit by any of the trillions of distributed agencies among the cells of the body, ultimately endangering the persistence of the individual. Cancer is so threatening to life because cancers advance through manipulating the *same* molecular discourses used in the regulatory and creative power of the functional, material body.[54]

Would thinking about organismal biology with Foucault's conceptual vocabulary yield empirical progress in biology? I am not sure yet, but I do think that Foucault's explicit emphasis on the local distribution of self-regulatory power, and the authorless purpose of power make clear contributions to frame problems in physiology and developmental and evolutionary biology. Ultimately, I think that Foucault's concept of distributed regulatory power can tell us more about canalization, homeostasis, evolutionary complexity, and embodied innovations of complex multicellular phenotypes than does Dawkins's concept of selfish genes. The continuity of the themes of regulation, constraint, conflict, and creative innovation between the humanities and developmental and evolutionary biology is striking indeed, and it relates directly to fundamental questions of the origin and maintenance of sexual reproduction, the origin of multicellularity, morphogenesis, evolutionary innovation, and cancer biology.

Physiology and Immunity

Genetics and development are not the only performative phenomena in the biological sciences. For example, at the heart of organismal physiology is the principle of homeostasis—the maintenance of physical or biochemical steady states, or "set points," within the body, its organs, and cells. Homeostasis is maintained through negative feedback loops between environmental and cellular conditions, and gene expression, organ function, and organismal behavior. Homeostatic systems operate like a thermostat—turning on the heat when the environment is too cold, and turning on the air conditioning when the environment is too warm. Homeostatic responses include panting, sweating, or moving into the shade when it is too hot; shivering or moving out of the wind when it is too cold; and the raising of your heart rate in response to exercise.

Homeostasis maintains appropriate internal conditions by sensing variation in critical features of the external and internal environment—such as temperature, or oxygen and sodium concentration—and using the genetic, hormonal, and neurological communication systems to alter the genetic expression, physiological state, or behavior of the organism in an appropriate, compensatory manner. Like development, the performative features of homeostasis are both hierarchical—involving responses at multiple levels of organization in the body—and discursive—involving genetic, hormonal, and neurological communications systems that are themselves historically contingent, arbitrary, and context dependent. Following Austin, the communication of information about the conditions of the body is representational, or constative, communication, but enacting negative feedback loops to respond homeostatically is a performative expression or doing.

Like development, physiological homeostasis cannot be described exclusively as a consequence of changes in gene expression. Some homeostatic physiological responses are built into the biochemical networks of enzymes and biomolecules themselves. These networks can adjust to changes in environmental conditions nearly instantaneously, instead of the minutes required for changes in gene expression to have physiological effects. In yeast metabolism, for example, changes in gene expression function only to reinforce, or enhance, the plasticity of the phenotype's homeostatic response rather than initiating or guiding it.[55]

Another performative dimension of organismal physiology is the immune system. Indeed, immunology is such a huge scientific and intellectual challenge precisely because of the complexity of the molecular discourses that underlie immune-system function.

The immune system includes innate defenses, which fight infections nonspecifically, and adaptive, or acquired, immunity, which fights specific infections and is influenced by the life experience of the individual. Adaptive immunity is a complex capacity that evolved early in our vertebrate ancestors. The adaptive immune system identifies, targets, and eliminates novel pathogens—like viruses, bacteria, or single-celled protists—that invade and infect the body. Disease pathogens are constantly evolving new ways to evade the host's immune defenses and to successfully infect new hosts, creating an ongoing existential challenge to all organisms. In response to these ever-evolving, novel threats, vertebrates have evolved a particularly robust adaptive immune system that recognizes *novel* pathogens by the distinctive, molecular-scale, physical features of their surfaces (called antigens), and produces antibodies and immune cells that bind selectively to these alien targets and eliminates them. The adaptive immune system then remembers these pathogens virtually *forever*, so that the body can mount an even more vigorous and timely immune response in case those pathogens ever return again. (For further description of the performativity of the acquired immune system, see appendix 2.)

Neurobiology and Psychology

The continuum of performative biological processes also extends through the neurological, behavioral, psychological, and social behavior of the organism. Although the topic is too vast to cover here, I want to make a few preliminary observations about the implications of a performative model of the phenotype for neurobiology, neuropsychology, psychology, and social behavior.

Neurobiology investigates the anatomy, molecular biology, physiology, and behavior of a networked type of highly reactive cells, called neurons, that communicate with each other using a combination of rapid waves of cell membrane depolarization and cell-to-cell molecular signaling through tightly interlocked junctions called synapses. Like the paracrine signals that function in other aspects of cellular development and physiology, the intercellular messages among neurons involve molecules released into the intercellular space by one cell and received by coevolved, binding receptors on the membranes of another cell. But neurons are special because their synapses are anatomically and physiologically specialized to achieve these communication events with great specificity, rapidity, and consistency. Synapses are anatomical-physiological intra-actions. The nervous system combines billions of networked neurons sharing trillions of rapid-fire synapses. By acquiring information about the external and internal environment from special sensory neurons, which are sensitive to physical stimuli and physiological conditions, and interfacing with the muscles and organs of the body, the nervous system integrates and coordinates the behavior of the animal body.

How the activities of the material nervous system generate individual sensory experience, consciousness, psychological experience, the capacity for language, mood, and personality is, of course, the subject of intense research and debate within neuroscience, psychology, and developmental psychology. As in genetic, developmental, and evolutionary biology, neurobiology and psychology include both committed reductionist and emergentist programs of research. Without delving into that debate, we can see clearly that there can be no such phenomena as perception, cognition, consciousness, psychological experience, language, mood, or emotion except through active *doings* by the individual, iteratively realized through physical and social interactions with the material and social environment. In other words, the brain is not a network of genetically determined cognitive modules that can be booted up to function as a mind. The brain develops the capacity to function by actively *doing* its functions. The psychological agency of the individual is necessary to achieve the capacity for any nonreflex cognitive functions. Like other complex phenotypic features, the development of individual cognition and psychology involves the same familiar elements of iterative intra-action, agency, hierarchy, and becoming through discursive doing.[56]

Furthermore, neurobiology and psychology are obviously manifestations of the structured, discursive, molecular intra-actions among billions of networked cells. The agency of human consciousness emerges from the hierarchical intra-actions of genes, molecules, neurons, ganglia, and other higher-order neuroanatomical structures within the physiological/behavioral context of the body. Even modern pharmaceutical efforts to intervene on the psyche to treat depression, addiction, obsessive-compulsive disorder, and other maladies function explicitly as neuromolecular discourse modulators. Selective serotonin uptake inhibitors (SSRIs) act biochemically to alter the physiology of intercellular discourse among serotonin-signaling neurons by extending the active durations of their specialized paracrine signaling expressions—that is, the number and duration of molecules of serotonin in these twenty-nanometer-wide, intercellular synaptic spaces.

The performative gender theories of Butler and others are deeply rooted in the psychoanalytic theories of Sigmund Freud and Jacques Lacan. But a performative concept of the organismal phenotype provides a deeper framework for an explicitly performative view of neurobiology, cognition, and developmental psychology.

Sexual Selection

Sexual selection arises from heritable variations in mating or fertilization success. Sexual selection has two modalities—competition within a sex and choice between sexes, referred to as intrasexual and intersexual selection, respectively. Intrasexual selection is usually described as competition among Males for control over mating opportunities with Females, or for scarce resources that Females require for reproduction, and it is associated with the evolution of large Male body size and weapons of aggression, such as antlers, horns, and spurs. However, Female-Female competition is common, and can be a major force in the evolution of some species. For example, in the polyandrous Wattled Jacana (*Jacana jacana*), a lily-trotting Neotropical shorebird, the Female defends a nesting territory from other Females, and those Females with high-quality nesting resources can attract multiple Males as mates. The Males conduct all of the parental care—building nests, incubating the eggs, and caring for the chicks—by themselves within the Female's territory. Female Wattled Jacanas are on average 48 percent heavier than Males (creating a strongly bimodal frequency distribution in body mass), and have evolved wing spurs on their wrists which they use as weapons in fights with other Females.[57] Intrasexual selection is by no means limited to Males.

Sexual selection by mate choice is associated with the evolution of ornamental displays, and it encompasses mate choice by Females, mate choice by Males, and mutual mate choice by all individuals. Typically, sexual selection by mate choice is strongest, leading to elaborate ornaments in species with Female-only parental care and no pairing behavior beyond copulation. In my 2017 book *The Evolution of Beauty*, I explore an explicitly aesthetic perspective on mate choice and sexual ornament, the differences between arbitrary and adaptive mate choice, the evolutionary consequences of sexual coercion, and the implications of these evolutionary processes for sexual autonomy, human evolution, and human sexual diversity.[58]

The performativity of mating displays is obvious; sexual displays are so often referred to as *performances* because they are communicative actions conducted for an audience of evaluating observers. Displays stand out from other behavior of an organism for their stereotyped form and their highly contextual social function.

However, the performativity of sexual display becomes even more profound when one examines the mechanism of sexual selection by mate choice in more detail. Almost every sexual display involves a coevolved correspondence between a display trait and a mating preference.[59] Displays and preferences correspond because they have shaped one another over evolutionary time through iterative performances, and autonomous social choices within populations of organisms. Mating preferences involve individual subjective judgments, and mate choices are actions enabled by the aesthetic social agency of animals.

This coevolved correspondence between preferences and traits is a premiere, behavioral example of communicative, sexual *intra-action*. Sexual displays have no function in the world except through the subjective evaluations of other conspecific individuals. Likewise, mating preferences exist solely through their potential correspondence with coevolved sexual displays. Like the biophysical correspondence of endocrine and paracrine signals and their receptors, the correspondence between display traits and mating preferences is frequently arbitrary in form, and the result of unique, contingent histories of intra-active sexual discourse.[60] Creating explicit intellectual connections between mate choice, aesthetic coevolution, and performativity provide new opportunities for embodied, posthuman performance studies.

In previous work, I have proposed that human and biotic art are forms of communication that coevolve with their evaluation. In this definition, it is the aesthetic intra-activity of art that differentiates it from other sensory/cognitive experiences.[61]

What Is Not Performative in Biology?

It is also important to ask explicitly what is *not* performative in biology. The answer is lots of things. The key feature of performative biological phenomena is the presence of coevolved intra-action, or discourse. Consequently, the lack of intra-action provides a clear criterion for the absence of performativity. For example, natural selection is not, by itself, performative. Selection does not necessarily involve an intra-active discourse or the mediation of agencies. Temperature and humidity, for example, can be strong sources of natural selection, but these abiotic environmental conditions do not have agency as evolved biological individuals like genes, biomolecules, tissues, organs, or organisms do. Environmental temperature is not involved in a discourse with organisms. Furthermore, not all biotic interactions are intra-actions. For example, exploitative competition among individuals for the same resource is not by itself performative; whether the early bird, or an even earlier bird, gets the worm may determine who lives, dies, or reproduces, but this process does not involve intra-actions between them. Interference competition, in which individuals interact aggressively over the same resource is not necessarily performative either, but could also involve social communications among species that become so.

Likewise, predation is not, by itself, performative. Being preyed upon is not the result of an individual of a prey species pursuing its agential choice; it is simply bad luck. Whether a cheetah chasing a gazelle is successful in preying upon them may *even* depend upon their differential motor performances. It may even involve coevolution between predator and prey, with cheetah predation selecting for faster gazelle, which selects for faster cheetah, and so on. But the predation by cheetahs of gazelles and the coevolutionary arms race that it inspires do not involve an inter-agential discourse. The speed of the gazelle also functions in fleeing from lions, wild dogs, or other terrestrial predators, and has a definable function outside of its interactions with predators. However, some responses to predation may be performative. For example, in response to the observation of a terrestrial predator, various species of gazelles will conspicuously leap into the air as they run away. This unusual display, called stotting, is thought to have evolved to communicate to the predator that it has been observed, and perhaps that the individual is healthy enough to evade capture. The evolution of stotting raises a potentially performative wrinkle into predator-prey interactions.[62]

Why Performative Biology Now?

The performative view of biology proposed here reframes organismal development and evolution by placing genetic individuality and environmental context at the foundations of the scientific explanation of the phenotype. When we view the organism as a performance of itself, and developmental mechanisms as molecular discourses among many intra-acting agencies within the body, we refocus scientific inquiry to capture the full complexity of the intellectual challenge of biology.

Some humanists and biologists may think that performative biology stretches the concept too far, perhaps beyond recognition. It is important to note, however, that Barad's material performativity already extends the concept beyond the social context of gender in humans to encompass the physical intra-actions of subatomic quarks, leptons, and bosons. So, biological performativity is not the first posthuman application of the concept, and it fits squarely in the intellectual space between Butler and Barad.

The term *performance* is already in common use in many biological contexts to refer to variations among individual organismal phenotypes or genomes in ways that align with how the concept has been defined by queer and feminist philosophers, as well as how it has been applied in other realms of knowledge making. In the same sense that engineers refer to high-performance cars or instruments, biologists refer to biomechanical investigations of fish swimming, lizards running, and birds flying as research on organismal *performance*. Likewise, physiologists refer to the osmoregulatory *performance* of fishes that migrate between fresh and salt water, or of kangaroo rats that can thrive in the desert without ever drinking liquid water. Biologists are already used to describing biomechanical and physiological variations among phenotypes as varieties of performance, far outside of the social and dramatic connotations of the word. The framework I am proposing extends this common understanding to demonstrate that organisms are indeed performance all the way down.

The concept of evolution itself as a drama performed by players of various historically contingent roles also has a long and illustrious history in evolutionary biology. In 1965, the pioneering community ecologist Evelyn Hutchinson entitled an influential collection of essays about the history of life *The Ecological Theater and the Evolutionary Play*.[63] Although Hutchinson did not elaborate on this metaphor further in his text, the clear implication of the "evolutionary play" is that organisms are individual performances in a broader drama of survival, fecundity, and reproduction. It is notable, I think, that Hutchinson did not name his book, say, *The Ecological Arena and the Evolutionary Wrestling Match*, which certainly would have been a better fit to the virile, deterministic, "survival of the fittest" conception of evolution prevalent in the mid-twentieth century. However, Hutchinson was a very cultured man (one chapter of this book is entitled "The Naturalist as Art Critic"), so it is possible that he merely preferred the more erudite setting of the theater. However, I think Hutchinson accurately perceived that evolutionary history is more than mere grinding, deterministic competition (even though he is most famous for his work on how competition structures ecological communities). Hutchinson's idea of ecology and evolution as a *theatrical discourse* captures the open-ended, creative, innovative, and emergent complexity of eco-evolutionary feedback in evolutionary process.

The concept of biological performativity that I am advancing here simply takes seriously these previous usages of the performance in biology to explore the generation of the phenotype—the material development, physiology, function, psychology, and behavior of the organism.

Having explored the concept of performativity in gender (chapter 3) and outlined a performative theory of the phenotype (chapter 4), I will bring these two concepts together in the next two chapters to explore the embryological development of the human sexual phenotype.

CHAPTER FIVE

How Do Our Sexual Bodies Develop?

Perhaps no scientific concept of sexual development is more deeply engrained in biology than the idea of "sex determination."[1] This way of thinking about sexual development assumes that sex is a binary fact about the genes, genome, chromosomes, and cells of the body, and that biological development is the process of materially representing that prior genetic fact. The continued use of the term *sex determination* reinforces the impression that science involves, or endorses, biological or genetic determinism. For decades, feminists and anti-racists have criticized "genetic determinism" in biology, and expressed concerned that it implies certain individuals are immutably superior or inferior to others. Some biologists defend against being labeled as genetic determinists by countering that the phrase does not acknowledge that biologists have long recognized an active role of the environment in phenotypic development (see appendix 3). However, if biologists really want to credibly claim that they do not endorse "genetic determinism," they can start by no longer using the phrase *sex determination*, or the related phrase *sex reversal* when their assumptions about "sex determination" prove wrong (see chapter 2).

A fundamental insight from the performativity of the phenotype I have been arguing for is that no complex feature or cluster of features, can be "determined" by any single gene, singular underlying cause, or definable feature of that organism's genotype. Although the sexual phenotypes of humans, and many other animal species, do exhibit strongly distinct, bimodal distributions in reproductive morphologies, the existence of such phenotypic differences cannot create, stipulate, or define essential qualities to the genes, chromosomes, genomes, or cells of individuals. In other words, the idea of a sexual binary obscures a much more dynamic set of phenomena.

Sex is an historical constellation, or cluster, of iteratively realized, co-occurring anatomical and physiological homologies that have evolved over evolutionary history through selection for sexual reproduction (see chapter 2). In humans, for example, the bodily co-occurrence of paired ovaries and fallopian tubes, uterus, vagina, labia, and clitoris are a frequently observed cluster of morphological structures that facilitate reproduction in specific and structured ways involving ovulation, menstruation, implantation, pregnancy, birth, and lactation. Likewise, the embodied presence of paired testes and vasa deferentia, prostate and other accessory glands, penis and a scrotum facilitate reproduction in different, corresponding, and coevolved ways through the production and ejaculation of sperm. The common co-occurrence of these reproductive anatomies in humans, other mammals, and most tetrapods provides evidence that the phenomenon has an ancient history and is deeply enduring. However, a bimodal distribution in reproductive anatomy and physiology within species does not justify the existence or assignment of an inherent, individual sexual essence or identity to the zygote or genome prior to, or independent of, the sexual phenotype's material realization. As we will see, even strong canalization of the development of reproductive capacities does not imply that the genome, a fertilized egg, or particular chromosomes have a sex.

In the next two chapters, I will describe how the reproductive anatomy and physiology of individuals are better understood as embodied enactments, or performative achievements, by those individuals. Accordingly, sexual development is best understood as a process of *individuation*—the realization of reproductive possibilities by the differentiated, multicellular, hierarchically complex body of an individual. Recall from chapter 2 that differentiation does *not* refer to the origin of sex differences—that is, differences between Female and Male—but to the process by which the cells and parts of the body become differentiated from each other, and from their earlier developmental states of cellular and anatomical uniformity. Likewise, the development of sexual difference refers to all variations in the gender/sex phenotype without reference to a priori categories (binary or otherwise). Phenotypic performativity transforms the biology of sexual development from the genetic determination of a preexisting binary essence into a variable, individualized, differential becoming.

Before diving into the biological details, it is important to acknowledge that the science of sex and sexual development can never be completely independent of cultural concepts of gender. The beliefs of scientists themselves are omnipresent in the experiments they create to investigate the biology of sex. This observation is one reason why feminists have advocated the combined, non-dualist concept of gender/sex to capture the entwined social and biological, cultural and evolved components of the extended sexual phenotype. However, to understand and investigate the materiality of human reproductive capacities, we will need a common language, or at least a shared conversation, to describe the biological and cultural components and influences on our sexual selves. In this chapter, I will present a summary of the current understanding of the molecular and genetic processes involved in embryonic development of the human sexual phenotype. In the next chapter, I will discuss how variations from this generalized account further reveal the performative properties of human sexual development.

Having criticized the flaws of "canonical accounts" of development (see chapter 4), I will also present my own summary or outline of the process of sexual development of humans. However, this consciously *performative* telling of human sexual development will not be a canonical or normative account. Like a synopsis of a play, my account does not aspire to be a full, standard, or definitive representation of any specific performance, but an introduction to understanding the dynamics of these performances generally. This summary provides background for description of variations in this process in chapter 6.

The following two chapters will include substantial detail from biological and biomedical literature. Although a lot of this research was designed and conducted in ways that implicitly rely on an assumption of an essential, individual sexual binary that I reject, the analysis of these cultural influences is another sort of work. Excellent examples of such research include works by Richardson, Fausto-Sterling, Hird, Jordan-Young, Karkazis, Wilson, Gill-Peterson, Willey, and others.[2] Rather, my goal here is to discuss findings from the biological literature on sexual development as statements about sexual difference (i.e., pure individual sexual variation) rather than evidence of sex difference (difference between defined sex categories). By discussing the development of human gender/sex phenotype in substantial molecular and anatomical detail, rather than with just a few brief examples, I hope to document my assertion that our sexual bodies are indeed *performances all the way down*, and to encourage, by example, future queer feminist analyses of the human phenotype in similar molecular detail. I will continue to use the words *Female* and *Male*, and a host of terms to refer to co-occurring features of sexual anatomy and physiology. However, as outlined in chapter 2, these terms refer *not* to definable, binary categories but to historical clusters of iteratively recurring reproductive homologies.

The Role of Chromosomes

According to textbooks, sex in mammals is "determined" by the "sex chromosomes," with Females being XX, and Males XY. For example, one respected, contemporary introductory college biology textbook begins the topic of sexual development with a section titled "Sex Determination by Chromosomes," which describes examples of the "sex chromosomes" of mammals and other organisms. A subsequent section is titled "Sex Chromosome Abnormalities Revealed the Gene That Determines Sex," which further states that "some women are genetically XY but lack a small portion of the Y chromosome. Some men are genetically XX but have a small piece of the Y chromosome attached to another chromosome." Ironically, the statements in the second section contradict the bold statements in the previous one. Yet this contradiction is not discussed.[3] "Performance all the way down" is a way of exposing this explanatory shell game. There is no level of phenomena at which we can ultimately state that sex is "determined," because individual sex can only be materially realized.

Chromosomes are subunits, or packages, of the nuclear genome housed in every cell.[4] The typical human genome includes twenty-three pairs of chromosomes, or forty-six total. Typically, each individual inherits one chromosome of each of the twenty-three pairs from each parent through the egg and sperm that combine to create the genome of the zygote. Twenty-two of these pairs of chromosomes are identical in size and structure, and may differ only slightly in DNA sequence. But one pair of human chromosomes—called the X and Y chromosomes—are vastly different in size and shape. In fact, the majority of genes on both the X and Y chromosomes have no function in sexual differentiation at all. Nor are there features of either the X or Y chromosomes that can solely "determine" the sex of the body. Moreover, the vast majority of genes that play a role in human sexual development are not found on either the X or Y chromosome! For this reason, there is an emerging consensus that it is erroneous to refer to them as "sex chromosomes" at all.[5]

In mammals, because XX individuals have two X chromosomes and can only produce gametes (in this case, eggs) that contain X chromosomes, Female mammals are referred to as *homogametic*. In contrast, XY Male mammals can produce gametes (in this case, sperm) that may contain either an X or a Y chromosome, and are referred to as *heterogametic*. Typically, all individuals inherit at least one X chromosome from their XX parent. XX individuals inherit X chromosomes from both parents, whereas XY individuals have one Y chromosome which they can only inherit from an XY parent.

Because all XY individuals also have one X chromosome, the sexual development of XX individuals proceeds using only those genes that are common to *all* individuals. In contrast, the development of XY individuals can also deploy those few genes exclusive to the Y chromosome, which will be found *exclusively* in XY individuals. In other words, in mammals, the development of homogametic Females employs a *genetic common denominator* shared by all individuals. However, the unique genes of the Y chromosome can act as a potential genetic *difference maker* in mammalian sexual development. This does not mean that the difference the Y chromosome can make is the binary difference between Male and Female. Or that the differences between Male and Female are encoded on the Y chromosome. In fact, the SRY gene is the *only* gene critical to mammalian sexual development that is located on the Y chromosome; all other genes that participate in this process are located on autosomes, or non–X or Y chromosomes, that are present in all individuals. Because some of the genes on the Y chromosome are unique in the genome, their presence can initiate a distinct cascade of developmental events that can lead to reproductive morphological differences. In the language of our revised, performative phenotype landscape (figure 4C), a genetic difference maker on the Y chromosome can have critical influence at early stages of the development toward a phenotypic peak in sexual morphology. Genes can influence the developmental path of the individual, but genes cannot define them.

As I will discuss in more detail in chapter 7, there is nothing inherently "Male" about sexual development involving a genetic difference maker, like a Y chromosome or the mammalian SRY gene. In birds, for example, Males are the homogametic sex (ZZ), and Females are the heterogametic (ZW) sex. Instead of requiring a distinct, genetic sexual difference maker, sexual development in birds is initiated in a dosage-dependent manner by the gene DMRT1 on the Z chromosome. Typically, individuals with two copies of DMRT1 develop as Male, and those with one copy develop as Female.[6] Despite developing from the genetic common denominator shared by all individuals, Male birds develop testes, produce sperm, and in some species, like ducks and ostriches, penises that are homologous with those of mammals. Likewise, sexual differentiation in many reptiles is initiated by variation in temperature of the eggs during development, yet these reptile species share many homologous genital and gonad structures with mammals.

Why are the textbooks so confused about sexual development? One answer comes from historian Sarah Richardson, who suggests that during the second half of the twentieth century scientific investigation of mammalian "sex determination" became equated with the role of the Y chromosome in the development of Maleness.[7] In accordance with particular culturally conditioned values, likely stemming from who was doing the research, Male sexual development was depicted as an active, assertive, highly nuanced process justifying detailed investigation, whereas Female sexual development was characterized as a simple, passive, "default" process that is less interesting or worthy of focused investigation.[8] In such accounts, only the Y chromosome and its unique genes—and therefore Maleness—are assigned any agency, or causal power. Although these biases still exist (as in the textbook excerpted above), we will see that recent genetic research has uncovered new details that reject this passive or default view of ovarian or Female development.

On Gene Nomenclature

Before diving into the molecular details of sexual differentiation, I want to introduce, and apologize in advance for, the awkward and idiosyncratic gene nomenclature in biology. The names of genes for many structural proteins—like the collagens in skin and cartilage or hemoglobins that carry oxygen and carbon dioxide around in your red blood cells—come from the names of the proteins themselves, which were discovered long before the genes that encode their amino acid sequences. Genes for enzymes that catalyze physiological reactions in the body were often named after their chemical functions, like alcohol dehydrogenase, which, as you can tell, removes the hydrogen bound to an alcohol molecule. But the names of most genes for developmental paracrine signaling proteins, receptors, and transcription factors were coined by genetic research labs when their existence could only be inferred from the morphological effects of mutations to these genes or from diseases whose etiology implied their existence. Subsequent discoveries of other homologous developmental genes were typically named as variants of these earlier names.

Genes discovered through mutation experiments were often named for the specific morphological effects of their absence. Understanding their molecular functions would take years more research usually by multiple different laboratories. For example, the diverse "hedgehog" family of extracellular paracrine signaling proteins was given its name in 1980 by Christiane Nüsslein-Volhard and Eric Wieschaus following their observations that mutations that functionally eliminate (or "knock out" in experimental lingo) this gene in *Drosophila* fruit flies result in a headless, spiny thorax that looks like a curled up hedgehog. They named other *Drosophila* development genes for other odd morphologies that their knockouts produced, including patch, gooseberry, Knirp (German for "tiny tot"), and Krüppel (German for "cripple").[9] When first coined, these gene names were relevant only to a few specialized researchers in *Drosophila* genetics. Unexpectedly, however, these same genes turned out to be fundamental to the developmental pathways shared by nearly *all* multicellular animals, including humans. The genes with these casually odd fruit fly names became topics of intense research, and Nüsslein-Volhard and Wieschaus went on to win the Nobel Prize for Physiology and Medicine in 1995.

When new genes that were evolutionarily related to hedgehog were discovered later, they were given modifying monikers—like the paralogous vertebrate genes Sonic Hedgehog, Indian Hedgehog, and Desert Hedgehog. As I mentioned in chapter 4, Sonic Hedgehog was named after a computer game character. But Indian Hedgehog and Desert Hedgehog were both named in honor of living species of wild hedgehogs—*Paraechinus micropus* and *P. aethiopicus*, respectively—even though these genes were not studied in those species. Lab biologists viewed the naming of new genes as a lighthearted opportunity to add levity and memorable marketing value to their laborious routines. Problematically, some genes that were given idiosyncratic or humorous names were later discovered to be related to significant human diseases.[10] Although the guidelines of the Human Genome Names Consortium, which regulates gene names and abbreviations, prevents the coinage of offensive or pejorative gene names, they apparently do not prevent gene names that are silly or humorous.[11] There are no clear criteria over how this consortium decides whether gene names like Indian Hedgehog or Lunatic Fringe are offensive.

To complicate matters further, biologists use italic and roman fonts to differentiate the names of the gene itself and its protein product, respectively. So, *hedgehog* is the name of the gene, and hedgehog is the name of the protein that the gene encodes. Furthermore, the human version of any gene is typically capitalized, while all other species are not. So *Sonic Hedgehog* refers to a human gene, whereas sonic hedgehog is used to refer to the homologous protein of a mouse, chicken, or other nonhuman vertebrate. Lastly, all the genes are given an official abbreviation. For example, *Sonic Hedgehog* is abbreviated as *SHH*.

At the risk of offending human geneticists, I will simplify my presentation here by referring to most genes by their capitalized, roman font abbreviations—that is, by the names of the human proteins they encode—regardless of whether I am referring to this gene/protein in a human or other animal. If they are not mentioned in the text, the full names of the genes will be given in an endnote at the first mention of the gene.

How Do Gonads Differentiate?

Until six weeks after conception, human embryos are sexually indistinguishable. By this time, however, embryos have developed "bipotent" gonads that have a distinct gonad identity, but have the capacity to differentiate further into *either* ovaries or testes (figure 6). Located by the incipient kidneys near the position of the mature ovaries, the bipotent gonads have achieved a unique state of gene expression from all other tissues of the body. They have also sequestered the germ line cells that will

6. Molecular signaling pathways in gonad development

Forty-four days after conception, cells in the undifferentiated, bipotent gonad (*top*) can develop toward ovary (*bottom left*) or testis (*bottom right*) identity and morphology in response to different gene signaling pathways. Arrows depict excitatory, paracrine signaling pathway molecules involved in (*gray*) bipotent gonad development, (*white*) ovary development, and (*black*) testis development. Dashed lines depict inhibition of gene expression between ovary and testis development pathways. Illustration by Rebecca Gelernter; redrawn from Baetens et al. (2019) and Ono and Harley (2016).

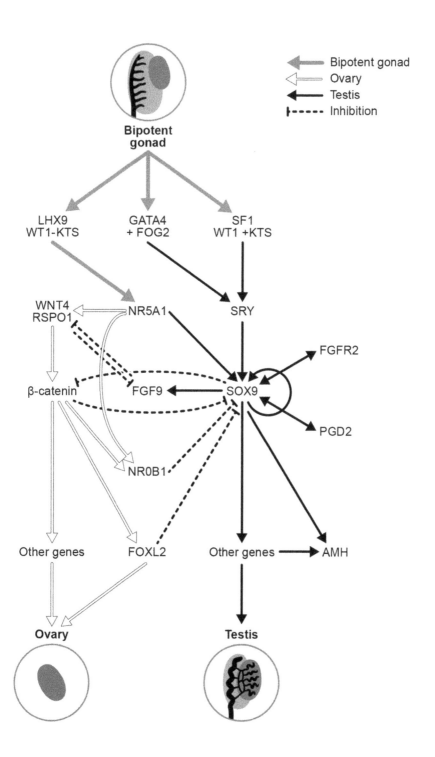

become eggs or sperm, but they have yet to develop further into the more specific, differentiated cells of ovaries or testes.

On day forty-four after conception, a few cells in the center of the genital ridge of the bipotent gonad of XY embryos typically begin to express the gene SRY—one of only seventy-five functional genes left on the greatly reduced human Y chromosome. The name SRY is short for "sex-determining region on the Y," which is a scientific-cultural holdover from the days when: (1) testis development *was* considered synonymous with sexual development itself, (2) when sexual development was conceived as a process of "determining" an essential binary fact about the individual, and (3) when this gene was known only as a broad stretch of real estate on the Y chromosome.[12] SRY is a transcription factor—a protein that is actively transported from the cytoplasm of the cell into its nucleus, and binds to specific, coevolved binding sites on the DNA where it can then influence gene expression (figure 7). When SRY binds in an appropriate place and manner to the SOX9 enhancer, it can promote, or upregulate, the expression SOX9—another transcription factor that plays vital roles in vertebrate sexual differentiation *and* development of the vertebrate skeleton, heart, kidneys, and brain.[13] The transcription factor gene NR5A1 encodes Steroidogenic Factor 1 (SF1) which also binds to the SOX9 enhancer near SRY's binding site (figures 6–7).[14] Together they can trigger the expression of SOX9.

In the bipotent, embryonic gonad, SOX9 acts as an excitatory signal that further enhances expression of SOX9 in that cell, and by other neighboring gonad cells through positive feedback loops (figure 6). SOX9 does this by upregulating the expression of two extracellular, paracrine signaling molecules—FGF9 and PGD2—which can diffuse to other nearby cells in

7. SRY transcription factor function

Top: Before it can be translocated into the nucleus of a bipotent gonad cell, SRY protein in the cytoplasm must be phosphorylated and acetylated at specific sites, and then bind to calmodulin (CaM) and importin-ß (Imp-ß). *Middle:* Once inside the nucleus, SRY protein unbinds from calmodulin and importin-ß, then binds to a specific site in the SOX9 enhancer to promote the expression of SOX9 (see also figure 13). *Bottom:* After expression, transcription factor protein SOX9 can then bind to its own enhancer and further upregulate its own expression. Illustration by Rebecca Gelernter; redrawn from Kashimada and Koopman (2010).

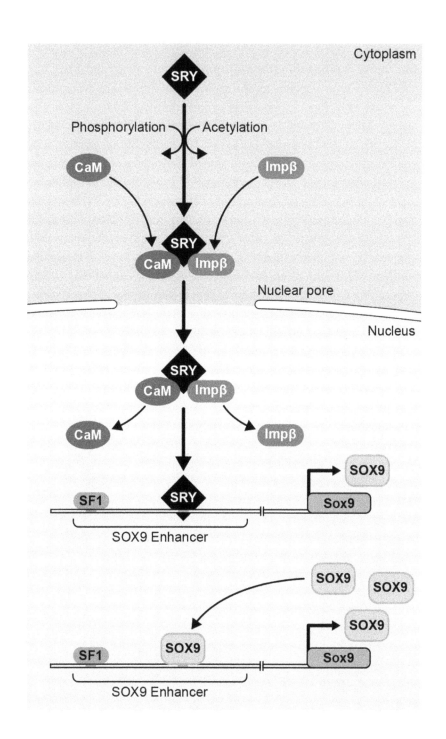

the gonad and bind to specific, corresponding receptors on their surfaces.[15] Binding of FGF9 and PGD2 can stimulate other intracellular signaling pathways that further enhance the expression of SOX9 in those new cells.

In the 1939 super-camp, classic movie musical *Babes in Arms*, the teenage Mickey Rooney encourages Judy Garland and their dejected pals by declaring "Hey! Let's put on a *show*!" Likewise, the organization of the undifferentiated gonad cells is encouraged to assume a new, coordinated, shared developmental identity as testis cells by way of a positive reinforcement of an initial signal coming from very few cells.[16] SRY expression begins with just a few young "Mickey Rooney" cells located in the center of the genital ridge of the gonad of XY individuals, and expands outward in a wave of recruitment mediated by the SOX9, FGF9, and PGD2 signaling pathways. This positive feedback loop creates a rapid, exponential increase in the number of SOX9-expressing cells over the next day, first converting the cells of the genital ridge and then spatially expanding over the entire bipotent gonad to assume testis cell identity.

What turns on SRY expression in the first place? Little has been established in detail. However, because the cells of the bipotent gonad do not "know" in advance whether the SRY gene is present or not, SRY upregulation is likely accomplished by genes that already function in the development of the bipotent gonad itself. It is known that two isoforms of the transcription factor WT1 play important roles in the origin of gonad identity, and in facilitating SRY expression when that gene is present. Isoforms are different proteins made by alternative splicings of the messenger RNA from a single gene before it is translated into a protein. In this case, WT1–KTS lacks three amino acids at the end of exon 9, and functions in initial gonad development. WT1+KTS includes those three amino acids, and functions in stabilizing SRY messenger RNAs, and preventing their degradation before they can be translated into a functioning transcription factor protein (figure 8). In other words, WT1 does not itself turn on SRY expression, but a specially spliced version of the gene makes SRY signaling possible by keeping its signal from being molecularly swamped before it can act. GATA4 and its cofactor FOG2 (i.e., friend of GATA2) may also be required for the initiation of SRY signaling (figure 6).[17]

As we will discuss in greater detail in chapter 6, the reliance of SRY on numerous other upstream and downstream molecular players means that it is not the "master switch" of testis development or of sex determination, as it is so frequently described. Nor is SRY expression either necessary or

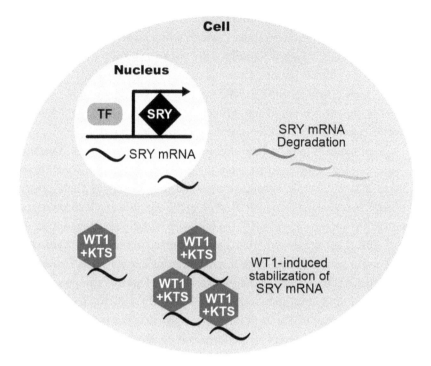

8. SRY mRNA stabilization

Upper left: In the nucleus, the transcription of SRY messenger RNA (mRNA) is promoted by an unknown transcription factor (TF) and released into the cytoplasm. *Upper right:* In the absence of the WT1+KTS protein in the cytoplasm, SRY mRNA rapidly degrades. *Bottom:* The protein WT1+KTS can stabilize SRY mRNA, so that it can be translated into SRY protein and function as a transcription factor in testis development (see figures 7 and 13). Genetic mutations to WT1 or failure to make the correct WT1+KTS protein isoform can disrupt or prevent SRY function and testis development. Illustration by Rebecca Gelernter; redrawn from Baetens et al. (2019).

sufficient for testis development. Other genetic variations can produce a fertile Male sexual phenotype in its absence. However, its role as a potential genetic difference maker comes from its status as the only gene on the Y chromosome with any role in early mammalian sexual development.

Soon after SRY and SOX9 activation, SOX9-expressing gonad cells begin to develop the morphological features characteristic of testes by activating additional gene pathways (figure 6). It is notable, however, that a cell's gene regulatory "decision" to differentiate morphologically always precedes its morphological differentiation. SOX9-expressing gonad cells begin to develop into Sertoli cells, which will nourish developing sperm, and Leydig cells, which will produce the hormone testosterone that facilitates further differentiation of various tissues around the body. Once the Sertoli cells originate, SOX9, NR5A1, and the GATA4-FOG2 complex enhance the production of anti-Müllerian hormone (AMH), which functions in reproductive tract development (see below; figure 6).[18]

What happens in gonad development of XX embryos? First, I want to assure you that I did not first trace the development of the XY embryos before XX embryos in continuation of the long history of phallocentric research in sexual development. Rather, in the presence of a Y chromosome with a functioning copy of SRY, the initiation of testis development occurs first. This is not accidental. To make a difference in sexual differentiation, SRY must initiate a trajectory toward testis development *before* ovarian identity starts to develop. Because ovarian development can proceed using genes found in *all* individuals—including XY individuals—the SRY genetic difference maker must intercede *before* that process starts if it is to have any effect. In terms of our performative developmental landscape (figure 4C), SRY cannot initiate a trajectory toward the development of testis identity once the bipotent gonad has initiated a developmental trajectory toward a different stable developmental peak.

Typically, around day forty-nine after fertilization, in the absence of functioning SRY signals, the still undifferentiated cells of the bipotent gonad ridge begin to express the paracrine signaling protein WNT4 which diffuses nearby, binding to the Frizzled (FZD) receptor on the membranes of neighboring gonad cells (figure 6).[19] The binding of WNT4 to FZD initiates a molecular signaling cascade inside the receiving cells that stabilizes molecules of β-catenin (an important transcription cofactor) so that they can move into the nucleus, bind to transcription factors (including LEF1) that then bind to specific sites in the DNA, and upregulate gene expression.[20] In other words, β-catenin signaling function is facilitated not by turning on β-catenin expression, but by temporarily disrupting the ongoing background process of β-catenin protein degradation. (Note, yet again, that this is *not* how computer programs or algorithms work.)

When it moves into the nucleus of bipotent gonad cells, β-catenin can enhance the expression of FOXL2 (figure 6).[21] FOXL2 is in the Forkhead box family of pioneer transcription factors, which are capable of binding to specific sites in the dense chromatin of regions of the DNA that are tightly coiled, opening up those regions of the chromosome, and exposing new promoter and enhancer sites for *other* transcription factors to bind and have a potential gene regulatory effect. In the gonad, FOXL2 stimulates the production of the ovary-specific growth factor follistatin and other genes. Meanwhile, RSPO1 is expressed in parallel with WNT4, and also enhances FOXL2 production.[22] In response to this new wave of gene expression, the previously bipotent gonad cells assume a specific ovarian identity, and begin to differentiate anatomically into granulosa and theca cells. Granulosa cells maintain the germ line cells that will become ova, and theca cells will later produce the hormone estradiol that contributes to sexual development in other tissues of the body in adolescence and afterward.

In addition to facilitating ovary development, FOXL2 expression also actively *inhibits* the expression of SOX9 in the developing gonad, actively blocking the potential development of testis-specific cell identity and gene-expression pathways (figure 6). This SOX9-blocking action of ovarian development is another reason why SRY expression and testis development must be initiated *before* ovarian development begins. If SRY misses its early developmental window of opportunity, SOX9 will be suppressed from being expressed later on, and will not get another chance. Correspondingly, in adult mouse testes cells, the continued expression of the transcription factor DMRT1 is necessary to maintain Sertoli cell identity, and to block their development into ovarian granulosa cells.[23] A similar function may occur in adult human testes.

These networked molecular intra-actions provide critical evidence that ovarian development is not a mere default, but an active and assertive process in itself that involves both the realization of ovarian identity and cell types, and the active inhibition of testis development. Similarly, the maintenance of testis identity is an ongoing and active genetic process, involving the inhibition of transformation to ovarian identity. Congruent with Günter Wagner's model of molecular character identity networks, the ovarian and testis identity signaling networks sustain their own expression and suppress the enactment of alternative cell identities by disrupting alternative identity networks.[24] These fundamentally bipotent possibilities remain an inherent capacity of gonads. In chapter 7, we will see that certain species of fish change individual sex during their lives, having evolved to actively exploit this inherent potential for gonad identity transformation.

Reproductive Tract Development

After gonad differentiation, the next events in sexual development involve the development of the anatomical tubes that form the internal reproductive tracts. Unlike gonad development, in which one bipotent organ can differentiate alternatively into a testis *or* ovary, the mammalian embryo begins with two *different* sets of precursor reproductive tracts—the Müllerian ducts and the Wolffian ducts. Individuals typically develop one set of these tubes into the reproductive tract, and get rid of the other (figure 9).[25] The testis and ovary are sexual homologs that develop from the same antecedent structure, but in mammals Male and Female reproductive tracks are true anatomical alternatives that are not homologous to each other. (As we will see in chapter 6, this means that both sets of ducts may be absent or present at the same time.)

In the absence of hormonal signals produced by the testes, during the seventh week after conception, the lower sections of the paired Müllerian ducts begin to fuse medially, and expand to form the uterus, the cervix, and the inner portion of the vagina, while the upper, unfused sections of Müllerian ducts develop into the paired fallopian tubes leading to each ovary (figure 10). Simultaneously, in the absence of hormonal signals produced by cells in the testes, the Wolffian ducts will degenerate and disappear. The muscles of the uterus and vagina are developed by seventeen weeks after conception.[26]

In contrast, after day forty-four, Sertoli cells in developing testes begin to produce anti-Müllerian hormone (AMH) which signals to cells of the

9. Human reproductive tract development

Top: Up to the seventh week after conception, both the Müllerian ducts (*gray*) and the Wolffian ducts (*black*) are present. Development of the Female or Male reproductive tract typically involves elaboration of one pair of these ducts and atrophy or elimination of the other. *Lower left:* Over the next two months, the Female reproductive tract develops by the central fusion and expansion of the Müllerian ducts to form the inner vagina, the uterus, and the fallopian tubes. The Wolffian ducts (*open lines*) degenerate (see figure 10). The outer portion of the vagina develops from the urogenital sinus. *Lower right:* During the third month, the Male reproductive tract develops from the Wolffian ducts in response to testosterone, and the Müllerian ducts (*open lines*) degenerate in response to anti-Müllerian hormone produced by the testes (see figure 6). Illustration by Rebecca Gelernter.

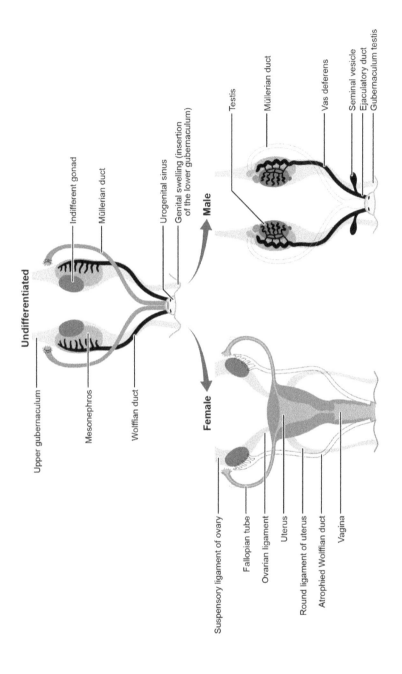

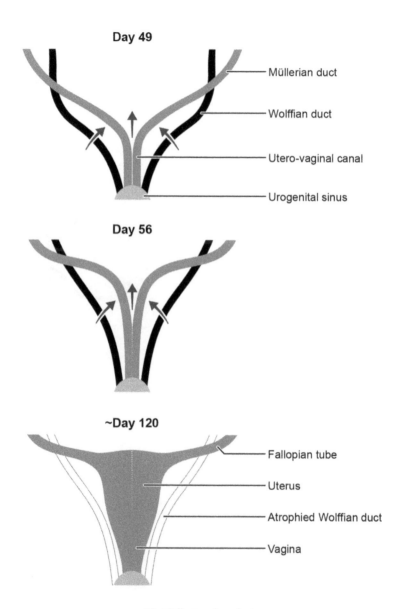

10. Müllerian duct fusion

Around day forty-nine (*top*), the paired left and right Müllerian ducts begin to fuse where they join the urogenital sinus. Over subsequent weeks (*middle and bottom*), the fusion proceeds upward to create the uterus and the inner portion of the vagina. The unfused left and right Müllerian ducts become the fallopian tubes. The outer portion of the vagina is derived from a compartment of the urogenital sinus. Illustration by Rebecca Gelernter.

Müllerian ducts to degenerate. Later, during weeks eight to eleven, in response to the hormone dihydrotestosterone that is intracellularly converted from testosterone (produced by Leydig cells in the testes), the paired Wolffian ducts begin to develop into the epididymis, the vasa deferentia, the seminal vesicles, and the ejaculatory ducts. Apparently, when the first vertebrate testes needed tubes to conduct sperm from the mature gonads out of the body, our ancient chordate ancestors pulled an evolutionary MacGyver maneuver, and co-opted some nearby excretory tubules from the primitive kidney—voilà, the Wolffian ducts were pressed into the service of vertebrate Male reproduction.

Occurring after the initiation of differentiated gonads from bipotent precursors, the development of our internal reproductive tracts demonstrates another possible path to materially realizing distinct reproductive possibilities—developing two redundant structures and selectively knocking one of them out.

Genital Development

The next event in sexual development is the differentiation of the external genitalia. As evolutionary biologist Neil Shubin explains in *Your Inner Fish*, our ancient vertebrate ancestry is manifest in each of our bodies. Non-mammalian vertebrates—including fishes, amphibians, reptiles, and birds—have only one ventral body orifice—the anus. The anus opens into a single internal body cavity, called the cloaca, onto which the digestive, urinary, and reproductive tracts all flow. In contrast, human bodies typically have either two (anus, urinary meatus) or three (anus, vagina, and urinary meatus) ventral body orifices. How does this variation arise? Like our fish ancestors, all embryonic mammals begin genital development with a single, cloacal cavity. Then, the distinct digestive, urogenital, or urinary and vaginal tract openings develop by the eversion and partitioning of the original cloacal space (figure 11).

In humans, the first event—the partitioning of the anus and digestive tract from the urogenital sinus—takes place around day thirty-two after conception when the embryo is still in a sexually undifferentiated stage. At this point, the opening of the cloacal cavity is surrounded by thickened margins, called the cloacal folds, with a distinct protuberance at the front, called the genital tubercle (figure 11). Then, the urogenital sinus is separated from the rectum by a growing wall of tissue, which is organized by Sonic Hedgehog expression in the cloacal endoderm.[27] This wall of tissue

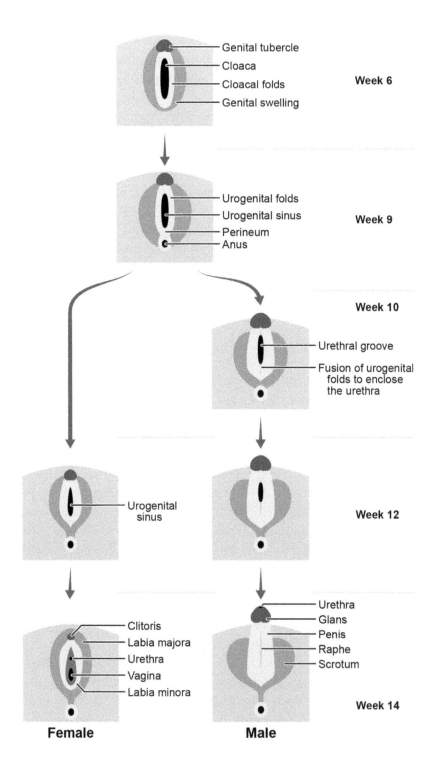

fuses together across the cloacal space to separate the external openings for the digestive tract and the combined urogenital sinus. This fusion, or raphe (pronounced *RAY-fee*), creates the perineal space between the anus and the genitalia.

Like the ovaries and testes, the external genitalia of humans grow from bipotent, precursor tissues—the urogenital folds, the genital tubercle, and genital swellings—into differentiated but sexually homologous possibilities. At this early stage, the button-shaped genital tubercle is distinguished from the neighboring urogenital folds by the expression of HOXD13, which is also expressed in the presumptive hands and feet at the same stage (see figure 16).[28] The formation of the genital tubercle is also organized by SHH expression in the nearby cloacal endoderm.[29] Around ten weeks after conception, depending on the presence or absence of testosterone and its derivative dihydrotestosterone, or DHT, the genital tubercle, the urogenital folds, and surrounding genital swelling will begin to differentiate into either a penis and scrotum, or clitoris and labia.

During weeks ten to eleven after conception, the cells of the undifferentiated urogenital folds express the paracrine signaling protein DKK,[30]

11. Development of human external genitalia

From weeks six to fourteen after conception (drawings not to scale). Through week nine, individuals are not sexually differentiated. **Week 6**—Cloacal folds develop around the cloacal opening with the genital tubercle in the front. **Week 9**—Partitioning of the cloaca into the rectum and the urogenital sinus, as well as the formation of the anus and the perineal raphe. Urogenital folds and genital swellings expand. *Female left:* **Week 12**—Genital tubercle forms the clitoris. Urogenital folds and genital swellings develop into the labia minora and majora, respectively. **Week 14**—Partitioning of the urogenital sinus into urethra and vagina with separate openings. Outer quarter of vagina, which is derived from the urogenital sinus, fuses internally to the Müllerian-derived, inner vaginal canal. Labia majora extend and connect anteriorly to form the mons pubis above the clitoris. *Male right:* **Week 10**—Urethral groove forms on the ventral surface of the genital tubercle between the urogenital folds. Lengthening of the genital tubercle and the urogenital folds begins to form the penis. **Week 12**—Fusion of the urogenital folds begins over the urethral groove starting at the base of the penis, proceeding forward. **Week 14**—Completion of urogenital fold fusion to create a tubular urethra. Fusion of the genital swellings along the midline to form the scrotum, extending the perineal raphe forward. Illustration by Rebecca Gelernter.

which locally suppresses WNT signaling in the urogenital fold. In the absence of WNT signals, the urogenital fold and genital tubercle will develop into the labia and clitoris, respectively. Later, between weeks twelve and fourteen, the descending inner vaginal tract, which is formed by the central fusion of the Müllerian ducts, extends outward, fuses to, and opens onto the urogenital sinus or the remaining cloacal space (figure 11, lower left pathway). Subsequently, the urinary tract differentiates completely from the vagina, creating separate, external body orifices for the digestive, urinary, and reproductive tracts. The extension of the upper portion of the vagina, which is derived from the Müllerian ducts, and its fusion to the lower part of the vagina, which is derived from the urogenital sinus, are active developmental processes involving cell-to-cell signaling and morphogenesis mechanisms that have been grossly understudied, and are still largely undescribed.[31]

In contrast, after week ten, testosterone produced by the testes can suppress the expression of DKK in the cells of the urogenital folds and genital tubercle, initiating their development into a penis and scrotum. Specifically, suppression of DKK signaling alleviates DKK's suppression of WNT, resulting in initiation of endogenous WNT signaling by cells of the urogenital fold and genital tubercle, which contributes to the differentiation of these tissues into a scrotum and penis. Recall that WNT belongs to the same paracrine signaling pathway that was critical to ovarian development only a few weeks earlier in embryogenesis. The ironic evolutionary co-option of the same paracrine signaling pathway critical to ovarian development for penis development demonstrates the inherently arbitrary, flexible, context-dependent nature of the "meanings" of molecular signals in organismal development.

To be effective in the embryonic development of the penis and scrotum, however, the small quantity of testosterone molecules produced by the testes at this age must be converted within the cells of the urogenital fold and genital tubercle into dihydrotestosterone, or DHT, by the enzyme 5α-reductase. The hormone DHT can elicit a much stronger gene upregulatory effect than testosterone can.[32] By ten weeks after conception, in response to DHT, a stripe of Sonic Hedgehog (SHH)–expressing cells forms along the ventral surface of the genital tubercle creating the urethral groove, which acts as a signaling center to organize the expansion, elongation, and tubular formation of the penis (figure 11).[33] The tissue of the left and right urogenital folds become integrated into the shaft of the penis. Migration of cells from the tissue between the rectum and the urogenital sinus into the penis expands its thickness, which displaces the stripe of SHH-expressing cells toward its center where they will later become the lining of the enclosed penile urethra. SHH signaling by the urethral groove cells also enhances the proliferation of the surrounding penis cells. By around fourteen weeks after conception, SHH signals from the urethral groove cells organize expansion of tissues on the left and right sides of the penis to fuse along the ventral midline to form the tubular urethra (figure 11). This tubulation process begins at the base of the penis and extends outward to form the urethral opening, or meatus. Lastly, in response to DHT during these weeks, the genital swellings on either side of the left and right urogenital folds elongate, and then fuse together along the central midline to create the scrotum, which will later contain the descendent testes. Thus, in Males, the raphe continues forward from the perineum along the genital midline of scrotum and the lower surface of the tubular penis.

Meanwhile, internally the ejaculatory ducts and ureters retain their embryonic connections to the urogenital sinus, so that Male mammals typically have only a single common external orifice for both urinary and reproductive tracts. Around week twelve after conception, the prostate gland, the bulbourethral glands, and the urethral glands begin to develop from protrusions of the urethral epithelium, also in response to DHT hormone derived in those cells from circulating testicular testosterone.

Lastly, between three months after conception to soon after birth, the testes migrate from their original internal position near the kidneys forward and down into the scrotum (figure 12). This migration is accomplished by the mechanical force exerted by the shortening of the lower gubernaculum ligaments, which originate on the gonads and insert on the genital swellings. As the gubernaculum shortens, the testes slide forward and down into the inguinal canals, and out through the abdominal wall, pulling the two vasa deferentia along behind them. The testes descend into expanding out-pockets of the peritoneal lining of the body cavity, which are called the "vaginal processes." Referring to these paired invaginations of the peritoneum, this historic name is confusing because these paired extensions of the peritoneal body cavity are not homologous with the vagina, which is comprised of the fused Müllerian ducts and cloacal endoderm. However, the idea that the testes are "birthed" into the fused labia majora that form the scrotum through the "vaginal" passages in the abdominal wall is a little piece of anatomical poetry that is, I think, well worth conserving. In

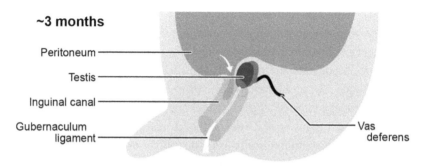
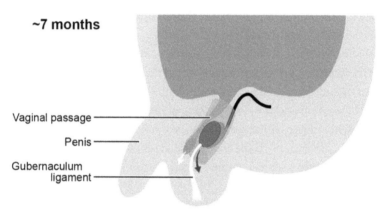
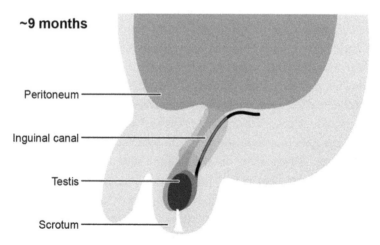

response to the mechanical force imposed as the gubernaculum shortens and the vaginal processes extend, the testes descend completely into the scrotum.[34]

The upper gubernaculum (or the suspensory ligament) also functions in displacing the ovaries medially. In their mature state, the lower gubernaculum ligaments connect the uterus to the tissue at the base of the labia majora, where they may function in Female orgasm.

Post-embryonic Sexual Development

Gender/sex development continues after birth in adolescence and throughout adult life. Although tracing the performativity of gender/sex development from birth through adulthood is beyond the scope of this book, I do want to establish a few connections between this performative view of molecular genetics and this vast literature on human gender/sex development.

After birth, human gender/sex development proceeds by hormonal, behavioral, and psychological mechanisms in interaction with the cultural and material environment. Of course, this process encompasses a myriad of important topics from molecular endocrinology to environmental stress, diet, religion, culture, and more. (Anne Fausto-Sterling and colleagues provide recent, feminist conceptual and empirical reviews of these scientific, psychological, and cultural literatures.)[35] In this section, I want only to comment on the role of hormones in post-birth gender/sex development.

In adolescence, gender/sex development is influenced by hormones secreted by the gonads, pituitary, adrenal glands, and other endocrine glands, which have been the subject of intense study for the past century.

12. Descent of the testis

At around **three months** after conception, an out-pocketing of the peritoneum (*white arrow*) extends into the inguinal canal to form the vaginal passage. By **seven months,** the vaginal passage elongates, and shortening of the gubernaculum ligament (*black arrow*) draws the testes forward, down, and into the inguinal canal (transparent tube). At **nine months** (or after birth), the testis passes through the abdominal wall and emerges from the inguinal canal into the membrane-lined space in the scrotum created by the extended vaginal passage. Illustration by Rebecca Gelernter.

These hormones, including estradiol, testosterone, prolactin, progesterone, and others have receptors distributed in cells of various tissues of the body. Their developmental and physiological effects are not determined by the hormones themselves, but by the cells that receive these signals, and the cascade of molecular and physiological events in those cells downstream from those receptors.

As in embryonic sexual development, hormone action in adolescent and adult bodies is also performative. The phenotypic effect of any hormone is contingent upon, and shares causal parity with, a diverse array of upstream and downstream metabolic pathway enzymes (which synthesize many hormone molecules from dietary precursors), the coevolved receptors that specifically bind those hormones, cytoplasmic modifiers, intermediary signaling pathway molecules, cofactors, and coevolved DNA-binding sites. The performativity of hormonal action—marked by the mediating and modifying effects of numerous other molecular agencies within the body—contributes greatly to the complexity of understanding their biological activities.

This complex intellectual territory has provided a particularly rich opportunity for cultural biases and concepts to influence the course of science. In *Testosterone: An Unauthorized Biography*, Rebecca Jordan-Young and Katrina Karkazis analyze the impact of the accumulated culturally based assumptions about the bodily functions of testosterone—a body of lore they refer to as "T Talk"—on testosterone research on human fertility, violence, power, risk taking, parenting, athleticism, and more. They document the long legacy of confirmation biases on analytical choices in hormone research that have been used to support concepts of categorical sex difference.[36]

For example, Jordan-Young and Karkazis describe the analytical method of "p-hacking," or conducting numerous statistical tests using different complex combinations of variables from a single data set in search of a significant result that will support a specific hypothesis. They also document how testosterone researchers have alternated frequently among different incomparable physiological measures of testosterone expression in search of "significant" results, and how the dosages used in experimental tests of testosterone action are frequently well outside of natural ranges in concentration. Overall, Jordan-Young and Karkazis find that the effects of testosterone on a diversity of classic phenotypes and behaviors associated with masculinity—including violence, power, risk taking, parenting, athleticism—are surprisingly inconsistent, variable, and contingent on other factors. In short, practically every simple or blanket statement one can make about the effect of testosterone on complex features of adult phenotype—including aggression, muscle mass, and body strength—is unsupportable. From a performative perspective, their findings demonstrate how individual organisms are not simply shaped by hormones; rather, individual organisms *use* their endocrine systems to shape and achieve their phenotypic ends.

Jordan-Young and Karkazis also document the critical and unexpected roles of the so-called sex hormones in human reproductive physiology. For example, recent biomedical research documents that testosterone plays a critical role in follicle maturation, which precedes ovulation, and is an *indispensable* contributor to Female fertility in humans. Equally surprising, anti-Müllerian hormone, which is used during development of XY embryos to eliminate the Müllerian ducts, is vital to management of folliculogenesis in sexually mature XX adults.[37] Furthermore, research also documents that estradiol plays a critical role in the maintenance of Sertoli cells in the testes, and is *indispensable* to Male fertility. Outside of reproduction, estradiol also plays an indispensable role in the maintenance of bone calcium density in both Females *and* Males. These findings demonstrate a diversity of functions for these so-called sex hormones—long associated with canonical masculinity and femininity—in sexual and nonsexual functions of *all* human bodies. Their roles in sexual development and reproductive physiology are only one of their evolved functions, and their reproductive roles are not exclusive to one sex or the other. This is another reason why it is erroneous to refer to estradiol, testosterone, and anti-Müllerian hormone as "sex hormones" at all.

Hormonal intra-activity has the unusual property that empirical questions about the concentrations and mechanisms of hormone action lead immediately to more questions about the mechanisms of hormone action. The body's homeostatic regulation of hormone-expression levels (i.e., "thermostatic" negative feedback) can confound both the investigation of observed natural variation in hormone expression, and the experimental manipulation of hormone action. The level of expression of hormone receptors frequently covaries inversely, or in a compensatory fashion, with hormone-expression levels. Such linkage acts to "turn up the volume" on weak hormonal signals, or muffle the effective impact of stronger hormonal signals. So, higher or lower hormone-expression levels are not necessarily associated with greater or less phenotypic impact. Establishing differences in hormone concentration is *not* equivalent to establishing differences in phenotypic impact of those hormones. Furthermore, experimental manipulations of hormone levels often fail to show phenotypic effects because the body can compensate by adjusting other components in the hormonal signaling pathway to *maintain* the same level of function. Accordingly, hormonal manipulation experiments often fail to create significant results without unnaturally large dosages that overwhelm the body's homeostatic endocrine regulatory system. However, the phenotypic effects of unnaturally large megadoses of hormones have little to tell us about the actual function of these molecules in the body.

The role of hormones in sexual development and reproductive physiology and behavior have made them a particular focus of research on human sex differences. As I discussed in chapter 2 ("Sex Difference versus Sexual Difference"), the existence of statistically significant quantitative differences between average traits of individuals assigned to Male and Female sex classes does not necessarily demonstrate bimodal distribution of trait values among people. If quantitative differences in a trait between the sexes are not even close to diagnostic, then why are they being investigated as such?

None of these criticisms of hormonal research on sex differences—that is, between classes of individuals assigned as Male or Female—should be construed as arguments for the absence of sexual difference—individual variations in phenotype influenced by variations in hormone action. Nor are they arguments in favor of the social construction of individual gender/sex. Rather, these critiques suggest that future research should focus more productively on understanding individual varieties of hormone action—a research question that arises from a systems-level focus on the performativity of the phenotype. Hormones are not the causes of the adult sexual phenotype. Rather, individual bodies employ their endocrine systems in the realization of their phenotypes.

Sexual Development Summary

Sexual development involves a process of realizing different reproductive possibilities from an embryonically uniform and undifferentiated state. In the case of gonads and genitalia, human sexual development proceeds by the differentiation of bipotent homologous precursor tissues toward different functions, whereas human reproductive tracts develop from anatomically distinct precursors. Sexual development typically involves the maturation and differentiation of one set of ducts, and the atrophy or active elimination of the other. In all cases, sexual differentiation is realized through inter- and intracellular molecular signaling using paracrine and endocrine pathways. The role of these genetic communication and regulation pathways is to organize, spatially coordinate, and materially realize the differentiation of reproductive organs, tissues, and cell types.

None of the individual genes, signaling pathways, or hormones can be cited as a singular material cause of sexual differentiation. None can "determine" the sex of the individual. Rather, the process of sexual development results from aggregate intra-actions of numerous signaling pathway molecules, cells, tissues, and developmental signaling centers. The process can be best understood as an inter- and intracellular molecular discourse among diverse, hierarchically distributed agencies.

This chapter has focused on the complex genetic networks that contribute to the creation of differentiated gonads, reproductive tracts, and genitals. In the next chapter, we will explore the anatomical and functional consequences of variation in these molecular intra-actions which further reveal the performative dynamics of sexual development.

CHAPTER SIX

Variations in Our Sexual Development

In his twentieth-century introduction to the first publication of the nineteenth-century autobiography of Herculine Barbin, identified as a "hermaphrodite," philosopher Michel Foucault asks:

> Do we truly need a true sex? With a persistence that borders on stubbornness, modern Western societies have answered in the affirmative. They have obstinately brought into play this question of a "true sex" in an order of things where one might have imagined that all that counted was the reality of the body and the intensity of its pleasures.[1]

In this chapter, we will investigate Foucault's provocative question through an exploration of the "reality of the body," and the variability in human sexual development in terms of the performative framework.

In chapter 5, we explored the molecular, genetic, and physical processes involved in the development of the human sexual phenotype. The investigation of embryonic sexual development we have pursued makes clear that individual sex is not determined by any particular gene, chromosome, or hormone. Nor is it a fact etched in our genes, or determined by chromosomes, hormonal profiles, or transcriptomes (the description of all the genes expressed by a tissue or cell type at a given time). In other words, sex is not an innate property of individual genomes or embryos waiting to be represented in the material world, but a suite of reproductive possibilities realized, or enacted, *by* those individuals through development. The sexual phenotype is an active, assertive developmental achievement by the organism.

This reframing is critical to understanding the substantial variability in human sexual development. Anatomical variation in human sexual anatomy *is* strongly, bimodally distributed—that is, the vast majority of human individuals develop the anatomically and physiological capacities to sexually reproduce either through the production and delivery of sperm, or through the production of eggs, pregnancy, lactation, and so on. This bimodal distribution is evidence of the strong canalization (i.e., evolutionary reduction of variation) of sexual development, but it is not sufficient to imply, support, or create an essential (i.e., inherent and indispensable) scientific fact about the sex of a genome, a zygote, or its phenotypic fate. Understanding variation in the development of sexual anatomy and physiology scientifically confirms Foucault's suspicion that there is no such thing as the "true sex" of an individual. Again, to paraphrase Simone de Beauvoir, the scientific data on variation in sexual development show that to be a woman, or to be a man, is to have become one. By rejecting the a priori assumption of two constitutively or inherently distinct biological sex classes, we reframe the traditional discussion of sex difference—the differences between binary sex categories—to the broader analysis of sexual difference—the full gamut of realized variations in sexual phenotype independent of imposed identity categories.

In this chapter, we will focus on how variations in the molecular, cellular, and tissue level enactment of the sexual self contribute to variation *within* as well as between the predominant modes of reproductive anatomy. By thinking about bodies that may be quite different, we learn anew that all bodies develop by the *same* process of individualized self-realization.

Of course any discussion of variability in human sexual phenotype raises the complex history of how Western and other societies have conceived of sex and its variations. Historian Thomas Laqueur documents that prior to the Enlightenment, sex in Europe was not thought of as a binary pair of opposites, but as a single feature varying in its degrees of expression.[2] The Roman anatomist Galen theorized that Male and Female genitals were the same but inverted relative to each other—the penis was an inside-out vagina, for example—a conception that continued well into the seventeenth and eighteenth centuries. Laqueur argues that the transition to a binary model of sex happened *despite* scientific discoveries of sexual homology, sexual variation, and other scientific advances, not *because* of them. Instead, the conceptual transition to a binary concept of individual sex was driven by cultural rather than scientific criteria.

Scientific concepts of individual humans as inherently Male-and-Female continued into the early twentieth century, and were present in works of Sigmund Freud, Carl Jung, and early sexologists. Forty years before Foucault's philosophical questioning, the pioneering Canadian physiologist and endocrinologist Frank Lillie wrote, "There is no such biological entity as sex" and lamented that "we have been particularly slow . . . in divesting ourselves not only of the terminology but also of the influence of such ideas." Likewise, in 1940 developmental-evolutionary biologist Conrad Waddington wrote that "the alternative between maleness and femaleness is not so definite."[3]

As Myra Hird, Anne Fausto-Sterling, and others explain, the scientific insistence on a *real*, or essential, sex of the body didn't arise until the second half of twentieth century with the advent of experimental genetics, and the medical technology to take control of human sexual variability through surgical and hormonal treatment—what Sarah Richardson calls the "geneticization" of sex. The history of the medical treatment of human sexual variation has been investigated by Myra Hird, Suzanne Kessler, Anne Fausto-Sterling, Katrina Karkazis, Georgiann Davis, Jules Gill-Peterson, Beans Velocci, and others. In aggregate, their analyses describe how physicians and scientists sought to reinforce, indeed create, binary sexual clarity from human variability, insisting on fundamental sex differences that had been thoroughly rejected by developmental biologists and physicians only a few decades before. The medical treatments developed in the late twentieth century involved projecting onto the infantile body the cultural requirements of function in heterosexual intercourse. In this way, cultural expectations of gender deeply impacted scientific concepts and medical treatments of embodied human sexual variation.[4]

Variability in human sexual phenotype is not the result of mistakes, or errors, in binary sex determination; rather it is evidence that there is no inherent, individual sexual binary to begin with. Learning about variation in sexual embodiment teaches us about the unique, emergent nature of each of our own individual bodies. We are *not* products of lawlike, genetic determination mechanisms, but iterative, individualized bodily self-enactments situated in specific biological and social environments. We are *performance all the way down*. The purpose of this chapter is to show that those molecular mechanisms were actually discovered *through* the biomedical investigation of the queer bodies and lives in which they vary. A broader scientific and cultural understanding of the performativity of human sexual development can contribute to a safer world for all, including infants born with sexually non-canonical bodies.

Terminology and the Framing of Embodied Sexual Variation

The cultural, scientific, and medical histories of the scientific vocabulary used to categorize and diagnose variations in human sexual anatomy have been investigated in detail by others, so I will only briefly summarize that history here.[5] In Western cultures, sexually indeterminate, non-canonical, or anatomically variant individuals were historically referred to as hermaphrodites, implying a mixture of Male and Female traits, or an intermediacy between Male and Female. In the nineteenth century, the term *hermaphrodite* was taken up by Western medicine to describe individuals with "ambiguous" genitalia and sexual physiology. With the development of a concept of differentiated gonads as the causes of sexual development in the early twentieth century, the new term "pseudohermaphrodite" was introduced into the medical literature to describe individuals with gonad morphologies (i.e., "genuine" sex) that differed from their expected genital anatomy or other bodily features (i.e., apparent sex). For example, *female pseudohermaphrodite* was used to refer to an individual with ovaries, a penis, and a scrotum.

In the late twentieth century, physicians concluded that the terms *hermaphrodite* and *pseudohermaphrodite* had unwanted sexual, even mythological associations. The term *intersex* was proposed and adopted into biomedical usage to describe individuals with variable genitalia, or other unexpected forms of sexual development or infertility.[6] By the twenty-first century, however, discomfort with the term *intersex* arose in various contexts for a variety of reasons. Some patients and families thought the term had unnecessarily sexual connotations or pejorative associations with variations in sexual orientation. (How infants would grow up to participate in heterosexual relations has been a persistent biomedical and parental concern in thinking about their bodily sexual variation.) Meanwhile, a new generation of survivor/activists who had been subjected to genital surgeries without consent as infants and children began to organize against the medical characterizations of sexually variable bodies as pathological, and many adopted intersex as a powerful, new political and sexual identity.

In 2005, a conference of pediatric endocrinologists, urologists, medical geneticists, and surgeons convened in Chicago to reconsider the standards of care and proposed to characterize variations in sexual development and anatomy as "disorders of sexual development" (DSD). Katrina Karkazis and Georgann Davis have each argued that this was a political maneuver to dissociate and insulate the medical community from the unfortunate history of abusive medical treatments of many intersex patients in the twentieth century, while maintaining their social and intellectual control over how to think about variability in human sexual anatomy.[7] The "disorders of sexual development" framework maintains, and reinforces a fundamentally pathological conception of genital variation for which biomedical technology provides the necessary treatments.

Most recently, in recognition of the fact that many developmental variations in sexual phenotype are not medically pathological in any way beyond their rarity, a team of medical geneticists and pediatric endocrinologists led by Dorien Baetens at Ghent University have proposed that the term *disorders of sexual development* should be replaced by *differences in sexual development* (also DSD). This proposal constitutes a biomedical reframing of sexual variation away from sex difference—pathological variations from a priori binary expectations—toward sexual difference—the full spectrum of individually realized variations in reproductive anatomy and physiology. When I use the term *DSD* hereafter, it is to refer to *differences*, not disorders, in sexual development.

The history of continued revision and instability of the terminology used to refer to human sexual variability document ongoing unease with the socially destabilizing fact of sexual variation and the political struggles by affected individuals themselves, families of variable children, and physicians over control of how we think about these phenomena. The result has been a "euphemism treadmill," similar to changing vocabulary used to refer to race, ethnicity, sexual orientation, gender, and other socially divisive categories, as concepts and language respond to the social forces generated by the biomedical community, parents of impacted children, and adults subjected to various medical treatments as infants and children.[8] At stake is whether anatomically variant individuals, their parents and families, or the biomedical community will determine how we view and think about sexual variation itself. Who should speak for the interests of the infants themselves?

Moving beyond Pathology

An important goal of exploring developmental variation in human sexual phenotype is to push back upon, and overcome, the history of biomedical control over the topic, and to recognize that many differences in sexual development are not in any way pathological to the health of the individuals themselves. Although some physiological conditions can pose life-threatening challenges to individual health, the correlated impacts of these syndromes on sexual anatomy are not necessarily pathological themselves. Despite the fact that vertebrate animals, including ourselves, have evolved to develop adult bodies that facilitate specific reproductive possibilities, creating distinct reproductive differences from common, uniform, embryological precursors remains a developmentally challenging and persistent biological problem. Consequently, a consistent number of individuals simply defy binary classification, and most of these variations are neither harmful nor debilitating medical conditions that should necessitate infant and childhood surgeries.

By viewing development as a performative phenotypic landscape that the organism actively *ascends* as it grows and differentiates, the distribution of human sexual anatomy can be visualized as two primary peaks connected by a broad ridge of sexual possibilities with multiple, locally stable peaks (figure 4C). This broad ridge of sexual possibilities is populated by many extraordinary people. People that have been depicted in human arts and culture for millennia. People that become world-class Olympic athletes. People that become activists and academics. People that may be us or members of our families. People that can thrive and flourish in unique ways.

The existence of an expanse of human sexual possibilities also means that the morphological peaks in the landscape are flatter and broader than our cultural and biomedical expectations of Female and Male. Because human sexual development is complex, including differentiation of multiple bipotent tissues and the development of non-homologous alternatives (i.e., Müllerian and Wolffian ducts) by a complex combination of paracrine and endocrine signaling networks, the sexual development landscape includes multiple pathways to intermediate phenotypic peaks that lead to unexpectedly queer phenotypes. Our traditional cultural belief in—indeed, perhaps even cultural need for, and insistence upon—a strict sexual binary has distorted our scientific perceptions. This has led many people, including biologists, medical researchers, and physicians, to "tidy up" the situation by actively sorting variable people into binary categories, and pathologizing the rest. As lots of previous scholarship has shown, the medical approaches to the "tidying up" of this variation involve applications of gender theory disguised as objective science—deployments of cultural concepts about what Male and Female mean, what penises, clitorises, and vaginas are for, what sorts of individuals can be expected to function, thrive, and be happy. The reproductively nonfunctional, short vagina of an otherwise Male person would commonly be considered a pathological "intersex" condition, while the presence of nonfunctional nipples on Male human bodies is not. However, the rarity of the former does not make it anymore medically pathological than the latter. Infant and childhood surgical interventions to "normalize" differences in sexual anatomy may be motivated by a compassionate, though paternalistic and patriarchal, desire to help variable people "fit in" to social expectations, but such efforts necessarily reinforce the (often unconscious) gender concepts of the biomedical community, and can cause real and substantial harm to the capacity of these individuals to experience sexual pleasure as adults.[9]

The biomedical and scientific focus on differences between sexes has also obscured variability *within* sex categories. In other words, the sexual "normal" is more variable than we have allowed ourselves to realize. For example, anatomical variation in the position of the urethral opening—or urinary meatus—of the penis is a developmental condition referred to as *hypospadias*. However, as Anne Fausto-Sterling points out, the position of the urethral opening in men is itself highly variable. A biomedical survey based on a sample of five hundred men found that only 55 percent of them had the urethral opening in the what the researchers had defined as the "normal" position at the outset of the study—in the outer third of the glans.[10] As a consequence, surgeries to address the medical diagnosis of "hypospadias" may be conceived around a cosmetic ideal that represents only a slight majority of Males.

Like the size of Male penises, there is also tremendous variation among women in the absolute and relative sizes of the labia majora and labia minora. Yet, this extensive variation is poorly appreciated. Driven perhaps by the cultural trends of greater exposure to pornography and more extensive pubic hair removal, increasing numbers of women and girls are now motivated to pursue an "ideal vulva," with an apparently more "tidy" appearance. Aesthetic dissatisfaction with vulvar appearance may also be triggered by racial associations with the common darker melanin pigmentation of the labia minora that project beyond the labia majora in about two-thirds of women. Labiaplasty, or cosmetic surgical reduction of labia minora, is now among the fastest-growing plastic surgical procedures. Plastic surgeons themselves recognize that most women and girls considering these procedures have little knowledge of how variable the sizes, relative sizes, and appearances of labia really are, but there are few modern data describing this variation. Misperceptions about vulvar appearance may also be fed by the archaic anatomical names "labia majora" and "labia minora," which plainly imply norms about relative size that are not based on any description of natural variation. (Perhaps they should be changed to *medial* and *lateral*, or *inner* and *outer labia*?) Furthermore, because the inner labia (labia minora) are sexually homologous with portions of the shaft of the penis, variations in labial size may genetically covary, or be correlated with, variation in penis size. I want simply to note here that the extensive variation in labial morphology documents that differences in sexual development occur as much *within* traditional normative sex categories as between them.[11]

Of course, many individuals that would be characterized as biomedically pathological at birth can thrive and realize fulfilled lives of diverse personal and sexual possibility. In the early twentieth century before surgical interventions were common, the American urologist Hugh H. Young at Johns Hopkins portrayed a few such remarkable individuals, whom he described as "practicing hermaphrodites," or "individuals who alternately led the sexual lives of Males and Females." For example, Emma T. was a twenty-six-year-old African American who was raised as a girl and lived as a woman, but had both a vagina and a hypospadic penis. She was married twice and had frequent vaginal intercourse with her husbands, but she also maintained a series of more sexually satisfying relationships with women lovers with whom she had intercourse with her penis. In this brief portrait, Young described Emma's overall health as "splendid," and that she was "fond of reading." Emma T. told Young that "I feel sometimes that I should like to be a man." But when asked by Young whether she would like to be surgically "made into a man," she replied, "Would you have to remove that vagina? I don't know about that because that's my meal ticket." Emma explained that her husband supported her well, and if she left him she would have to support herself on her own. (The interview occurred during the height of the Great Depression.) "So," Emma concluded, "I'll keep it, and stay as I am."[12]

Even in Young's short clinical account of Emma T., we get a rare glimpse of an extraordinary individual living in a unique fashion in the broad space between our binary cultural expectations. Interestingly, Emma T.'s choices may have been facilitated by her race. In an historical analysis of the treatment of intersex and transgender patients at Johns Hopkins University in Baltimore in the early twentieth century, Jules Gill-Peterson documents that Black families and communities were quite accepting of intersex children, and skeptical of medical interventions that reinforced binary sex. In response, physicians characterized their beliefs as ignorant and irrational.[13]

This brief glimpse of Emma T.'s life contrasts strongly with the still common biomedical view that the birth of a child with "ambiguous" genitalia should be considered as an "endocrine emergency."[14] Biomedically diagnosed differences in sexual development are surprisingly frequent, arising somewhere between 1 in every 250 human births (for severe hypospadias), and 1 in 2,000–4,500 human births for other genital variations. The failure of one or both testes to descend into the scrotum, called cryptorchidism, occurs in about 3 percent of full-term and 30 percent of premature Male births, and has been characterized as the most frequent human "birth-defect."[15]

Although some differences in sexual development can be associated with painful, functionally compromising, or physically debilitating conditions that can properly be classified as medical disorders, most conditions recognized historically as "disorders" of sexual development are not medical problems to the children themselves, but social phenomena. As Kessler, Fausto-Sterling, Hird, Karkazis, Davis, and Gill-Peterson have documented, it is the strictly binary sexual expectations that have led variations in genital morphology to cause psychological, social, and emotional distress to parents, families, physicians, and communities, not the variations in and of themselves.[16]

Likewise, although it is often viewed as a pathology, infertility is not, in itself, a challenge to individual health, vigor, or life span. Whether infertility is perceived as a medical problem is an individual experience, which can be exacerbated by social and cultural expectations that reproduction is a necessary part of a normal, successful, or proper adult life. So, like anatomical differences in sexual development, infertility can be as much a social problem as a medical problem. Similarly, as we will see, some bodily variations may be appropriately characterized as asexual. In the human context, asexuality may be more frequently thought of as a sexual orientation, but some DSDs are characterized by the absence of reproductive anatomy including gonads, internal reproductive ducts, puberty, etc. The existence of such biological states do not imply, however, that asexuality is somehow an inappropriate sexual orientation for an otherwise fertile individual.

In the sections to follow, I review differences in sexual development associated with variations in chromosomes, genes affecting gonad development, and genes affecting reproductive tract development. I then examine sexual variations associated with noncoding genome sequences and environmental influences. I conclude the chapter by exploring the biomedical investigation of differences in sexual development as a queer science.

Chromosomal Contributions to Differences in Sexual Development

As we discussed in the previous chapter, chromosomes have played an important role in scientific history as the "causes," and "definition," of individual sex. It is therefore worthwhile to start with a discussion of the role of variations in chromosomes and chromosome number in sexual development. Some differences in sexual development arise from the absence of a chromosome, or from the presence of an extra chromosome. Such variations in chromosome number arise randomly in meiosis during the production of eggs and sperm which unite to form the zygote. X0 individuals have only one X chromosome and lack a Y chromosome, a condition called Turner's syndrome. These individuals develop ovaries but may have problems with fertility and kidney function. XXY individuals, a condition called Klinefelter's syndrome, may present as Male, but have small testes and are unable to produce viable sperm. However, not all X0 and XXY individuals are infertile. Furthermore, not all variations in X and Y chromosome number contribute to variation in sexual development. For example, XXX and XYY individuals develop as Females and Males, respectively, and may experience no fertility or health problems at all. So defining human Males and Females as XY and XX, respectively, is patently incorrect.

A quite rare variation in sexual development can arise from genetic chimerism, or the presence in a single body of both XX and XY cells. Genetic chimeras can arise either by the fusion of two fertilized eggs, or by the fertilization of a single egg by two sperm. The adult body becomes a mosaic of tissues with two different genomes. The result can be the development of both ovaries and testes, ovotestes (which are gonads that combine cellular, anatomical and physiological features of ovaries and testes, and usually do not produce fertile gametes), and variable genitalia. One patient was also noted for having irises of two different colors—brown and hazel—each associated with their XX and XY genotypes.[17]

Genetic Variations in Gonad Development

Differences in sexual development can also arise from variations in the many genes involved in sexual differentiation. A recent biomedical review identified twenty-nine and seventeen different genes that have been associated to date with differences in sexual development in XY and XX humans, respectively.[18] Just about every detail in the summary of genetic pathways of sexual development in chapter 5, and many more that we glossed over, has been discovered as potentially contributing to differences in human sexual development, and more such variants are being identified all the time. However, despite advances in genomic technologies, the molecular mechanisms of sexual development are so complex that the genetic variations associated with differences in sexual development are yet unidentifiable in nearly 50 percent of cases.[19] In this section, I will discuss developmental outcomes of variations in genes involved in the development and differentiation of gonads.

In an unfortunate reification of the idea of decisive Male power, SRY has frequently been referred to as "the master switch" of mammalian "sex determination." A recent Google Scholar search for the terms *master switch* and *sex determination* returned over one thousand peer-reviewed scientific papers.[20] But a simple look at the details of the process demonstrate that SRY is not the *master* of anything. Nor can it *determine* anything on its own. Rather, to influence gonad development, SRY must function through a network of coevolved intra-actions with many other molecular agents within the bipotent gonad (figure 6). Functional variations in SRY protein, or any of the other components it interacts with downstream, can prevent, disrupt, or delay the development of testes, leading to the development of ovaries or ovotestes.

Because SRY is among the very best-studied of all the genes involved in sexual development, it provides particularly good insight into the complex variety of contributions of genetic variation that can make variation in sexual phenotype. SRY gene sequence variations are also the most commonly observed genetic variation found in XY individuals that fail to proceed to puberty.[21] Most such individuals are usually considered to be Female at birth and are raised as girls.

The amino acid sequence of the human SRY protein contains several regions that are critical to its signaling function including: (1) its DNA-binding domain, (2) two nuclear translocation sites that bind proteins in the cytoplasm that usher the molecule through the pores in the nuclear envelope and into the nucleus, and (3) additional phosphorylation and acetylation sites (figure 7). To initiate SOX9 expression, the SRY DNA-binding domain must attach to the minor groove of a specific, four base pair sequence in the nuclear DNA—ACAA/TGTT—and induce a 60°–85° bend in the DNA helix (figure 13). (This event constitutes the most fundamental role of kink in Male gender/sex development.) The most common, loss-of-function variations in the SRY gene are mutations to this binding domain that either inhibit its binding to the DNA, or prevent it from bending the DNA after binding.[22]

Before binding to the DNA, however, transcription factor proteins like SRY must be translocated from the cytoplasm where proteins are assembled into the cell nucleus. Moving SRY into the nucleus requires the binding of calmodulin and transportin-β to their respective nuclear translocation sites on the SRY protein (figure 7). XY individuals with loss-of-function variations at either of these two amino acids may fail to develop testes, and proceed to develop ovotestes or ovaries, and Müllerian duct–derived reproductive tracts. In humans, translocation of SRY into the nucleus is further facilitated by acetylation of a specific amino acid (lysine) at a highly conserved site on the protein, which facilitates the binding of transportin-β. Meanwhile, in humans and other primates (but not in horses or mice), the DNA-binding function of SRY also requires the phosphorylation of specific amino acids well outside of the binding domain.[23] Like the grammatical declensions of a noun or the tense of a verb, each of these tiny modifications to the surface of the 204-amino-acid-long SRY protein is essential to its molecular communication function.

Even after SRY initiates the first few gonad cells to express SOX9, testis development can also be disrupted by genetic variations in the subsequent molecular players that act in the spatial wave of recruitment of additional gonad cells to express SOX9, including the paracrine signals FGF9 and PGD2 (figure 6). For example, loss-of-function mutations to either one

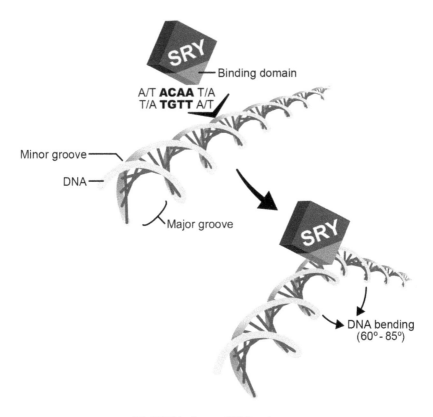

13. SRY binding to SOX9 enhancer

To promote SOX9 expression during testis development, SRY must bind to a specific DNA sequence—ACAA—in the minor groove of the DNA within the SOX9 enhancer and induce a 60°–85° bend in the DNA helix. Failure to bind to the SOX9 enhancer or failure to induce the appropriate kink in the DNA after binding will prevent SOX9 expression in the developing gonad, leading to the development of ovaries or ovotestes. Illustration by Rebecca Gelernter; redrawn from Kashimada and Koopman (2010).

of these genes, or to the genes for their receptors, can mute the positive feedback loop that enhances testis identity development. In other words, if nearby gonad cells cannot hear the first few Mickey Rooney cells declare, "*Hey! Let's put on a show!*"—a signal that is passed on and amplified downstream by FGF9 and PGD2 pathways—they will not be recruited to become testis cells. Loss-of-function variations in the FGF9 and PGD2 pathways can result in the development of ovotestes with testicular morphology in the center surrounded by ovarian tissue around the outside, anatomically preserving the extinction of the initial wave of SRY recruitment to testis cell identity. Thus, the absence of a transmembrane receptor for a specific paracrine signal constitutes an inability to attend to certain messages of other nearby cells. Managing the differential "inaudibility" of paracrine and endocrine signals is a critical element of developmental biology. Analogously, in *The Epistemology of the Closet*, Eve Sedgwick describes the role of willful incomprehension—the social inability to perceive and respond to queer idiom—within the nineteenth- and twentieth-century social discourse around same-sex sexual attraction.[24]

The initial balance between the pursuit of testis or ovary identity is so tenuous that genetic variations in multiple signaling pathways can also lead to the development of testes in XX individuals. For example, testis development can occur in XX individuals in response to mutations in RSPO1, NR5A1, or WT1, genes involved both in the early development of the bipotent gonad and in SRY upregulation (figure 6). Likewise, a lack of WTI+KTS function in XY individuals allows the ongoing degradation of SRY messenger RNA before it is translated into a functional protein, which can also prevent development of testes (figure 8).

Alternatively, a copy of SRY can be translocated to the X chromosome from the Y via rare crossing over during meiosis. The presence of a functional SRY gene on the X chromosomes can lead to a variety of possible outcomes, including fully fertile testis development and Male phenotype (the most common condition), a typical Male phenotype with infertility, testes with variable genitalia, ovotestes, or even one ovary and one testis. The variation in outcomes can depend upon the size of the Y chromosome segment translocated, and whether that segment includes all the appropriate regulatory binding sites for SRY.[25] (See also "X Chromosome Inactivation" below.)

The diversity of genetic pathways that are known to contribute to differences in human gonad development reflect the failures of the "master-switch" concept, the concept of genes as strict causal explanations of material bodies, and the broader concepts of "sexual determination" and "sex reversal." Any regulatory gene can only realize its phenotypic effects through intra-actions with a contingent network of diverse, coevolved molecular agents that all share causal parity. However, as I described in chapter 5, there is still an appropriate sense in which SRY can be a genetic difference maker—that is, SRY can play a distinct role in influencing the course of individual sexual development, and it appears to be among the earliest genes yet identified with such an impact. But the genes that initiate SRY expression are not yet known, indicating that previous intellectual satisfaction with the "master-switch" concept has contributed greatly to research on SRY function, and to our continued ignorance about SRY expression regulation.

X Chromosome Inactivation

Another factor that can affect the development of XX individuals with a functioning copy of SRY on an X chromosome is the phenomenon of X chromosome inactivation. X chromosome inactivation can have profound effects on the XX body. To compensate for the fact that the genes on each X chromosome have evolved to function two-thirds of the time in XX individuals and one-third of the time XY individuals, XX individuals must somehow adjust the quantity of expression (or "dosage") of genes on the X chromosome to match the levels achieved in XY individuals. In mammals, this is accomplished by inactivating (essentially, moth-balling) one of the two X chromosomes in every cell of the body by wrapping it in dense heterochromatin, making it inaccessible for transcription.

X inactivation occurs in the embryo at the blastocyst stage (about two hundred to three hundred cells), or about the time of implantation in the uterus. As the body grows, all cells inherit the inactivation state of their progenitors' cells, leading to a unique mosaic pattern of X inactivation throughout the tissues of the entire body. This phenomenon results in tortoiseshell and calico coat-color patterns in Female cats that have different alleles for fur color genes located on each X chromosome. The process is entirely absent in XY mammals; XY Male cats cannot be calico (though XXY, or Klinefelter, cats can be). In other words, XX individuals are genetic mosaics because their bodies contain two populations of cells that express genes from one X chromosome or the other. Pioneering geneticist and pediatrician Barbara Migeon has written a detailed, far reaching, and fascinating review of the mechanisms of X inactivation in humans and its complex consequences for women's health.[26]

The process of X inactivation at the blastocyst stage is an additional source of stochastic contingency, or randomness, in the development of *every* individual XX mammal. The pattern of X chromosome inactivation will even be entirely different, and independent for identical XX twins.

This means that XX individuals with a complete, active copy of SRY on one of their X chromosomes can have quite different sexual developmental outcomes depending upon the X inactivation state of the cells that give rise the bipotent gonads and other tissues of the body. These conditions may lead to individuals with an active ovary on one side of the body and an active, descended testis on the other, typically with a hypospadic penis.[27]

Other genetic mutations that lead to constitutive overexpression of SOX9 in XX individuals can also lead to testis or ovotestis development in the absence of SRY. Because SOX9 has many other functions in development of other parts of the body, these individuals may also have a host of associated and developmental differences unrelated to sexual anatomy and function.[28]

Genital and Reproductive Tract Development

After the differentiation of Leydig cells in testes and theca cells in ovaries, the gonads can assume their functions as endocrine glands producing hormones that circulate generally throughout the body. Interestingly, however, the role of hormones in embryonic development of the genitals and reproductive tract is notably distinct between testes and ovaries, and this has important impacts on the realized differences and possibilities in sexual development. (See also "Performative Scientific Hypotheses" in chapter 8.)

An important class of differences in sexual developmental are a consequence of the inability to receive or process embryonic steroid hormonal signals that play a role in reproductive tract and genital development in XY individuals. For example, genetic variations in the gene for 5α-reductase can prevent the conversion of testosterone into DHT. Because the embryonic development of the penis and scrotum require local conversion of testosterone into DHT by the cells of the genital tubercle, urogenital folds, and genital swellings, the absence of a functional 5α-reductase enzyme can prevent the embryonic differentiation of these tissues in XY individuals. At birth, such individuals typically present as Female with labia and small vagina but with a large clitoris or small hypospadic penis. But internally these individuals have undescended testes with complete vasa deferentia and ejaculatory ducts. Because their fully functioning testes produce anti-Müllerian hormone (AMH) during embryonic development, the fallopian tubes, uterus, and upper vagina derived from the Müllerian ducts are all absent. The small vagina consists of the outer portion derived from the urogenital sinus. At puberty, however, the testes of these individuals begin to produce much higher quantities of testosterone which can trigger renewed development of the penis, descent of the testes into the expanded labia majora to form a bifid scrotum—a pair of sacs separated from one another at the midline—and other adolescent Male characteristics including a lower voice, and additional facial and body hair (figure 14C). Although most of these individuals are completely fertile, insemination may require accommodations because of the hypospadic position of the urethral meatus at the base of the penis.[29]

Another class of differences in sexual development are related to the lack of functioning steroid receptors, referred to as androgen or estrogen insensitivity syndromes, that can disrupt the differentiation of tissues that rely on these hormonal signals.[30] Lack of a functioning "androgen receptor" gene produces complete androgen insensitivity syndrome (CAIS). CAIS individuals are XY with testes, but their urogenital folds and genital tubercle develop into labia and a clitoris, because these tissues will be insensitive both to testosterone and to DHT. However, their active testes still produce AMH, eliminating the Müllerian ducts and their derivatives.[31] Alternatively, genetic disruption of AMH production or reception in XY individuals can lead to retention of Müllerian duct derivatives, and to the development of fallopian tubes, uterus, and vagina along with mature testes.

AMH, 5α-reductase, androgen receptor, and other hormone receptors have decisive impacts on the development of the sexual phenotype of XY individuals even though they are far "downstream" of the action of SRY in the initiation of testis identity and development. Thus, we can see again the failure of the concepts of SRY as a "master switch," and of "sexual determination" as a switchable binary process, and the intellectual necessity within this awkward paradigm for a concept of "sex reversal."

In mice, the absence of a functioning estrogen receptor protein does not affect embryonic sexual development in either XX or XY individuals but reduces adult fertility of both. In humans, however, the absence of a functioning estrogen receptor (ESR2) can block or prevent the primary development of the gonads. Although the details are not yet understood, this finding implies an unknown role of estrogen signaling in the origin of the bipotent gonad.[32]

Disruptions in the fusion of the paired left and right Müllerian ducts

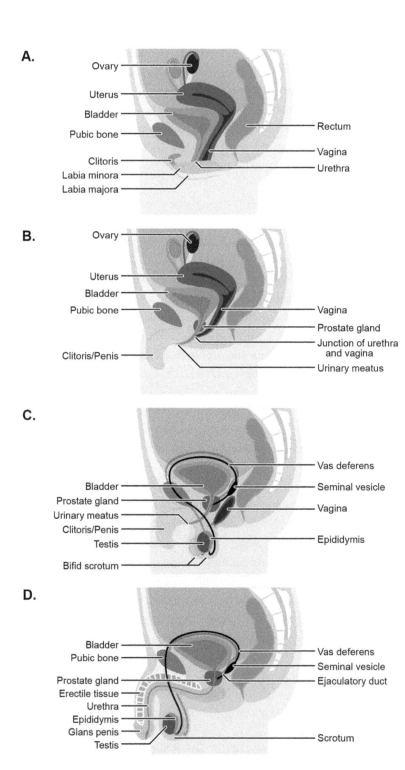

in XX individuals, possibly related to functional variations in HOXA13 and EMX2 transcription factors,[33] can result in left-right separations, or duplications, of the uterus and vagina. Because the central fusion of the Müllerian tubes begins at the base near the urogenital sinus and proceeds upwards (figure 10), incomplete fusion results in side-by-side anatomical duplications from the top and proceeding downward. There can be a septum dividing the top part of the uterus (uterus septus), or two uteruses connected with a single cervix (uterus bicornis), or two uteruses with separate cervixes (uterus didelphus, named after the genus of opossums which have evolved this morphology). Complete separation of the left and right Müllerian ducts may lead to completely duplicate, left and right vaginas connected to separate left and right uteruses, fallopian tubes, and ovaries. Although these conditions can be associated with some complications (especially menstrual disorders), many individuals experience no detrimental symptoms at all. In some cases, individuals can have intercourse, become pregnant, gestate, and give birth from either vagina and uterus.[34]

Occurring about once in every 125 live Male-presenting births, hypospadias is a quite common variation in penis development that involves variable position or size of the urethral opening on the ventral side of the penis, at the base of the penis, or even along the midline of the scrotum.[35] Because the fusion of tissues that form the extended raphe and urethral tube begins at the base of the penis and proceeds toward the tip, hypospadias

14. Developmental variations in human reproductive tract anatomy

A, "Typical" Female reproductive tract with separate external orifices for the urinary meatus and vagina. B, XX individuals with in utero exposure to testosterone (CAH) or other developmental influences may develop a large clitoris (or hypospadic penis) with the urinary meatus at its base and complete fusion of the genital swellings, extending the perineal raphe forward. The vagina may connect internally to the urethra, and a prostate gland may be present. C, XY individuals without a functioning 5α-reductase enzyme may develop a hypospadic penis with the urinary meatus at its base. Absence of midline fusion of the genital swellings can lead to development of a bifid scrotum with testes on either side of a small vagina, which is derived from a portion of urogenital sinus. D, "Typical" Male reproductive tract features a single external orifice, the urinary meatus, for excretory and reproductive tracts. The left and right vasa deferentia connect to the urethra internally just below the bladder at the location of the prostate gland. Illustration by Rebecca Gelernter; B and C redrawn from Young (1937, 380).

vary among successively more basal positions on the ventral surface of the penis or scrotum. Hypospadias has been associated with loss-of-function variations in five different developmental genes, though most cases are thought to result from multiple factors.[36]

In all adults, the adrenal glands produce testosterone and estradiol in physiologically necessary quantities. However, a genetically heritable disruption of adrenal gland function, called congenital adrenal hyperplasia (or CAH), can lead to the embryonic production of testosterone by the adrenal glands of XX individuals during the development of the external genitalia (weeks ten to fourteen after conception).[37] This condition can lead to a wide spectrum of variations in development of the testosterone-and-DHT-sensitive tissues of the urogenital sinus, urogenital fold, and genital tubercle. These possibilities may include combinations of a large clitoris or a small hypospadic penis with the urethral opening at its base (no distinct difference), a small external vagina (i.e., limited to only components derived from the urogenital sinus), or no external vaginal opening and large labia majora to form an empty, bifid scrotum (figure 14B). These individuals usually lack the vasa deferentia and seminal vesicles derived from the Wolffian ducts because of the absence of earlier testosterone during weeks eight to ten after conception. However, they may have a partially or fully developed prostate gland, which grows later from glandular tissue surrounding the urethra in response to DHT converted from testosterone. In some cases, the upper, Müllerian-derived portions of the vagina connect internally to the urethra near the prostate gland and before the urinary meatus at the base of the hypospadic penis (figure 14B). As adults, these individuals may menstruate out of their urethras.

Because mammalian sperm require lower temperatures to survive, the testes typically migrate from their original position forward, down, out of the body cavity, and into the scrotum (figure 12).[38] When testes do not descend completely into the scrotum it is called cryptorchidism. It is very common at birth, and has been associated with loss-of-function variations in at least seven different genes. Quite rarely, testis migration can lead to ectopic testes that may end up nearly anywhere in the lower body cavity. Ectopic testes may be a result of anatomical variations that block the path into the scrotum, or developmental variations in the gubernaculum ligament attachment.[39]

Again, because the glans penis and the clitoris, portions of the shaft of the penis and the labia minora, and the scrotum and labia majora are all sexual homologs of each other, these anatomical structures cannot be strictly defined as distinct, and they morphologically intergrade into one another. So, a large erectile clitoris *cannot* be technically distinguished from a small hypospadic penis. Likewise, there is no criterion for distinguishing between large labia majora and an empty bifid scrotum. Morphological variations in these structures reveal their shared developmental origins.

Noncoding Genetic Variation

So far, I have focused on the effects on human sexual development of loss-of-function mutations in the coding regions of various genes—that is, variations that affect the amino acid composition of proteins. However, this class of genetic variations accounts for only about 50 percent of all the biomedical diagnoses of genetic influence on differences in sexual development. These findings document a fundamental role for noncoding genetic variation in human differences in sexual development, including deletions or rearrangements of promoter or repressor regions where transcription factors bind or gene duplications that create new associations between gene copies and novel upstream promotors sites, and so on. Referred to by one group of biomedical researchers as the "dark side of the genome," noncoding genomic variations are more difficult to identify, more highly variable, and less well understood. However, the few that have been well characterized illustrate the fundamental roles of the coevolved binding sequences that function in intra- and intercellular molecular discourses to the material realization of the human sexual phenotype.[40]

Deletions, or other rearrangements, of the gene promoter regions—the sites where transcription factors bind—have been found to influence the functions of SRY, SOX9, DMRT1, DAX1, and other genes active in sexual development. For example, in one XY individual, a three-base pair deletion in the SRY promotor region led to an absence of gonad development. This individual's father had had multiple surgeries to treat hypospadias, implying a heritable range of possible phenotypic effects for this promotor region deletion in other genetic backgrounds.[41]

Because of its multiple functions in development, variations in regulatory sequences of SOX9 are associated with many different developmental syndromes, including campomelic dysplasia—a severe developmental disorder of the skeleton that is caused by disruption of SOX9 function. (About 75 percent of XY individuals with campomelic dysplasia also exhibit partially or completely Female-like body characteristics.) Deletions in different promoter regions upstream of SOX9 can result in either the development of testes in XX individuals, or the development of ovaries in XY individuals.[42]

Mutation or deletion of the transcription factor DAX1 is associated with congenital adrenal hypoplasia (see above).[43] However, when duplicated in the genome (where it can be associated with *different* upstream promoter sequences), DAX1 can also lead to complete or partial failure of gonad development in XY individuals. Normally, DAX1 expression in the developing ovary downregulates anti-Müllerian hormone production. So, extraneous DAX1 expression resulting from a gene duplication can lead to the retention and development of Müllerian tube derivatives in XY individuals—that is, testes associated with fallopian tubes, uterus, and vagina.

Loss-of-function variations in the SOX3 transcription factor gene have no effect on sexual development in humans.[44] However, SOX3 regulatory sequence rearrangements can result in the development of testes in XX humans. Comparative phylogenetic studies show that the mammalian SRY gene originally evolved from a duplicate copy of the ancestral version of SOX3. Although SOX3 typically plays no part in human sexual development, misexpression of SOX3 in the bipotent gonad cells can replace the function of SRY, and lead to the initiation of testes development in XX individuals. Thus, differences in sexual phenotype can arise not only by eliminating molecular signaling pathways that are typically involved in sexual development, but may also occur by the interjection of expression of other historically related genes that usually have nothing to do with human sexual development.

How Does the Environment Affect Sexual Development?

Performative gender theory focuses on the individual's developmental interactions with agents and influences in their social environment. Likewise, a performative theory of the sexual phenotype should also provide insights into the role of the environment on sexual development.

In placental mammals, the developmental environment consists of the womb in which gestation proceeds. The gestational environment can have multiple effects on human sexual development. For example, maternal adrenal gland tumors that elevate maternal testosterone production during pregnancy can affect the genital development of an XX embryo. This developmental trajectory parallels the possibilities observed in congenital adrenal hyperplasia (CAH) but without the additional physiological complications for the child after birth. Similarly, alcohol consumption and cigarette smoking during gestation have been correlated with cryptorchidism.[45]

Gestational environment may have other complex effects on gender/sex development of human offspring. Although the mechanism is unknown, there is strong evidence of a nongenetic, birth-order effect in the development of self-identified homosexuality in men with older brothers, but there is no such effect in women, or men without older brothers. Men born to parents that have gestated previous Male pregnancies are significantly more likely to grow up to self-identify as exclusively homosexual. This now well-documented phenomenon is particularly fascinating because it involves an influence on fetal development that depends upon the sexual phenotype of previous offspring gestated by that individual. Explanations for this nongenetic, environmentally induced variation focus on gestational effects. Recent data indicate that people who have gestated gay sons have higher concentrations of immune-system antibodies to specific neuroligin proteins (NLGN4Y1 and 2) that are encoded exclusively on the Y chromosome, and are hypothesized to function in fetal brain development. Neuroligins are extracellular proteins that bind to neurexin and play a role in synapse formation in the brain. Antibodies to neuroligins developed during prior XY pregnancies could circulate across the placenta and potentially influence fetal brain development of subsequent XY offspring, and the subsequent development of sexual orientation and gender/sex identity.[46] (For further discussion of parental-fetal intra-actions, see "Placental Performativity" in chapter 7.)

Another aspect of the embryonic environment is the potential influence of additional embryos in the uterus. An extreme example of in utero environmental effects on sexual differentiation occurs in cows, where only 1 in 250 births consists of twins. XX embryos with an XY twin can develop undifferentiated genitalia and no reproductive tract. These rare cows, called freemartins, were the subject of intensive research in the role of hormones in sexual development, and they contributed to the twentieth-century attempts to "define" sex as a result of "sex hormones." The altered development of the XX twin cow is apparently the result of hormonal exchange between embryos that share blood flow across their fused placentas. Specifically, the anti-Müllerian hormone produced by the testes of the XY twin can result in the degeneration of the entire Müllerian-derived reproductive tract of the XX embryo.[47] Such in utero sibling effects are likely to be of higher impact in species in which the majority of births are single offspring, but nothing is known about how species in which multiple births are the rule have evolved to prevent such interfetal developmental effects.[48] (This is a great example of a potential future research program in performative molecular-developmental biology.)

A very subtle in utero phenomenon is suggested by a recent study of mixed-sex human twins. A study of a huge sample of thousands of twin births in Norway reported that Females with a Male twin have a lower probability of completing high school or college or of being married, and have fewer children and lower lifetime income.[49] The results held up in a smaller sample of Female twins whose Male twin died in the first three months of life, implicating in utero hormonal effects rather than family environment. Although the mechanism is also unknown, the authors propose that exposure to fetal testosterone from the Male twin may influence the embryonic development of the Female twin, and later social and sexual development. Alternatively, blood exchange across a shared placenta can also produce microchimerism, in which blood stem cells of the twin are held permanently in the bone marrow of both twins. Around 65 percent of dizygotic twins share some placental fusion that would facilitate such interactions.

Another recently recognized source of environmental influence on fetal sexual development is exposure to environmental chemicals, industrial pollutants, and pharmaceuticals that are released, discarded, or excreted into the environment. One important class of environmental pollutants is endocrine-disrupting chemicals that can bind to steroid or other hormone receptors in the human body, and influence embryonic development.[50] There is controversy about whether endocrine-disrupting chemicals from environmental pollution have contributed to an increase in differences in human sexual development, including hypospadias and precocious puberty, in recent decades. However, the capacity of many human-created environmental pollutants—including polychlorinated biphenyls (PCBs), bisphenol-A (BPA, in polycarbonate plastics and epoxy resins), dioxins, pesticides, and fungicides—to bind to steroid receptors and affect sexual development is well established in mouse and frog models. In the United States, the frequency of hypospadias *doubled* between 1970 and 1997. The rapid increased incidence of hypospadias cannot be explained by genetic risk factors alone.[51] The increase in incidence is evidently not due to enhanced surveillance because the ratio of mild to severe hypospadias has remained unchanged. (Enhanced surveillance without a change in the rate of incidence would increase the relative frequency of mild or marginal examples of hypospadias.) It is interesting to note that these endocrine-disrupting environmental chemicals function by modulating the roles of steroids in developmental discourse of the fetus or child.

Pharmaceuticals can also have a potent effect on embryonic development. For example, exposure to diethylstilbestrol (DES) during pregnancy, a medicine that was prescribed for decades during the mid-twentieth century to prevent miscarriage, was linked to the later development of vaginal adenocarcinoma and testicular cancer in offspring. Finasteride is a drug that is used to treat Male pattern baldness, and the prostate enlargement that can precede prostate cancer. Finasteride functions by blocking the 5α-reductase enzyme from converting testosterone to DHT. However, DHT plays an important role in the development of the penis and scrotum, and the descent of the testes. Consequently, pregnant individuals are advised to avoid any exposure to finasteride, including other individuals who are taking it, during pregnancy because the drug can disrupt external genital and reproductive tract development of XY embryos.[52] Experimental studies also indicate that the artificial estrogens from birth control pills can disrupt sexual development in frog embryos, and that environmental exposure to molecules excreted in sewage waste water can contribute to population declines in wild amphibians.[53]

Queer Science

The term *queer science* can have multiple references, but here I want to focus specifically on the scientific study of *queer* bodies—those anatomies and physiologies that vary from normative expectations. I have no desire to label anyone as queer against their wishes, but I want to raise this general perspective as a counterweight to the biomedical reflex to pathologize sexual anatomical variations from social expectations.

For over 150 years, the scientific study of queer bodies has advanced our understanding of the breadth of variations in sexual development. Since the discovery of DNA, however, this area of laboratory and clinical research has largely been deployed to reinforce the normative expectations through which these bodies *become* queer—are marked as problematic, pathological, and in need of medical intervention and correction.

In this chapter, we have explored some of the variations in human sexual anatomy that can develop in the phenotypic space between and around the so-called canonically Female and Male. Most of the scientific research that has contributed to this account has explicitly involved the study of queer bodies, patients, and lives that deviate from normative, binary sexual expectations. The roles of variation in chromosomes, gene regulatory pathways, and environmental chemical influences on gonad, genital, and reproductive development and physiology have been identified through clinical investigations of individuals with variable genitalia, delayed puberty, or other embodied sexual differences. Such biomedical research involves the analysis of individual anatomies, chromosomes, genomes, and developmental and sexual histories. Once a sufficient sample of individuals with related genetic variations and bodily conditions have been identified and their genomes are sequenced, then hypothesized mechanisms for these variations can be investigated using transgenic mice, which have been engineered to have similar genetic variations to those observed in the sample of humans. These genetically engineered mice are then grown, anatomically characterized, and further investigated with pharmaceutical or other experimental interventions to test the proposed developmental mechanisms. Thus, *queered* strains of mice are grown and experimented on to investigate the development of queer human bodies.[54]

For example, the role of SOX9 in development of human Males was first implicated through genetic analysis of XY patients with campomelic dysplasia. The role of the SRY DNA-binding domain in upregulating SOX9 and the other functions of SOX9 development were then tested and established in genetically engineered, knockout mouse models.

The clinical and laboratory context of this research anonymizes and obscures the queerness of the individual bodies and lives that motivate it and make it possible. But the research largely pathologizes them. Greater scientific and cultural appreciation of the performativity of the human sexual phenotype can reroute the scientific-cultural feedback loop that has reinforced traditional individual binary gender/sex norms and concepts.

A classic example of research that needlessly reinforces the concept of a sexual binary comes from the discovery of the role of DHT in genital differentiation. The function of 5α-reductase enzyme in the production of DHT from testosterone was first identified in the 1970s by endocrinologist

Julianne Imperato-McGinley and colleagues investigating cluster of thirty-eight sexually unusual individuals born over three generations in the isolated, rural town of Salinas in the Dominican Republic.[55] These individuals had a distinct cluster of anatomical features at birth—a large clitoris or small hypospadic penis, large labia, and a small vaginal opening—followed at puberty by the development of an enlarged penis, a lower voice, expanded pubic, facial, and body hair, enhanced muscular development, and the descent of the testes into a bifid scrotum. Known in their community as *guevedoces* (meaning "penis at twelve"), or *machihembras* ("men-women"), these individuals were typically raised as girls, though later awareness of the phenomenon led some families to raise some of these children as boys.

The biomedical researchers who studied the *guevedoces* viewed them as an "experiment of nature" in mistaken sex assignment, which would provide them with the opportunity to examine the relative impact of social rearing and in utero hormonal exposure on gender/sex identity. Of course, the research assumed that there is a definitive fact of individual sex and gender to which all individuals can be correctly, or incorrectly, assigned. They reported that seventeen of the eighteen individuals that had been raised "unambiguously" as girls changed to a Male gender identity following puberty. They concluded that "it appears that the extent of . . . testosterone . . . exposure of the brain in utero, during the early postnatal period and at puberty has more effect in determining male-gender identity than does sex of rearing," and that their finding "demonstrates that in a laissez-faire environment, when the sex of rearing is contrary to the testosterone-mediated biologic sex, the biologic sex prevails if the normal testosterone-induced activation of puberty is permitted to occur."[56]

The study raises numerous questions (many of which are outside the scope of this book). How can these researchers separate the impact on individual gender identity of testosterone exposure to the brain from the embodied, adolescent changes in genital morphology, body hair, voice, and musculature? To what extent are these individuals' gender identities shaped by internal feelings or external cultural expectations? Is the name *guevedoce* a response to individual choices, or a cultural construct to guide and enforce such choices? Is there anything "laissez-faire" about being raised in a strongly gendered, *macho* culture of a rural, twentieth-century, Caribbean town with ambiguous genitalia and an unusual family sexual history?

In this context, however, the most salient feature of the researcher's argument is their implicit, unquestioned assumption that there is an actual fact to the matter of individual sex—that the *guevedoces* are *essentially* Male, that their condition led to mistaken sex assignment at birth by their rural parents, and that the embodied changes at puberty and gender re-identification are a reassertion of this binary truth. The researchers used the scientific study of the extraordinary lives of the Dominican *guevedoces* to simply reinforce their concept of binary sex as a *fact* about individual lives. But it does not have to be this way. In identifying these individuals as *guevedoces*, the members of the Salinas community may be demonstrating greater capacity for understanding and living with their individuality, and their anatomically spectral position, than the outside scientists viewing them as "experiment[s] of nature."

Conducted before the discovery of SRY, SOX9, WNT, or any of the other paracrine signaling pathways discussed here, the researchers applied their concept of gonads and hormones as strict physical *causes* of sexual differentiation and gender identity. Yet, this chapter documents the nearly unlimited number of caveats and ad hoc exceptions that must be appended to any statement about the causal effect of specific genes, hormones, or chromosomes in the development of the sexual phenotype. The statement that "SRY is the master switch of mammalian sexual determination" must be exhaustively footnoted to account for variation in SRY's binding domain, or a three-base pair deletion in the promoter region upstream of SRY, or a duplicate copy of SRY on the X chromosome, or WT1+KTS function, or SOX9's binding domain, or SOX9 genes' promoter region, or SOX3's promoter region, and so on. Neither a Y chromosome, an X chromosome, SRY, SOX9, WNT, testes, ovaries, testosterone, nor estrogen are *sufficient* to cause Maleness or Femaleness, because everyone of their roles in sexual development is contingent upon other, additional agents in their pathways of molecular action. Because other genetic or environmental variations acting downstream of their activities can mimic their functions, many of these agents are not even *necessary* causes of specific events in sexual development either. For example, SOX9 promoter mutations can induce completely functional testis development in the absence of SRY.

Eventually, any hypothesis about the genetic causes of sexual differentiation must dangle with conditional clauses, caveats, and ad hoc exceptions, like so many charms on a gaudy bracelet. In *The Structure of Scientific Revolutions*, Thomas Kuhn argued that the use of ad hoc exceptions and conditions to maintain and prop up the explanatory power of a scientific paradigm creates an intellectual burden under which the paradigm ultimately collapses and fails. The idea of specific genes, hormones, or chromosomes as the causes of individual sex determination is such a failed paradigm.[57]

The scientific investigation of queer, sexually exceptional bodies and lives *has* contributed greatly to our understanding of sexual development. Genetic and environmental variations *do* matter to the process of sexual development. But what do all these data really mean about sex? About the sex of individual people—infants, adolescents, and adults? About "the sexes" in general? What has science really learned from all this queer variation? What should we understand about sex as a result? What alternative paradigm is available?

The performative view of the phenotype as the material realization of the individual self provides an intellectual framework that can encompass all the genetic, hormonal, and biological details of development without making erroneous essentialist assumptions and ad hoc exceptions, or forcing reductive causal explanations. Each individual life begins with an event of sexual reproduction—what I referred to in chapter 2 as the reproductive bottleneck in the human life cycle. Every embodied individual is realized through a material becoming that involves the regulated expression—or performative enactment—of their inherited genomic resources within a specific environment. The iterative process of the development of human lives provides no opportunity and no need for the input of any preexisting fact or essence, and no means to establish or express an a priori intention about the outcome of that process. Individual sex cannot be "determined" during development because there is nothing outside of the process of development itself to do so. There is no descriptive binary blueprint, no prior parental intention, no developmental program, and certainly no sexual identity certificate issued at fertilization. When it comes to individual sex, there can be no scientific fact of the matter prior to, or independent of, its individualized realization. If there is no individual genomic or zygotic binary fact of sex, then there is no such process as "sex reversal" (see chapter 2).

The historical persistence of a reproductive bottleneck going back hundreds of millions of years in our evolutionary history has imposed strong natural and sexual selection, which has resulted in substantial canalization, or limitations on variation, in sexual phenotype. (In the next chapter, we will explore this evolution further.) As a result, most bodies conform to a bimodal distribution of reproductive possibilities. But, as we have seen, there are also plenty of variable possibilities both within and between historical, cultural sexual norms.

Given the extensive canalization of sexual development, many people may persist in thinking, "Isn't it obvious what sex is? Why should we question the existence of something that seems so clear most of the time? Why does it matter what we think sex is?" Whether individuals have an inherent, biological binary sex matters because essentialist categories of binary sex have been, and are, used to justify medical, surgical, cultural, educational, and social norms and policies that function to enforce, and indeed to create, a binary "fact" of individual sex to the detriment of many individual lives. The scientifically unsupportable concept of individual, binary sex does real harm to real people.

As Kessler, Fausto-Sterling, Karkazis, Davis, and Gill-Peterson have shown, many surgical interventions to the bodies of infants and children with "ambiguous" genitals are *not* done to treat painful or problematic medical conditions but to make these infants conform to the anatomical "norms" associated with their presumed, "real" individual sex.[58] These medical judgments involve detailed, implicit cultural assumptions about what genitalia are for and what it means to be a man or a woman—that is, implicit gender theory cloaked in scientific objectivity. Is the proper function of penis to urinate while standing up? Is a clitoris that is large enough to intromit when erect incompatible with heterosexuality, and therefore pathological? Are the variety and intensity of pleasures that such organs may afford beyond what medicine considers permissible or culturally acceptable?

Perhaps the least justifiable, common infant surgical intervention is the reduction of a large clitoris/hypospadic penis in XX individuals, to conform to normative anatomical expectations for Females. These surgeries could be classified as cosmetic because their only goal is to achieve a "Female appropriate" genital appearance, but they involve removal and major rearrangement of highly sensitive, erectile tissue. Because many of these surgeries are done in infancy, profound interventions into the sexual futures of these infants are being made before these individuals can give any consent. Research shows that the scars—physical, emotional, social, and sexual—of infant genital surgeries can remain into adulthood, and that the individual consequences of trying to conform to binary cultural expectations can be severe.[59]

Of course, the parents and families of the approximately one in two thousand individuals born with major differences in sexual development are often not prepared for such surprises at birth, and they face confusing, agonizing, and often lonely decisions about what to do. Parents may also face a host of other complex medical decisions and health management requirements for infants with conditions that bring additional medical complications. Their life experiences have likely provided them with little preparation for thinking about differences in sexual development, and they may bring strongly normative sexual expectations to these decisions. Moreover, physicians and surgeons who have been trained to believe that there is an actual biological fact of individual sex may see their role as helping these children conform to this "fact" and live more "normal" lives. Attitudes of medical authority figures have a strong impact on parental decisions.

But we must ask ourselves: What we are correcting? Why shouldn't a fertile, self-identified woman have an erectile intromittent clitoris/penis? What is wrong with having a self-identified Male having a penis, a bifid scrotum, and a vagina? Why must a child (or adult) conform to binary gender/sex expectations? The rarity of these individual phenotypes is not by itself a reason to classify them as pathologies, medical conditions, or syndromes requiring correction, repair, or reconstruction. Some genital surgeries may be justifiable. For example, surgery to create an external opening for a vagina with an internal junction on the urethra can restore sexual and reproductive functions that would otherwise be unavailable to that individual. Surgeries for major hypospadias may expand the capacity for insemination during intercourse. But such surgeries can be conducted at later ages when individuals can make informed decisions. Major infant genital surgeries that enforce purely cosmetic norms may not be scientifically or medically justifiable before the individual develops their own concept of themselves, and is capable of giving consent.[60] It would be wise, if not legally required, to give those with an erectile clitoris/hypospadic penis the opportunity to experience and understand these anatomies for themselves before they are given the option to be surgically altered.

The ethical confidence of physicians and surgeons to conduct genital reconstructions on infants has been justified by the scientific assumption of an essential binary sexual fact about human lives. Rather than regard differences in sexual development as pathological failures of a binary sexual development, we should realize that the developmental variations in the sexual body are actually evidence that the individual sexual binary does not exist in the first place. That "we" includes biologists, physicians, parents, the public at large, including the children themselves.

"Serious talk about sexuality is... inevitably about the social order that it both represents and legitimates," writes historian Thomas Laqueur.[61] Along these lines, I argue that conversations about the science and treatment of differences in sexual development cannot be isolated from broader culture questions about gender/sex. The biomedical practices of treating differences in sexual development can no longer assume binary heterosexual conceptions of human normality as justification for invasive, life-changing surgeries on infants and children.

CHAPTER SEVEN

How Evolution Generates Sexual Variability

Even though the anatomical homologies we share with the sexual bodies of other vertebrate animals are ancient and have deep histories, the mechanisms by which those bodies are materially realized are truly diverse and dynamically evolving. Attending to patterns of diversity and evolutionary instability in the development of the sexual phenotype reveals many new consequences for thinking about sex and sexual becoming. In short, biological evolution extends the performative continuum we have been exploring into deep, historical time. In this chapter, we see that the mechanisms of sexual development evolve so dynamically because the evolved solutions to the challenge of sexual development create additional new challenges that contribute directly to the *generation* of queer bodily sexual variation.

The word *evolution* may refer to a breadth of different phenomena of change in various disciplines and contexts. In biology, evolution refers to changes in the frequencies of genetic variations within a population, to inherited changes in phenotype within a population, to speciation and extinction, and to diversification among species and higher groups. Although individual development is a phenomenon of biological change, however, development is *not* by itself a form of evolution. In anthropology, the humanities and some areas of zoology, evolution can also refer to cultural change and diversification, which may take place so rapidly as to include change within an individual lifetime. Astronomers even refer to the evolution of black holes.

In this chapter, I want to expand the idea of queer biology that I have pursued in the context of sexual development to ask a broader question: In what ways can biological evolution itself be appropriately and productively described as performative? I propose that the iterative materialization, or development, of complex phenotypes (i.e., hierarchies of homologies) with selective genetic evolution and historical contiguity (i.e., genetic inheritance, selection, and drift) create distinctively and uniquely performative phenomena on the longer, historic, and comparative scales of biological evolution. In other words, the emergent phenomenon of evolution is deeply impacted by the performativity of organismal phenotypes.

Stephen Jay Gould famously considered whether, if one replayed the history of life, one would observe a similar pattern and distribution of species, morphologies, ecologies, and adaptations. To up the stakes, Gould rephrased the question as, "Are we humans inevitable?" He answered with a resounding *No*, inspiring an intellectual debate that is still ongoing.[1] The idea that natural selection is a strong deterministic force that would contribute to the repetition of similar adaptations in different replayings of the history of life has long held sway among adaptationists. To them, history doesn't really matter, its relevance erased by the deterministic force of adaptation. I have been arguing, expanding on Gould's position, that the performativity of the phenotype—the actions and intra-actions of numerous agencies between the levels of genetic variation and the embodied, realized phenotypic diversity within populations—ensures that all replays of the history of life will be strikingly different. (It is not accidental that Gould was also deeply interested in development.) Consequently, the performativity of phenotype, its intra-active emergences, profoundly affects Earth's biodiversity *and* its history.

Like the performativity of gender, the performativity of the phenotype carries, in Judith Butler's words "the double-meaning of 'dramatic' and 'non-referential.'"[2] On one hand, the phenotype is an individual enactment realized in a particular environment through communicative interactions (i.e., molecular and social discourses) among multiple, hierarchically nested agencies. Simultaneously, the regulated expression of genomic information is itself nonreferential, or nondescriptive; it is not a representation of a prior essence but an expressive action—an enactment by the cell. In this way, gene expression is a molecular performative, an action that achieves its meaning in the moment and manner in which it comes into being. In this way, individual phenotypes are both material realizations of a historically derived, constitutive, genomic, organismic individuality in a specific environment, *and* iterative becomings through material-discursive action. The performative enactment of the phenotype is manifest in the contingent, phylogenetic (or genealogical) history of evolutionary change.

The Evolution of Sex

In *Bodies That Matter*, Judith Butler writes, "Nature has a history, and not merely a social one."[3] Indeed, a major focus of evolutionary biology is to uncover that material, biological history—the myriad of events and the complex processes that have contributed to the planet's current biodiversity and, in particular, to ourselves. Since the term *sex* has so many connotations and meanings in different contexts, I want to unpack some of the many details the term invokes, so that I can document their deep histories. This will involve summarizing the series of origins, events, and innovations over the last couple of billion years that incrementally and cumulatively contributed to the complexity of the biological phenomena that are packed into the concept of "sex" in humans.

Biologists use hierarchical branching structures called phylogenies to represent the histories of evolutionary relationships among species and higher groups of organisms (figure 15). Phylogenies are empirically derived, but they are not simply filing systems, statistical frameworks, or diagrams of convenience. They are our best attempts to reconstruct the explicit history of evolutionary diversification of biological lineages themselves.

The shape of phylogenies depicts the branching structure of proposed evolutionary relationships through the relative recency of common ancestry of species and higher groups of organisms. The branches depict lineages propagating and persisting through time. The nodes, or branching points, represent historic speciation events through which ancient lineages originally diversified, giving rise to multiple descendent lineages. Phylogenies also provide an analytical framework for understanding the history of evolutionary events in the past, and investigating the processes that were involved in them. Because the history of life is a tree, we can reconstruct the order of events in the evolutionary past by comparing organisms descended from different branches or trunks.[4] In this way, phylogenies provide both evidence of homology between historically related structures found among different species, and of convergent evolution of similar but historically independent, or analogous, structures in different

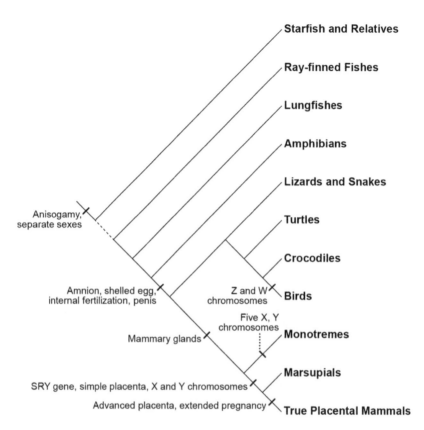

15. Simplified phylogeny of vertebrate animals depicting the evolution of novel reproductive traits

Anisogamy and separate sexes (dioecy) evolved before the most recent common ancestor of mammals and the starfish and its relatives (echinoderms). Numerous reproductive innovations evolved in the most recent common ancestor of reptiles and mammals (amniotes) including the amnion, shelled egg, internal fertilization, and the penis. Mammary glands evolved in the most recent common ancestor of living mammals. The SRY gene, a simple placenta, and the X and Y chromosomes evolved in the most recent common ancestor of marsupials and true placental mammals. An advanced placenta and extended pregnancy evolved in the most recent common ancestor of true placental mammals, including humans. Monotremes convergently evolved five X and Y chromosomes. Birds evolved Z and W chromosomes. There were many independent evolutionary origins of X and Y chromosomes and Z and W chromosomes in ray-finned fishes, lizards, and snakes (not shown). Dashed lineage indicates where the tree has been simplified (many lineages not shown). Illustration by Rebecca Gelernter.

lineages, such as the membranous finger-wings of contemporary bats and extinct reptilian pterosaurs.

To understand the evolutionary "genealogy" of the many components of human reproductive biology, we must investigate the history of sexual innovations on the phylogeny of life. At its most basic, sexual reproduction produces offspring that differ genetically from their parents and from each other. However, organisms have evolved multiple ways to accomplish this. Bacteria engage in various kinds of sexual recombination, which may involve the transfer of plasmids (circular pieces of DNA that act something like bacterial chromosomes), conjugation (a kind of cell-to-cell mating with genetic exchange), or homologous recombination, in which a gene copy is taken from a second individual to replace the first individual's copy of the same gene. Even viruses can "sexually" recombine—sort of—if an individual is simultaneously coinfected with two different strains. These processes are the simplest forms of sexual reproduction, because the new individuals incorporate genetic variation from another individual.

The evolution of true sexual reproduction using meiosis, or reduction division, evolved over a billion years ago in the ancient, common ancestor of all organisms with nucleated cells (eukaryotes), including all protists, plants, fungi, and animals. Meiosis creates gametes that each contain one-half of the parental genome. Gametes are subsequently combined by syngamy, or fertilization, to create a new individual that inherits half of its genes from each parent. Sexual recombination contributes to the efficiency of adaptation, and thus sex can enhance the persistence and evolutionary success of sexual species and clades.[5]

At multiple times in different lineages of plants, fungi, and animals, gametes later evolved to be differentiated into small, motile gametes that we call sperm, and larger, less motile gametes we call eggs. Because this condition (called anisogamy) evolved multiple, independent times in the history of life, not all "sperm" and "eggs" are homologous. Rather, eggs and sperm are basically a great evolutionary idea that was "discovered" over and over again. And anisogamy is not necessarily a fixed state. Some green algae have evolutionarily switched back and forth between isogamy and anisogamy, depending on the environments they live in.[6] If big, free-swimming gametes are more vulnerable to predation, then natural selection can *eliminate* anisogamy, and thereby differentiated individual sex. Such species have sex, but these individual organisms do not have a sex. There is also a complex history of the subsequent origins of separate bodies specialized for reproducing using different gametes.[7] In our evolutionary lineage, however, the origin of sexually differentiated bodies with ovaries or testes probably goes back at least 550 million years to the most recent common ancestor of humans and starfish (figure 15).

Ancestrally within the vertebrates, eggs and sperm were released into water, and fertilization was external. But this aquatic mode of reproduction limited animals to reproducing in open water or very humid environments. Several related reproductive innovations evolved in the most recent common ancestor of mammals and reptiles that allowed them to conserve water and reproduce independently on land (figure 15). One of these evolutionary innovations was the amnion—an embryonic membrane that surrounds and protects the embryo—which evolved in the most recent common ancestor of the vertebrate group, called amniotes which includes all reptiles and mammals. The amnion allowed for the evolution of a leathery, or shelled, egg that further protects the developing embryo from the challenging environment.[8] Because the amnion and egg shell must develop *after* fertilization, terrestrial reproduction in amniotes required additional evolutionary novelties to achieve internal fertilization—namely, the penis, which also originated in the most recent common ancestor of all reptiles and mammals. (Similar innovations for terrestrial reproduction evolved convergently in various groups of invertebrates and even in seed plants.)

Mammals themselves have evolved a number of additional reproductive novelties including specialized sweat glands—called mammary glands—for feeding infant young (figure 15). The most recent common ancestor of all mammals laid leathery eggs, as the contemporary monotremes, platypuses and echidnas, do today. Internal gestation evolved in the common ancestor of marsupials and true placental mammals. In marsupials, the fertilized egg is retained for only a short period of development, and the embryo continues to develop after birth in the mother's pouch. In true placental mammals, including humans, the embryo is retained for a longer period of development during which it is nourished and physiologically maintained through the placenta (see below).

Sexual reproduction has repeatedly been lost in different lineages of fishes, lizards, and invertebrates, but not in any mammals or birds. Parthenogenesis, or Female-only reproduction, involves Females laying eggs that develop into full adults. Although several variations exist, offspring are essentially clones—or they may inherit slight recombinations—of the uniparental genome. In some parthenogenetic lizards, Females require sperm to enter the ova to induce their development, but the DNA from these sperm are not incorporated into the genome of the embryo. In other words, sperm are still biochemically required to initiate the self-enactment of the embryo, but the sperm's genetic contributions are irrelevant. These parthenogenetic species are essentially sexually commensal with Males of the closely related lizard species that they must copulate with in order to reproduce.

These biological details are important to understanding how the diversity of sexual reproduction systems and sexual phenotypes is a product of the ever-iterating materialization of reproductive possibilities.

Evolutionary Variability of Sexual Development Initiation

Variability among species in the mechanisms used in the development of any homologous structure demonstrate that phenotypic features that are in a fundamental way the "same" can be materially realized in remarkably different ways via different molecular/discursive paths. This observation provides powerful additional support for the performative understanding of the sexual phenotype, and the disconnection of sexual anatomy from any individual, biological, or genetic essence. The evolved variability in the developmental mechanisms of homologous sexual organs among different species provides striking evidence of the absence of an essential, individual sexual binary.

Given the prevalence of sexual reproduction in multicellular animals, and the fundamental importance of reproduction to life itself, one might think that sexual animals would have quickly evolved a single, reliable mechanism to create sexually differentiated bodies, and then stuck with it—an evolutionary version of "*Set it and forget it*!"[9] Actually, the most surprising detail about mechanisms of sexual differentiation is how variable they are among species and higher groups of organisms. This variation constitutes a form of evolutionary performativity, and it arises from the inherent instability and variation-generating capacities of the various mechanisms for creating sexual differences in phenotype. In other words, the performative continuum in sexuality extends all the way back to the origin of bodily sexual differentiation itself, and persists dynamically in all its instances today.

Despite the great antiquity in our phylogenetic history of sperm, egg, testis, and ovary, vertebrate animals have evolved, and continue to evolve, an astounding diversity of ways to initiate and realize the development of egg-producing and sperm-producing bodies. As geneticist Ursula Mittwoch wrote in 1971, "One of the puzzling aspects of the genetics of sex determination [sic] is the apparently haphazard distribution of sex chromosomes between the sexes in different classes of animals."[10] For example, sexual differentiation in birds is initiated genetically with distinct chromosomes like mammals but in the opposite fashion. Male birds are homogametic, having two Z chromosomes, while Female birds are heterogametic, having one Z and one W chromosome. ZZ birds have testes, sperm, and even (in ducks and some others) penises that are homologous with those of humans and other mammals. Among insects, flies (Diptera) and grasshoppers (Orthoptera) are generally XX/XY, but moths and butterflies (Lepidoptera) are ZZ/ZW.

Another interesting evolved variation in sexual development is the extent to which embodied sex is organized by gonadal hormones, as in mammals, or by additional somatic or cell-autonomous mechanisms. Jason Ioannides and colleagues have shown that ZZ chickens with CRISPR knockouts of one copy of DMRT1 develop ovaries. However, the secondary sexual traits in plumage and skin still develop as ZZ Male.[11] This result demonstrates that avian somatic tissues have the capacity to develop some sexually differentiated traits autonomously based on their genomes alone, rather than in response to gonadal hormones. It would be tempting to conjecture that these avian cells, unlike those of mammals, have a definable sex. However, "cell-autonomous" sexual differentiation is just a finer-scale performative genetic enactment of a phenotypic state by individual cells, which will be subject to all the same influences of intra-active cellular contingency—as exhibited by the different developmental outcomes in the gonads and skin of experimental chickens. It's performance in every direction.

One of the extraordinary consequences of the combined gonadal and cell-autonomous sexual development mechanism of birds and butterflies is the extremely rare appearance of gynandromorphs—individuals that are bilaterally Male and Female.[12] Occurring much less frequently than one in a million, gynandromorph birds develop from the combination of ZZ and ZW cells in the same embryo. In 2019, the sighting in Erie, Pennsylvania, of a gynandromorph Northern Cardinal (*Cardinalis cardinalis*), with striking Maserati-red plumage on its right side and dusty brown plumage on its left, made national news.[13]

Similar to birds and mammals, the XX/XY and ZZ/ZW genetic sexual differentiation systems have each evolved multiple times independently in different lineages of fishes and lizards. A particularly unusual variant of chromosomal sex differentiation is found in the ancient lineage of egg-laying mammals, the monotremes of Australia. The echidna, or spiny anteater, has heterogametic Males like other mammals, including humans. But instead of simply having single X and Y chromosomes, echidnas have five X and five Y chromosomes. Individuals with $X_1X_1X_2X_2X_3X_3X_4X_4X_5X_5$ chromosomes develop ovaries, lay eggs, and lactate. Individuals with $X_1X_2X_3X_4X_5Y_1Y_2Y_3Y_4Y_5$ chromosomes develop testes, a (highly unusual, branched!) penis, and produce sperm.[14] The Xs and Ys line up to form two long chains during meiosis. All echidna eggs include $X_1X_2X_3X_4X_5$, but echidna sperm can include either $X_1X_2X_3X_4X_5$ or $Y_1Y_2Y_3Y_4Y_5$. Oddly, the gene sequences on the echidna Xn and Yn chromosomes share little sequence homology with the X and Y of marsupial and true placental mammals. Rather, they appear to be more similar in sequence composition to avian Z chromosomes! This finding means that the genetic mechanisms of sexual differentiation using differentiated chromosomes and Male heterogamy evolved *independently* in the monotremes, and in the common ancestor of the marsupials and true placental mammals.

Genetically initiated sexual differentiation does not even require differentiated chromosomes. Paleognathous birds—including ostriches, emus, tinamous, and their relatives—have genetic sex development initiation without differentiated Z and W chromosomes. Although still poorly understood, sexual differentiation in the lab model organism zebrafish (*Danio rerio*) is influenced in a complicated, quantitative fashion by the dosage of a number of genes at many locations in the genome, in the absence of differentiated chromosomes, and with some additional influence from external environment (temperature) and internal interactions (number of oocytes).[15] This enduring mystery strikes me as outstanding precisely because the zebrafish is one of the best-studied model organisms we have.

In many fishes and non-avian reptiles, sexual differentiation is not related to *any* genetic differences among individuals. Rather, it is initiated entirely by environmental cues, especially temperature during embryonic development. In these animals, sex is an environmentally performative state because individual reproductive anatomy and behavior are a result of the ambient temperature of the developing embryo. Interestingly, the actual relationship between the temperature of development and the sex of the embryo is itself variable, and continues to evolve within and among species of fishes and reptiles. For example, in turtles, Males may be produced at lower temperatures and Females at higher temperatures, or Males at intermediate temperatures, and Females at both warmer and colder extremes. Thus, even the environmental cues that initiate the sexual developmental trajectory are arbitrary, historically contingent, and evolvable.

Obviously, in certain habitats, temperature and other environmental cues can change, or become highly variable, "engendering" big swings in the primary sex ratios of wild populations. For example, in response to global climate change (warming, locally), the populations of loggerhead sea turtles (*Caretta caretta*) in the Canary Islands, which is the third-largest breeding population in the world, is now producing 85 percent Females.[16] Based on current climate-change predictions, the Canary Island population will be producing 99 percent Female hatchlings by 2100, causing a major new impediment to sea turtle conservation. Of course, this environmental change is creating extremely strong selection for the production of a greater number of Male sea turtles, so it is at least possible that the sea turtle sexual development system can evolve to evade this environmental challenge. However, given the longevity and reduced population sizes of these turtles, evolution may not be able to occur fast enough to avoid population extinction.[17]

Chromosomal mechanisms of initiating sexual development were long hypothesized to be an irreversible evolutionary trap—that is, once you have evolved a specific one, you can't lose it.[18] However, phylogenetic analyses and genetic research have now demonstrated that this is common in some animals. For example, geckos have evolved between seventeen and twenty-five *independent* evolutionary transitions between ZZ/ZW, XX/XY, and temperature-dependent sexual differentiation systems. The geckos alone account for over half of all the evolutionary transitions in sexual development initiation mechanisms known in all lizards and snakes. There appears to be something specific to their biology and history that makes them particularly susceptible to evolutionary instability in sexual differentiation initiation mechanisms.[19]

Environmental and genetic mechanisms of sexual development initiation can coexist, and evolve adaptively, among populations of a single species. In snow skinks (*Niveoscincus ocellatus*) on Tasmania, sexual development is initiated by temperature at low altitudes, where between-year variance in temperature is low and early birth date gives advantages to

Females. However, at higher altitudes, where between-year variance in temperature is high and there is little advantage to early-born Females, sexual development is initiated by multigene genetic variation (like zebrafish). Environmental sexual development initiation permits the evolution of optimum sex ratios at low altitudes, while genetic sexual development initiation prevents extremely skewed sex ratios at high altitudes.[20]

In other lizards, it has been shown that temperature-dependent sexual development initiation can reassert control over genetic sexual differentiation even after the evolution of dimorphic chromosomes. If some individuals of the homogametic sex (XX Females or ZZ Males, depending on the species) can be environmentally induced to develop into fertile members of the other sex, the result can be a single population with homogametic members of both sexes, the loss of "sex" chromosomes, and the re-evolution of temperature dependence. For example, in one population of the ZZ/ZW bearded dragon (*Pogona vitticeps*), some ZZ individuals at high temperatures developed into fertile ZZ Females (instead of becoming Male). These ZZ Females are entirely interfertile with ZZ Males.[21]

Similarly, it has been shown experimentally that ZW embryos of the half-smooth tongue sole (*Cynoglossus semilaevis*; Cynoglossidae) can develop as Male when raised at higher temperatures (28°C). Interestingly, these ZW Males epigenetically modify (i.e., methylate) gene sequences on the W chromosome in their sperm so that ZW offspring with a paternal W chromosome will *also* develop as Male, regardless of developmental temperature.[22] This is a rare instance of epigenetic modification producing transgenerational effects on the fundamental features of sexual development. It is unknown for how many generations this effect can persist.

Thus, comparative biological evidence demonstrates that environmental cues can *reassert* themselves to become sexual development difference makers, and *eliminate* the role of genetic difference makers and differentiated chromosomes. Even after the origin of differentiated chromosomes, the advantages of entirely environmentally induced sexual differentiation initiation mechanisms can lead to the re-evolution of temperature dependence.[23]

Venturing outside of the vertebrate animals, we encounter even greater diversity in sexual development initiation mechanisms.[24] Twelve percent of all animals, including bees, ants, and thrips, have haplodiploid sexual development. In these organisms, Females are diploid—meaning that they have two matching sets of parentally inherited chromosomes. Males, however, are haploid, and receive only one set of maternal chromosomes. Females develop from fertilized eggs, and Males develop from unfertilized eggs, so the frequency of Males in the population is determined entirely by the physiological decisions of fertile Females.

Social environment can also influence sexual development. Bonellid marine worms disperse as tiny pelagic larvae, and settle on the ocean floor. If they cannot detect (i.e., smell) another individual nearby, they develop as Females. If they detect a Female nearby, they develop as Males.[25] Interestingly, Male bonellid worms remain tiny, and encyst in the body of the Female where they become a symbiotic, sperm-producing internal organ of the Female.

In many insects and other arthropods, sexual development can be influenced by an inherited parasitic disease—an intracellular parasitic bacterium *Wolbachia* that is only passed down to offspring in the cytoplasm of the egg. To increase its fecundity and evolutionary success, many lineages of *Wolbachia* have evolved to molecularly *redirect* the sexual development mechanisms of their infected hosts in various ways toward egg production, allowing the parasite to propagate more efficiently. Thus, *Wolbachia* have evolved to reach out and manipulate the discursive molecular development mechanisms of the sexual phenotypes of their host species for their own adaptive advantage. Extreme *Wolbachia* infection rates can lead to severely skewed sex ratios within a population, and threaten it with extinction.[26]

In many plants, sexual development can be influenced by mutations in the genomes of the organelles in the cellular cytoplasm—the chloroplasts and mitochondria. Because these organelles are inherited from the ovum exclusively, these "cytoplasmic sterility mutations" have frequently evolved to produce Male sterility, thereby selfishly advancing their own success. The result can be the development of individuals with exclusively Female flowers within a population of individuals with both fertile Male and Female flowers, a breeding system referred to as gynodioecy. The evolution of cytoplasmic sterility genes is an important intermediate stage in the evolutionary origin of plants with completely separate sexes (dioecy), like papaya and gingko.[27]

In summary, the mechanisms for initiating sexual differentiation are evolutionarily unstable, and subject to various evolutionary perturbations and environmental influences. This diversity is a direct consequence of the iterative, performative enactment of each sexual body.

Why Sexual Differentiation Mechanisms Are Generatively Queering

Despite the ancient origin of separate sexes in many lineages of animals, a diversity in sexual differentiation mechanisms has continued to evolve. But why are sexual developmental mechanisms subject to so much evolutionary instability? Here, we will explore how the fundamental challenge of developing sexually differentiated bodies from otherwise common genomes and embryonic morphologies is so profound and enduring. We will see that all of the evolved solutions to accomplish sexual differentiation create various, new evolutionary challenges, each of which contributes directly to the *generation* of queer bodily variation that disrupts normative binary expectations. Evolution canalizes sexual development, but it also *creates* new conditions through which canalization is constrained.

The evolution of genetic sexual differentiation initiation mechanisms can evade some of the variability that results from environmentally induced sexual differentiation (as in sea turtles and snow skinks, above). However, once genetic initiation of sexual development has evolved, it can create new evolutionary challenges that *generate* phenotypic instability and ultimately variability in sexual phenotype.[28] In other words, evolution itself can be *generatively queering*; natural selection creates the conditions for the *persistence* of nonbinary, bodily variations.

A source of selection for chromosomal differentiation can arise during gamete production, when the arms of homologous pairs of chromosomes have the opportunity to "cross over," leading to the exchange or rearrangement of DNA sequences *between* homologous chromosomes. Recombination between homologous chromosomes gives rise to new—either advantageous or inferior—genetic combinations that natural and sexual selection must sort among. Although every individual inherits half of their chromosomes from each parent, crossing over provides an opportunity to mix it up, recombine portions of homologous pairs of chromosomes inherited from each parent, and create new combinations *before* they are passed on to the next generation in the gametes. This opportunity for novel variation is one of the great advantages of sexual reproduction—the possibility of producing offspring that are evolutionarily more successful than themselves. This may be the original adaptive advantage of meiosis and fertilization in the first place.

However, crossing over and genetic recombination of chromosomes can create problems for newly evolving genetic mechanisms of sexual development. Imagine a new gene variation that can initiate a difference in sexual development evolves on one of a pair of homologous chromosomes—like SRY evolving from a duplicate copy of SOX3 in the most recent common ancestor of marsupials and placental mammals.[29] Immediately, strong selection will arise to *suppress* crossing over during meiosis between these homologous chromosomes in order to prevent the new sexual development gene from becoming scrambled up or transferred to the other chromosome, which would disrupt the sexual difference-making function of this newly evolving gene. These sudden disadvantages to crossing over will select for the evolutionary inversion (i.e., flipping end to end) of the arm or section of the chromosome that carries the new sexual development gene, which can prevent these DNA sequences on the two chromosomes from lining up precisely during meiosis, and discourage crossing over.

Although suppression of crossing over limits certain developmental-evolutionary problems, it also creates *new* ones. Crossing over is the only way to dissociate deleterious variations in DNA sequence from advantageous variations on the same chromosome. As a result, once crossing over is suppressed, the remaining genes and regulatory sequences on the heterogametic (Y or W type) chromosome will gradually become *loaded* with functionally compromising mutations that they cannot eliminate.[30] This problem creates a novel source of natural selection to entirely eliminate from the Y or W chromosome those genes whose functions have been compromised by newly acquired mutations. Protecting a sexual development initiation gene from recombination shuts down the only mechanism available for elimination of problematic mutations! This process explains why the tiny human Y chromosome has only about 75 remaining functional genes, compared to the more than 800 functioning, protein-coding genes on the homologous human X chromosome.[31] Because the X and Y chromosomes started out identical in size, some 725 genes have been eliminated from the human Y chromosome since its origin in our most recent common ancestor with kangaroos.[32]

For most of the genes on the mammalian X chromosome, which have nothing to do with primary sexual differentiation, XY individuals can function perfectly well with one copy of the X. However, for the genes that function in the initiation of differences in sexual development, like SRY on the Y chromosome, there is nowhere else to go—no refuge from the relentless, unavoidable advance of compromising mutations. But what do these mutations do? In placental mammals, SRY mutations contribute directly to the frequency and diversity of differences in sexual development. Recall, for example, that the various functional elements in the human SRY protein—the large DNA-binding domain—have *two* nuclear translocation protein-binding sites, an acetylation site, *and* a phosphorylation site (figure 7). New genetic variations at any of these sites can alter the function of SRY, and prevent the initiation of SOX9 expression that is involved in testis development. Likewise, mutations that interfere with the capacity to bind to the SOX9 enhancer and induce an appropriate bend in the DNA can prevent the initiation of testes identity development. This is why variation in SRY variants are the single most frequently identified genetic factor in major differences in sexual development in humans. These variation-creating mutations on the Y chromosome are a direct consequence of the evolutionary cul-de-sac created by the evolution of a rarely crossing-over, reduced, mammalian Y chromosome over one hundred million years ago. In other words, the evolution of the Y chromosome *is the generative source of* queer variation in mammalian sexual development.[33]

This topic has not escaped scientific scrutiny, but it is usually described quite differently. For example, in a paper with the ironic title "Sry: The Master Switch in Mammalian Sex Determination," molecular-developmental biologists Kenichi Kashimada and Peter Koopman write, "Despite its dramatic biological role, SRY is a fragile and partly debilitated gene" that exists in a "degraded state" on "the rapidly degrading Y chromosome."[34] Juan Carlos Polanco and Koopman further state, "Instead of the robust gene that one might expect as the pillar of male sexual development, SRY function hangs by a thin thread."[35] (These researchers never explain why one should expect Male sexual development to require the support of a "pillar" in the first place.)

In their detailed investigation of the molecular role of phosphorylation in the translocation of SRY into the nucleus in humans and other primates, Yen-Shan Chen and colleagues write that SRY's "transcriptional activity lies near the edge of developmental ambiguity," and they consider "the *anomalous nonrobustness* of the male [sexual development] program" as an urgent evolutionary question.[36] Like the trope of the fragile male ego, XY chromosomal sex differentiation in mammals has led inevitably to an intrinsically precarious mechanism for testis development.[37] The language biologists use to describe this situation expresses anxiety and surprise that the essentialist genetic sexual binary rests on such insecure foundations. SRY is sounding less and less like a "master switch."

The relentless buildup of deleterious mutations in SRY creates natural selection to *compensate* for mutations that compromise its function, which leads directly to the dynamic evolution of the increasingly baroque biochemical mechanisms to regulate SRY signal maintenance, translocation, and binding function. For example, the role of WT1+KTS in stabilizing SRY mRNAs before translation in the cytoplasm is an adaptive compensation for the evolution of mutations that degraded their stability. Likewise, the roles of calmodulin and transportin-β, two different nuclear translocation protein-binding sites—an acetylation site *and* a phosphorylation site—are evolutionary compensations for prior mutations that degraded SRY's translocatability into the nucleus. The result is the ever-expanding evolved diversity in the structure and function of the SRY gene between humans, mice, and other mammals.[38]

In *Through the Looking Glass*, the Red Queen tells Alice that, in contrast to her "slow sort of country, . . . here it takes all the running you can do, to keep in the same place."[39] Likewise, the elimination of crossing over between differentiated chromosomes creates a constant and inevitable buildup of mutations on the reduced (Y or W) chromosome, and a compensatory evolutionary struggle to maintain the precarity of sexual development initiation. Like the Red Queen, these genetic signaling networks must jog or even run, evolutionarily speaking, simply to stay in the same place, simply to maintain the development of different sexual possibilities within otherwise similar bodies. All of the baroque details of managing the straightforward transcription factor function of human SRY (figure 7) have evolved in order to fix compromising genetic changes that have arisen by unavoidable accumulation of mutations to Y chromosome. For example, if mutations that cause mRNA instability arose in any autosomal gene, those mutations could be eliminated by recombination and natural selection. But when such mutations arise in SRY, there is no recourse but to evolve some *other* end around—like the unlikely evolutionary recruitment of a new isoform of WT1—to stabilize the newly fragile SRY mRNAs (figure 8). Of course, each of these new evolved solutions relies on the recruitment of the discursive intervention of other agents, which makes the whole system more vulnerable to new sources of genetic variation (e.g., future mutations that effect the function of WT1+KTS and newly compromise SRY).

Overall, this process contributes to the development of queer bodily variation *within* species, and evolutionary diversity *among* species. This is why SRY is so strikingly variable in sequence and structure across mammals, and is still the most commonly identified genetic influence on DSD. There are especially notable differences between humans and mice, which makes the mouse an inconveniently inappropriate model system for the study of Male sexual development in humans. For example, the mouse SRY gene lacks one of the phosphorylation sites that is necessary for nuclear translocation of SRY in humans; it also has a large, novel glutamine-rich region and an interceding "bridge" region that human SRY lacks. The features of these molecules that coevolve with other members of the network are precisely those that function in their molecular intra-actions.[40]

A radical solution to this evolutionary conundrum is to get rid of the Y chromosome entirely, and evolve a "dosage"-dependent XX-X0 sexual differentiation system in which Males have only one X chromosome and Females have two. This solution has evolved in wood lemmings (*Myopus*, Cricetidae). The mole voles (*Ellobius* species, Cricetidae) and Ryukyu spiny rats (*Tokudaia osimensis*, Muridae) have convergently evolved an X0/X0 system in which all individuals have only one X chromosome, and sexual development is initiated by new autosomal gene variants.[41] Recently, evolutionary geneticists Matthew Couger, Polly Campbell, and colleagues have discovered that the unassuming creeping vole (*Microtus oregoni*, Cricetidae), native to forests of the American Pacific Northwest, has gone unimaginably further than wood lemmings, mole voles, or spiny rats. Not only have creeping voles lost the Y chromosome, but Females possess only *one*, distinct X chromosome—XM (where M is for Maternal). Males have both a maternally inherited XM chromosome, and a novel *paternally* inherited version of the X chromosome, called XP, making them XMXP. This bizarre chromosomal, evolutionary two-step—involving the *loss* of the Y and the *creation* of a new Male version of the X—has involved the assimilation of multiple genes formerly unique to the lost Y chromosome—including SRY—into *both* new versions of the X chromosome. SRY has been duplicated into twenty-three total copies, including seven fully coding copies on XM and ten on XP. Another three genes formerly specific to Y chromosome are now found exclusively on the maternally inherited XM chromosome. Furthermore, the X inactivation dosage compensation machinery deployed in all XX Female mammals has evolved in creeping voles to control gene expression in XMXP Males! (No calico Males of this retiring little forest rodent have been yet identified, but lab scientists should be able generate them in captive strains.) Like the lost, ancestral Y chromosome, the novel XP chromosome, which is unique to this species, is already showing significant signs of genetic degradation from suppression of crossing over between the two versions of X. Lastly, it is known that Females always produce XM eggs, and that Males are still heterogametic, producing sperm with either an XP or no X chromosome, but it is still unknown how Males control this.

What set the creeping voles on this chromosomally revolutionary path? Like so many of the issues we have been exploring, it is currently unknown, but the reshuffling of the chromosomes involved in sexual development of the creeping vole may have resulted from the sexually disruptive natural selection that can arise from differential impacts of genetic variations on the survival and fecundity of the differentiated sexes. Of course, these recent findings raise numerous additional, yet unanswered questions about how sexual development proceeds in this species. As Couger, Campbell, and colleagues report, "Given the primary function of Sry [*sic*] in initiating the male developmental pathway, the presence of multiple copies of Sry on a maternally transmitted X chromosome is particularly surprising." But these stunning new findings further document the extraordinary diversity of mammalian mechanisms of sexual development.[42]

Although some researchers have proposed that the evolutionary reduction of non-recombining chromosomes means that the loss of the Y chromosome is inevitable in humans, this is not correct. Over five thousand species of mammal have evolved, and radiated since the ancient mammalian ancestor in which differentiated XY chromosomes first appeared, and the vast majority of these species have retained Y chromosomes. This overwhelming pattern of persistence of the reduced Y chromosome in most mammal species is equivalent to thousands of ongoing evolutionary experiments, each demonstrating that the plausibility of Y chromosome persistence, and refuting the idea that evolutionary loss of the Y chromosome is somehow inevitable.[43]

A similar evolutionary pattern of the chromosomal reduction found in placental mammals is observed in numerous different animal lineages. For example, the W chromosome found in neognathous birds is reduced in size and gene number, like the mammalian Y chromosome, and for the same reasons.[44] This shows that having a "rapidly degrading" chromosome involved in sexual development is unrelated to Maleness itself. Unlike mammals, however, no birds are yet known to have evolved a ZZ-Z0 sexual differentiation system. In fact, the ancient lineage of paleognathous birds (ostriches, tinamous, and their relatives) has entirely avoided reduction of the W chromosome by evolving mechanisms of Male-biased gene expression for all genes on both sex chromosomes.[45] So, there can be alternative evolutionary pathways to chromosome size reduction.

Another consequence of the origin of chromosomal dimorphism is the evolution of X chromosome inactivation in XX mammals (see above, page 171). This solution to the genetic dosage compensation problem generates a new opportunity for the introduction of random performative patterns of development into the phenotype of every homogametic Female mammal. Calico cats simply make visible the distinctive and random pattern of genotype-to-phenotype mosaicism that is an *inherent* and uniquely individual aspect of *every* XX mammal body. This individual-level, queer variability is a direct response to the ancient history of genetic sexual differentiation in mammals.

The variability generated by different mechanisms of sexual development initiation are evolutionarily unavoidable. Because sexual reproduction in animals requires generating anatomical and physiological differences in reproductive capacities among individuals that are otherwise genetically and developmentally equivalent, the process of sexual development is fundamentally fraught with variation-generating challenges. Creating stable sexual differences in anatomy, physiology, and behavior from bodily uniformity is inherently difficult for an organism to manage. All the evolved solutions to these challenges create other new challenges, which generate new sources of variation in sexual differentiation. The result is that fundamental and vital aspects of sexual differentiation and reproduction remain intrinsically variable. Because most mechanisms of initiating and completing sexual differentiation are subject to these constraints and inefficiencies, biological mechanisms of sexual development are themselves *generatively queering*. (Remember that queering refers to the disruption of the normative expectations through the introduction of variability from that norm.) Every evolvable mechanism to generate sexual difference from anatomical homogeneity will inevitably queer those differences.

Sexually Disruptive Selection

The challenges posed by the evolution of sexual bodies extend further to constrain adaptive evolution across the entire genome. The fact that every autosomal allele (or gene variant) must spend half its evolutionary lifetime in bodies developed from an XX genome or an XY genome creates unavoidable constraints on how finely tuned, or universally optimal, any particular gene can evolve to be. Referred to generally (and erroneously, in my opinion) as "intra-locus sexual conflict" or "sexually antagonistic selection," this phenomenon is more productively understood as sexually disruptive selection—the failure of natural selection to be able to optimize the survival and fecundity impacts of genetic variations that necessarily persist over evolutionary time in diverse sexual phenotypes. Consequently, natural selection in favor of a particular allele in one sexual phenotype can be constrained by its potentially negative contributions to survival and fecundity in another sexual phenotype.

For example, a study by Stephen Stearns and colleagues of genetic variations associated with biomedically relevant traits, including body mass, total serum cholesterol, blood pressure, height, and weight, in the Framingham Heart Study were under different selection pressures in men and women in ways that would clearly constrain future evolutionary responses to that selection. In other words, some specific genetic variants that foster cardiovascular health differ between men and women, contributing to the persistence of genetic diversity of these traits, and the persistence of suboptimal, health outcomes for both sexes. In this way, variation in sexual phenotype poses novel constraints to adaptive evolution, thus perturbing the adaptive expectations of evolutionary optimality.[46] Sexually disruptive natural selection favoring different alleles in each sex may also contribute to chromosomal differentiation, which may eliminate certain genes entirely from the genomes of one sex. However, this does not eliminate this adaptive inefficiency because the vast majority of all genes remain in autosomes that are common to all individuals. As a result, sexually disruptive selection may also contribute further to the instability of specific differentiated chromosomes once they have evolved (see creeping voles above).

This performative framework emphasizes that this phenomenon is not really about sexual antagonism or conflict, but evolutionary inefficiency, yielding novel evolutionary predictions. Sexually disruptive selection predicts that natural selected genetic mechanisms to control the development of cancer will be weaker at regulating cancers formed in sexually differentiated tissues and cell types. Because the genetic variations that contribute to cancers of sexually differentiated tissues are only exposed to natural selection approximately half of the time, genetic variations that contribute to these cancers are more likely to evade the body's evolved systems to control cancer and to persist in the population. As this hypothesis predicts, the two most frequently occurring human cancers are breast cancer and prostate cancer. (The annual number of cases of lung cancer in 2022 is only slightly lower than the number of prostate cancers, but the frequency of prostate cancer is more than twice as high as lung cancer because only half the population have prostate glands and everyone in the population has lungs.)[47]

Evolution of the Molecular Discourse of Sexual Development

We have examined the evolutionary consequences of variation in chromosomes, and how evolution of sexual development can be generatively queering. Now, we turn our attention to the evolution of genetic variations in the overall molecular discourse of sexual development.

In general, the downstream molecular mechanisms involved in vertebrate sexual development have not evolved as rapidly as the mechanisms that initiate sexual differentiation.[48] For example, the gene SOX9 plays an important role in testis development in many vertebrates from fishes to humans, indicating that some elements of the genetic mechanisms of sexual differentiation are quite ancient, and have been broadly conserved among many lineages since their origins. However, different genes have evolved in various independent lineages of XX/XY vertebrates to initiate SOX9 expression and the cascade of testis development. For example, in salmonid fishes, SOX3 plays no role in initiating testis development, but IRF9 does.[49] In Japanese rice fish, DMRT1 does the trick, but in the closely related Indian rice fish, SOX3 plays this role.[50]

Although the role of SOX9 in testis development across many vertebrates has been conserved since common ancestry, the novel gene SRY evolved through the duplication of SOX3 in our most recent common ancestor with marsupials to initiate SOX9 expression during testis development (figure 15). For a bipotent mammalian gonad cell to differentiate into a testis, SRY has to evolve to "mean" something new—"Turn on SOX9." But this novel way to enact this historically ancient discursive action can only occur through the coevolution of a new regulatory network to manage SRY's movement into the nucleus and its ability to bind appropriately to the enhancer site upstream of SOX9 in the genome.

Our understanding of the discursive nature of the coevolved gene regulatory networks can help us to think productively about the evolution of new ways to initiate the same old thing. In our ancient mammalian ancestors, SOX9 signaling in testis development was likely induced by environmental variation in temperature. The evolution of SRY as an initiator of sexual differentiation in XY embryos likely involved the derived *insertion* of its regulatory action at the *beginning* of the historically ancestral regulatory cascade in testis development.[51] By inserting its novel function at the beginning of the testis developmental cascade, this change to the initiation need not change other downstream details about the gene regulatory network.

However, similar evolutionary tinkering and replacement of various links in the sexual development gene networks have occurred in different mammal lineages.[52] In order to retain the continuity of inter- and intracellular developmental discourse, random changes in a structure of any paracrine signaling protein or transcription factor binding domains will require compensatory changes in the corresponding structures of protein receptors or DNA-binding sites they intra-act with, and vice versa.

To return to the linguistic analogy, the evolved differences in the structure of elements in the molecular-developmental discourse occur incrementally with a consistent maintenance of their downstream developmental meaning. In French, a verb is negated by placing *ne* and *pas* before *and* after it. The novel *ne... pas* double French negation system evolved culturally from the simpler, ancestral, one-word negation system found in other Romance languages. Likewise, primates have evolved to require phosphorylation *and* acetylation of SRY before nuclear translocation, whereas most other mammals get by with just the latter. Like the consistent meaning of negation across languages, these structural necessities occur to maintain a consistent developmental function and "meaning" of SRY within mammal evolution—"Turn on SOX9." Yet again, we see that an explicitly discursive, performative perspective on evolutionary developmental biology does not simply translate biological concepts into a humanities vocabulary—old wine in new bottles—but productively captures the complex roles of multiple developmental agencies in the evolution of development.

Why vary so much? Or, on the contrary, why conserve any genes at all? An important part of the answers is that the discursive properties of sexual development both facilitate and constrain evolutionary change. Like the coevolved correlation between linguistic statements and their comprehended meanings, genetic developmental regulatory systems cannot change so rapidly that paracrine signaling molecules and transcription factors do not bind functionally and selectively to the appropriate receptors or binding sites. If either changes too rapidly, the discourse fails.

For example, once when I was playing catch with my eldest son, Gus, sometime in the early 2000s, he tossed the ball wide and said, "My bad." I responded, "Huh?" because I had never heard that phrase before. Although slang like this evolves rapidly (probably because it functions culturally to distinguish speakers from their parents), language cannot change so rapidly that normal, mutual intelligibility is lost. Thus, just like the grammatical features that are shared by all Romance languages, many genetic details of vertebrate sexual differentiation systems have remained intact long enough for us to reconstruct their evolutionary histories.

At every level of the process, the coevolved molecular intra-actions involved in developmental gene regulation create endless opportunities for random changes that select for compensatory responses that maintain developmental function. Like the Red Queen, they must continue to evolve in molecular detail to compensate for changes among intra-acting molecular agencies, and maintain their consistent developmental "meaning" and discursive intelligibility. (See also appendix 7.)

However, these variability-enhancing properties have been consistently constrained by natural and sexual selection to maintain reproductive function. We can see this quite clearly in the presence of numerous inhibitory intra-actions between the gene signaling pathways for ovary and testis development (see dashed lines, figure 6). For example, β-catenin and SOX9 have evolved to mutually suppress each other during gonad development precisely because antecedent molecular events during gonad development were *insufficient* on their own to completely prevent the development of less reproductively functional morphological states (like infertile ovotestes). Yet again, we can see that SRY expression cannot "determine" anything—least of all the sex of an individual—if its transitory action must be continuously buffered against, or defended from, the effects of downstream events by other molecular and cellular agencies. The evolution of multiple inhibitory pathways in gonad development is evidence of the ongoing, evolved precarity of sexual differentiation itself.

These same issues apply generally to the evolution of any complex aspect of the phenotype. But curiously, sexual differentiation still stands out from other aspects of the body. As molecular-developmental biologists Amaury Herpin and Martin Schartl put it, "The variability and plasticity of the mechanisms that govern the development of the gonads is unmet by any other organ systems or tissues."[53] As we have seen, Yen-Shan Chen and colleagues are struck by the "anomalous nonrobustness" of Male sexual development. *Why* is this? The answer is that the fundamentally performative challenge of creating embodied sexual dimorphism out of embryonic uniformity is inherently unstable and subject to variation. These same phenomena and forces contribute to the generation and inevitability of queer, embodied sexual variation within species, including humans.

Evolution of Sexual Transition

Clownfish (Amphiprioninae) live exclusively in tropical coral reefs of the Indian and Pacific Oceans among the wiggly arms of colorful sea anemones. The clown fish and sea anemones have coevolved a symbiotic mutualism. The colorful arms of the anemones are armed with thousands of highly toxic, stinging cells that protect it and the clownfish from attack by predators. The clownfish have evolved a special mucous layer that protects them from the anemone's stings. Meanwhile, the clownfish eat parasites off of the surfaces of the anemones. Lastly, the clownfish's brilliant orange-and-white color patterns have evolved to send a negative, "*Don't Mess with Me!*" advertisement, signaling to potential predators that they are protected by the stinging anemone.

The clownfish's brilliant colors are effective warnings to predators when the fish is "at home" in its anemone, but they are disastrously *attractive* to predators if the fish is away from its protective home. Consequently, mature clownfish *never leave* the embrace of their anemone's slippery arms, which means that adult clownfish cannot safely move from one anemone to another. So, clownfish live in isolated colonies that include a reproductive Male and Female pair and a few, small, unrelated, nonreproductive Males. The Female is much larger than the Males, and produces large clutches of many hundreds of eggs. Males are *very attentive* parents, and they fan fresh, oxygen-rich sea water over their clutches of eggs to help the embryos develop successfully.

After hatching, the young clownfish rapidly disperse as tiny, microscopic, but not yet colorful(!), larvae into the ocean currents, where they travel independently, and ultimately settle down to join a new group of clownfish in another sea anemone, where they will stay and live out their entire lives.

So, what happens when the single clownfish Female dies? The once fecund colony suddenly becomes an isolated bachelor pad. Because it would be too dangerous for another adult Female to disperse to the colony from another anemone, there is only one choice. One Male—usually the largest, socially dominant, reproductive Male—grows even larger in size and *becomes* Female. The largest of the younger, nonreproductive Males then becomes a reproductive Male.

Clownfish biology raises a fundamental question about sex. Why produce sperm *or* eggs? Why not produce both? Or one, and then the other, at different stages of life? As clownfish demonstrate, different groups of organisms have answered these profound evolutionary questions differently. Although wholly absent from four-legged vertebrates, fishes and invertebrate animals have evolved many examples of sex-changing species, which zoologists have called sequential hermaphrodites. As one might imagine, the capacity for transformation between mature sexes creates its own new diversity to the sexual spectrum. There are species like clownfish that sexually mature first as fertile Males, and then transition to become fertile Females at a later date, which are called protandrous. There are also species that sexually mature first as Females, and then transition into fertile Males at later ages, which are called protogynous. Examples of protogyny include many of the colorful wrasses (Labridae) which frequent many of the same coral reefs as the clownfish. The evolutionary difference between the two developmental trajectories depends upon which sex has the greater reproductive success at larger body size, which is itself influenced by the ecology, behavior, and reproductive system in these species. A third variety, simultaneous hermaphroditism, includes species that can produce eggs and sperm at the same time. It exists as a permanent state in many corals, trematode worms, barnacles, snails, and many echinoderms, or as a temporary, transitional state during protandrous or protogynous transitions (e.g., a few fishes). Approximately 95 percent of all species of flowering plants are simultaneously hermaphroditic, producing fertile Male and Female flowers or flower parts.[54]

Although sexual development in tetrapod vertebrates typically leads to the elimination of other reproductive possibilities (at least so far), the radiation in sexual diversity among sequentially hermaphroditic fishes demonstrates an additional, evolvable possibility for the bipotent vertebrate gonad. For many species whose gonads are homologous to our own, sexual differentiation is neither an end point, nor an irreversible fork in the developmental landscape. Rather, sex can be a multiphasic process of becoming. Working from a single, homologous reproductive gland with bipotent possibilities, various groups of fishes have evolved the capacity to developmentally reengineer a mature gonad to produce the other type of gametes (or both simultaneously as a transitional state). It is a whole different kind of sexual life cycle.

As clownfish show, developmental decisions about whether to transition from one sex to another are often based on the *social* context in which the individual finds themselves. In other words, the difference between remaining Male or transforming into a Female will depend upon the social presence of a larger, dominant Female clownfish at your anemone. Consequently, sequentially hermaphroditic fishes have evolved to be highly sensitive to changes in visual and behavioral cues from the social environment (the size, sex, and behavior of other fishes), which then instigate neural and hormonal responses that stimulate the developmental transformation of their currently mature gonads into another state. Sexually transitioning Male fishes can begin to show changes in color and behavior within one day of the absence of the Female, and show substantial hormonal changes within one week. This rapid initiation of sex change in fishes is powerful evidence of the profoundly performative influence social environment can have on embodied sex, and a striking nonhuman example of the role of social cognition on the material sexual body.

Little is yet known about the molecular-developmental biology of sexual transitions in fishes, likely because most such molecular research is funded by biomedical sources that are usually quite uninterested in fish biology for its own sake. However, we have seen that mature ovaries in humans continue to produce the FOXL2, which inhibits the production of SOX9, throughout the lifetime (figure 6). Likewise, ongoing SOX9 expression in testes locally inhibits the expression of β-catenin, which functions in the development of ovarian identity and morphology. Thus, one ongoing function of being a mature human ovary is to prevent its conversion into a testis, and vice versa. Conversion of one gonad morphology to another in sequentially hermaphroditic fishes likely involves the input of social information from the central nervous system to control similar inhibitory feedback loops, followed by the conversion the network of gene-expression states to that characteristic of the other mature gonad type.

The genetic details of these sexual transitions in fishes are not yet understood to be biomedically relevant to humans. However, the rise of transgender medicine should contribute to new research on this topic, perhaps ultimately leading to medical treatments that could convert mature human gonads from one morphological state to another. Of course, humans have evolved more complex anatomical variations among individuals in internal and external sexual anatomy. All sex-changing fishes deliver their unfertilized gametes from their cloacae into the water, and thus do not have highly differentiated reproductive tracts or external genitalia. Clownfish don't have uteruses or prostates. However, in the long future of biomedical technology, it may become possible to transform between differentiated human gonad identities and morphologies, as some fishes do. Thinking about the evolutionary biology of sex performatively can provide new insights into understanding human sexual possibilities.

Evolution Is Incompatible with Sexual Essences

As we discussed in the introduction to historical ontology (chapter 2), all biological entities—from genes and proteins, to organelles, organisms, species, and other taxonomic groups—have the capacity, by virtue of evolution or aging, to evade or elude any definition of them that is based on any essential feature, trait, or characteristic. Again, "essential" refers to a quality that is both inherent and indispensable. Having four limbs might seem to be an essential feature of tetrapod vertebrates, but the evolution of snakes and caecilians demonstrates that limbs are neither an inherent nor indispensable feature of being a tetrapod. In short, evolved entities are historical individuals that do not have essences.

Yet, biological science continues to investigate sexual development as the material manifestation of some individual essence. As Sarah Richardson has shown, scientific concepts about the X and Y chromosomes and their functions have been greatly influenced by social concepts of femininity and masculinity in Western societies. Accordingly, the X chromosome has been referred to as sociable, controlling, monotonous, and motherly, while the Y chromosome and its genes have been "scientifically" described as active, macho, sexy, wily, clever, and dominate, but also degenerate, lazy, hesitant, and hyperactive. Similar statements have been made to characterize the gene SRY (see above). However, the evolved variability in the mechanisms of initiating and realizing sexual development demonstrates that concepts of Male and Female that are constructed upon essentialist notions about genes, chromosomes, hormones, or other biological features simply fail as accounts of organismal diversity and human biology.

The anatomical features that many would think of as "essential" to being Male or Female are much more ancient than the biological mechanisms by which these features are materially realized in the bodies of individual organisms, including humans (figure 15). For example, the molecular events that result in the development of testes and a penis in amniote vertebrates can be initiated: by variation in embryonic temperature (turtles, crocodiles, and many lizards); by possessing distinct genetic differences (XY mammals, and lizards); by lacking any genetic differences (e.g., ZZ ducks, and various other lizards); by simply lacking a full diploid complement of chromosomes (e.g., X0 Ryukyu spiny rats), by having a full diploid complement of chromosomes (e.g., XMXP creeping voles); or having an array of distinct chromosomes (e.g., echidnas, $X_1X_2X_3X_4X_5Y_1Y_2Y_3Y_4Y_5$). Likewise, the ovaries, uteruses, and mammary glands of mammals may develop in individuals that are chromosomally XX (most mammals), XM0 (e.g., creeping voles), or $X_1X_1X_2X_2X_3X_3X_4X_4X_5X_5$ (e.g., echidnas). Recall that the X_{1-5} chromosomes of the monotremes are *independently* evolved from the X chromosome of other mammals. One of the stunning conclusions from this observation is that mammalian mammary glands evolved *before* the origin of either X or Y chromosomes (figure 15). Thus, there is nothing about lactation or having a penis that is linked essentially to the possession of XX or XY chromosomes, or to any specific combination of genes or hormones.

Thinking more broadly, we see that no feature of an organism can be considered an individual essence if it can be alternatively initiated by one's genome, by developmental temperature, by the sexual composition of one's social group (e.g., clownfish), or by the presence or absence of an inherited parasitic infection (e.g., *Wolbachia* in insects).

One could argue that sexual reproduction is essential to the persistence of human beings, and other sexual species. However, that assertion does not mean that a *specific* mode of sexual reproduction—via sperm or eggs—is *indispensable* to a given individual organism. Even if reproduction were indispensable in general, either sexual possibility will encounter reproductive opportunities within a population. Thus, neither Maleness nor Femaleness, specifically, can be an essential feature of an individual genotype, embryo, or adult.

The evolutionary diversity of sexual development mechanisms establishes clearly that there is no essential fact of individual sex, only reproductive possibilities realized by individual bodies. Furthermore, the evolutionary and individual differences in sexual development—detailed in this and the previous chapters—show that all sexual phenotypes are performative states, not definable facts about individuals prior to, or independent from, their iterative individual realizations.

Norms and Innovation

Gender performativity provides both an account of the constraints on gender variation by cultural norms, and the process by which those same cultural norms can be transformed and expanded. In a precise parallel to this process, it is the regulatory and variation-generating properties of organismal development that facilitate evolutionary innovation and adaptive change in morphology. Both processes constitute transformations of the historic "norms" within a population or species—culturally "normative" expectations of gender in the former case, and the historic "norms" or modes of a frequency distribution in phenotypic traits in the latter. The fact that the phenotype of every zygote performs itself—working with information in its individual genome, and physically elaborating itself through interactions with its material and social environment—creates the potential for the evolution of both new phenotypic variations and genetic functions.

Biological variation poses two questions simultaneously; both "why is there so little variation?" and "why is there so much variation?" The limitations, or constraints, on phenotypic variation in response to genetic environmental variation—Waddington's canalization—is the result of two processes acting on different time scales. On historic time scales, natural and sexual selection on the function of the phenotype eliminates some genetic combinations from surviving and reproducing into the next generation. On the individual time scale, performative developmental processes themselves create biases among phenotypic possibilities. Like Foucault's concept of social power, selection and developmental constraints are both restrictive *and* creative, limiting *and* facilitating, operating negatively *and* positively.[55]

In each generation, sexual reproduction shuffles genetic variation into new, usually unique, individual zygotes that go on to perform themselves in the world. The genetic and physical mechanisms of organismal development create cell types, tissues, organs, and body parts that have their own emergent variational potentials within the organism. The specific ways in which the agency of genes, cells, signaling centers, tissues, and body parts gives rise to organismal complexity allows organisms to both iteratively generate normative bodies with surprising fidelity, *and* to exceed, defy, or transform those historic norms as well. It is the intrinsic variational potential created by the developmental mechanisms of the limb, the flower, and the feather that facilitates the explosive adaptive radiations in their forms among species of tetrapods, flowering plants, or birds and other dinosaurs, respectively.[56] (See also appendix 5.)

More specifically, we can see how variational potential inherent in the process of sexual development can contribute to evolutionary change. For example, we have seen that the fusion of the left and right Müllerian ducts that creates the upper vagina, uterus, and fallopian tubes in mammals begins where the left and right paired Müllerian ducts meet one another at the cloaca, and fusion proceeds upward like a zipper (figure 10). Limitations in the extent of Müllerian tube fusion lead to a specific pattern of organ duplication, but not others (see pages 173–75). Some possibilities—like two vaginas leading to a single uterus—are precluded because one cannot interrupt a temporally and spatially structured developmental process without having first started it. In this way, the process of development itself creates certain structured biases, or variational potentials, while closing down or excluding others. The individually realized variations observed in human reproductive anatomy are also indicative of the underlying structure of evolutionary potential *among* related organisms. As the name *uterus didelphus* indicates, this rare human anatomical variation is the evolved anatomy found typically in Female opossums (*Didelphus virginiana*).

Another example of the historic/developmental structure of phenotypic variation is presented by urethral tubulation in the mammalian penis. The medioventral fusion of the left and right sides of the developing penis, which converts the urethral groove into an enclosed urethral tube, begins at the base of the penis and proceeds toward the distal end. But hypospadias in humans is similar to the ancestral penis morphology that first evolved in the common ancestor of all reptiles and mammals. The penises of reptiles (including birds, crocodiles, turtles, lizards, and snakes) all lack an enclosed urethra, and have an open, ventral sulcus (an undifferentiated urethral groove), which functions as a tube for delivery of sperm during intromission. (Snakes and lizards actually have two sulci, because the original single penis has evolved into two paired hemipenises; but that is a story for another time.) Again, performative phenotypic variation among humans reveals details about the historic, phylogenetic variation realized evolutionarily among lineages of amniotes.[57]

Evolutionary innovations—such as flowers, feathers, and hair—are novel structures that are not homologous with any other antecedent or ancestral structures. In the evolution of amniotes, the genital tubercle, which develops into the sexually homologous penis and clitoris, is another example of an evolutionary innovation that originated in the most recent common ancestor of all reptiles and mammals. As we have seen, the evolution of the penis facilitated internal fertilization, which permitted the uncoupling of amniote reproduction from the water (as required in fishes and most amphibians), and facilitated the evolution of a fully terrestrial ecology. Although the genital tubercle is not homologous with any other body structure, it evolved through the co-option, or reutilization, of a modular network of interacting molecular signaling pathways that had previously evolved to control spatial differentiation of the limbs, hands, and feet.

Developmental biologist Martin Cohn and colleagues have shown that HOXD13 is expressed in a similar fashion in the genital tubercle *and* the developing feet and hands, and it has a similar function in the morphogenesis of all of these appendages (figure 16). Later in development of each, Sonic Hedgehog signaling creates a signaling center in the urethral groove of the *ventral* surface of genital tubercle, *and* in the zone of polarizing activity on the *posterior* margin of the developing hands and feet.[58] The latter is vital for the organization of the fingers and toes in all tetrapods. The co-option of this modular gene regulatory network from the limb bud is critical to the creation of the spermatic sulcus and the enclosed urethra in mammals.

During the evolutionary origin of the penis and clitoris, natural selection for reproductive innovation led to the redeployment—a molecular/evolutionary variety of discursive citation—of a preexisting network of coevolved molecular signals that had originated for the development of a different appendage—hands and feet with fingers and toes. In conclusion, the amniote genital tubercle is an evolutionary innovation without homology to any other body part, but the molecular module that was evolutionarily co-opted to create it *is* homologous with that evolved previously for the development of the tetrapod hands and feet. Like other evolutionary innovations in phenotype, including flowers, limbs, eyes, and feathers, the evolution of the genital tubercle provides evidence about the origins of evolutionary novelty in biology.

Placental Performativity

Thinking performatively in biology requires examining the role of the developmental environment in organismal becoming. In humans and other placental mammals, the developmental environment for embryos is created by the uterus and placenta.

The performativity of embryonic development is mirrored by the fascinating agential intra-actions of mammalian pregnancy itself. Specifically, embryonic development in humans is mediated by the placenta—a distinc-

16. An embryonic Male mouse (*Mus domesticus*)

At day 13.5 after conception, cells stained for the expression of the transcription factor HOXD13 appear in the developing hands, feet, and the genital tubercle. Although the clitoris/penis are not homologous with the hands and feet, the same modular molecular signaling networks involved in hand/foot development were evolutionarily co-opted (or reutilized) for genital morphogenesis. HOXD13 is also expressed during the development of the eyes. Photo courtesy of Marty Cohn.

tive mammalian fetal/maternal co-organ that provides a dynamic interface for nutritional and physiological support of developing embryos. The placenta's function and evolution provide special insights into the uniquely performative relation between the mammalian embryos and the gestational parent that provides the embryo's material developmental environment.

Recent research by my colleague Günter Wagner, his students, postdocs, and collaborators have established that the placenta and extended pregnancy of mammals have evolved through a modification of the maternal inflammatory immune response induced by the presence of the genetically foreign embryo inside the body.[59] To understand the evolution of the placenta and extended pregnancy in humans, we need to compare ourselves and other true placental mammals to our closest living relatives—the marsupials, or pouched mammals (figure 15).

In marsupials, pregnancy is brief—lasting less than a single menstrual cycle, or about fourteen days in the opossum. Indeed, the pregnancy is so short that the Female opossum does not even hormonally "recognize" being pregnant at all. For twelve of those fourteen days, opossum embryos develop in utero inside their egg coat (which is homologous to the reptilian egg shell), and they get little nutrition from the maternal physiology during that time. When the embryos "hatch" out of the egg coat *inside* the uterus, the embryos release protease enzymes that damage the uterine lining. Note that the brief uterine-fetal intra-action involved in marsupial pregnancy is initiated by the embryo's active efforts to *provoke* a response from the uterine tissue. The maternal inflammatory response to this uterine tissue damage is characterized by increased expression of the paracrine signals of the innate immune system including interleukins and other cytokines (IL1A, IL17A, IL19, TNR), and prostaglandins (PTGS2 and PGE2). These innate immune-system signals lead to: (1) the recruitment of white blood cells (or neutrophils) that migrate into the uterine lining, (2) contractions of the uterine muscles, and (3) capillary growth and an increase in vascular permeability, or swelling, of the uterus, which increases the flow of small molecules like sugar out of the uterine capillaries and into the fetal environment. Within two days after the internal "hatching" from the egg coat, marsupial embryos are born in a very undeveloped state. At birth, opossums do not yet even have hindlimb buds.[60]

Thus, birth in marsupials consists of a classic inflammatory reaction by the maternal uterine lining to the immune system challenge instigated by the genetically distinct embryo. During those last two days before birth and after "hatching," however, the marsupial embryo actually grows quite rapidly by absorbing glucose and nutrients acquired from the inflamed, leaky uterine capillaries. The birth of the embryos by myometrial muscle contractions only two days after implantation constitutes a classic innate immune-system rejection process.

In true placental mammals, by contrast, pregnancy endures longer than a single menstrual cycle. This extended form of pregnancy evolved through the elaboration of a complex placental organ that typically involves the growth of fetal cells *into* the uterine lining. Although marsupials have a rudimentary form of placenta, the interpenetration of maternal and fetal tissues creates a novel, collaborative, intra-active organ that is unique to true placental mammals.[61]

Wagner and colleagues have found that this advanced form of placentation evolved by the transformation, or rewiring, of the innate inflammatory stress response by the uterine lining, keeping those components of the inflammatory response that encourage sustained embryonic growth and development, and eliminating other components of the inflammatory response that would endanger the embryo's retention and successful development.

When a blastocyst—the tiny two-hundred- to three-hundred-cell-stage embryo—implants into the uterine lining of a true placental mammal, it stimulates the uterine expression of the inflammatory cytokines IL1, TNF, and others. These developing uterine cells then begin to express the prostaglandin PGE2 which fosters angiogenesis (i.e., blood vessel growth), and vascular permeability, or leakiness that will nourish the growing embryo. These responses are retained components of the classic inflammatory response that we observe in marsupials—think of how a sprained ankle turns both red and swollen. However, unlike marsupials, the cells of the placenta have evolved to *suppress* the expression of other cytokines, namely IL17A, which *prevents* the recruitment of white blood cells into the tissue that would attack the genetically foreign embryo. Unlike marsupials, implantation in placental mammals *lowers* the expression of prostaglandin PGF2A, which *suppresses* uterine muscle contractions.[62]

During implantation, the embryo actively *elicits* maternal inflammation by producing the paracrine signal IL1B (whose receptor is expressed in uterine lining), and receipt of this signal by the uterus initiates further inflammatory cytokine release. Uterine inflammation is in the interest of the implanting embryo, and the embryo has coevolved with the tissue of the uterine lining to stimulate it. The paracrine signaling systems that originally evolved to function among the immune cells of a single individual have been evolutionarily co-opted to function in intra-active developmental communication *between* two individuals—parent and offspring.

Once the placenta is developed, there is a long anti-inflammatory period of pregnancy during which the embryo can grow extensively. When the embryo is mature, the inflammatory process is reinitiated, inducing birth with fluid release and uterine muscle contractions. Thus, in true placental mammals, birth is accomplished by resuming the mother's original allogenetic inflammatory reaction, and proceeding with rejection of the foreign body through an increase in swelling and uterine muscle contractions.

In summary, the ancestral uterine immune response to a genetically foreign embryo that leads to early birth in marsupials was evolutionarily rewired in the ancestor of true placental mammals to retain those components—angiogenesis and vascular permeability—that are helpful to embryonic development, but eliminating others—white blood cell recruitment and uterine muscle contractions—that would endanger it. This evolutionary biology research further indicates that medical management of uterine immune inflammatory response may be critical to alleviating multiple problems of human pregnancy including difficulties with implantation that contribute to infertility, *and* difficulties in sustaining embryonic development for the full term that contribute to premature birth, both of which may be a response to *overactive* immune responses at inappropriate times.[63]

The evolution of the placenta is an example of the origin of a novel, collaborative, functional, material-social *intra-action* between parent and offspring. The gestating parent and the embryo are each part of the "environment" of the other. At implantation, these two agencies come into material-social contact, and engage in a coevolved molecular dialogue in which *both* are active agents.

The ancestral uterine immunological response to the marsupial embryo "hatching" is an exemplar of a physiological performativity—an expressive doing that constitutes a reaction to a provocative agent in the environment. The evolutionary flexibility of this inflammatory response facilitated the origin of an innovative reproductive possibility—true placentation with extended pregnancy. The performativity of the placenta is manifest hierarchically in its homeostatic physiological function, its intra-active developmental process, *and* the coevolved discourse between parental and fetal agencies.

Limits of the Binary Bottleneck

Extending from zygote to culturally realized gender, the performative continuum encompasses genetic, developmental, physiological, psychological, linguistic, sociological, and cultural processes in the iterative realization of each and every human being. However, at the initiation of this cascade of livable possibilities lies the historically persistent binary reproductive bottleneck. All human beings (so far) are the product of an egg fertilized by a sperm and gestated in a human womb. As a species of placental mammals, humans have evolved from ancestors with bodies that have historically differentiated to produce either eggs or sperm, to either have a uterus with the possibility for placentation and gestation of developing embryos or not, and to either have mammary glands for the possible production of milk that can nourish infant offspring or not. Although the lives of human individuals manifest extensive variation, the *origin* of each human life has always been, and remains, subject to this sperm-egg-womb constraint. Every human being has a binarily structured origin, and the persistence of this life-initiating bottleneck is a result of a strong history of natural and sexual selection on sexual reproduction.[64]

The existence and persistence of the binary bottleneck has been frequently misinterpreted as evidence of the scientific reality of an individual sexual binary. But the centrality of sexual reproduction to the origin of individual human lives cannot, by itself, dictate, determine, control, or override the mechanisms by which individual organisms are physically, iteratively materialized. Our species' long evolutionary history of sexual reproduction cannot create or insert a prior fact, essence, or intention of the binary sex of any gene, genome, zygote, or life prior to, or exclusive of, the process of its individual material becoming. The evolutionary history of sexual reproduction cannot veto the inherently performative process by which organismal bodies are iteratively materialized in the world. The canalized persistence and boundary-blurring diversity of sexual morphologies are both products of that *same* evolutionary history. The existence of the former cannot eliminate the reality of the latter. From our individual origins in the sperm-egg-womb nexus—made possible by embodied capacities necessary for egg and sperm production, fertilization, pregnancy, and gestation—we mature and realize a greater variety of sexually differentiated selves and possibilities than an essentialist binary concept of individual sex can contain.

Because individual sex has no status other than through its iterative realizations, sex cannot be an inherent fact about an individual or the body independent of that individual's bodily becoming. This further means that sex is not necessarily an essential fact about any individual life, but only a realized reproductive possibility within a life. As sequentially hermaphroditic fishes (which share homologous gonads with our own) demonstrate, the biological process of becoming can be continuous and ongoing until death, and many organisms have evolved reproductive lives that explore this reality.

Although no human life has evaded the sperm-egg-womb bottleneck at its origin, future advances in reproductive technology will almost certainly relax or eliminate the sperm-egg-womb constraint on human reproduction. Reproduction has already been extensively decoupled from social gender. Human dads with wombs, and moms with penises are already lived realities. Whether people choose to pursue future reproductive technologies or not, it is not difficult to imagine that humans will someday develop the technological capacity to decouple reproduction from many aspects of embodied sex.

CHAPTER EIGHT

The Future of Performative Biology

There is no reason to propose a new scientific concept or framework unless it does new scientific work.[1] The challenge to extending the cultural concept of performativity into the biological sciences is to show that a performative view of biology leads to new predictions or insights that are currently missing from biology. Here, I develop and present a broad framework for performative developmental and evolutionary biology, and a few examples of what performative biology can accomplish and what we will need to do to pursue the opportunities in the future.

Performative Scientific Hypotheses

New scientific hypotheses in a performative biology will not come by focusing solely on the traditional goals of demonstrating necessity and sufficiency of genetic causes of phenotypic variation in inbred lines or tissue cultures. Rather, progress will result from reframing our thinking about the phenotype as an *active doing*—a material realization by a hierarchy of biological agencies within the body in interaction with facilitating and constraining agencies and material and social influences in the organism's environment.

As an example, consider again the observation that adrenal disorders that result in high testosterone production during gestation can radically influence the sexual differentiation of XX embryos.[2] From a performative perspective, this observation leads us to ask, "How does the XX gestational environment differentially affect the development of mammalian XX and XY embryos?" As soon as we ask the question, we realize that mammalian gestation and embryonic mechanisms of sexual differentiation have not evolved in isolation of each other, but have *coevolved* through selection on the discursive process of individual embryonic development. As we have seen (chapter 7), the womb is more than just a gestational environment for the embryo; it is an intra-active interface between the fetal and parental agencies. Thus, mammalian sexual development must have evolved to accommodate, account for, and stabilize in response to, the predictable differential impacts of the XX gestational agency on the sexual development of different offspring.

The placental exchange of small molecules necessary for fetal nutrition and physiology cannot prevent the simultaneous transfer of hormones, growth factors, and other biologically active small molecules between the gestating parent and the offspring. Accordingly, a heterogametic gestational environment could include gene products not found in homogametic offspring, which could potentially affect their sexual development. Thus, from a performative perspective, it is not accidental that the gestating sex is homogametic, or XX. XX mammals have only those genes found in *all* individuals in the population—which would include all offspring. If Female mammals were heterogametic—that is, XY—then expression of the distinct, sexual difference-making genes on their Y chromosomes could potentially disrupt the development of their homogametic XX offspring.

The hypothesis that the gestating sex should be homogametic is a *performative* prediction because it explicitly reframes the inquiry into the development of the embryo as an active process of intra-action with the gestational environment. This prediction is congruent with the observation that Female birds are heterogametic, and that birds are the only major group of vertebrates that have *never* evolved live birth. Of course, there are many other potential reasons why birds have not evolved live birth, so this hypothesis should be best tested further by examining mechanisms of sexual differentiation in multiple other lineages of vertebrates with live birth.

Although more research needs to be done to establish sexual differentiation mechanisms in more live-bearing vertebrate groups, the performative prediction that the gestational sex should be homogametic is consistent with observations that *all four* lineages of live-bearing fishes for which data are available. Guppies (Poeciliidae), requiem sharks (Carcharhinidae), and hammerhead sharks (Sphyrnidae) all have XX-XY chromosomal sex differentiation systems[3] and live birth with homogametic Female gestation.[4] A powerful fourth test of this hypothesis comes from the exceptional case of seahorses (*Hippocampus*), in which the Male gestates the fertilized embryos in a special pouch and nourishes them with a vascular pseudoplacenta that supports the transfer of small molecules. Although there are still questions about the precise mechanism of sexual differentiation in seahorses (probably multigenic like zebrafish), a recently completed version of the *Hippocampus* genome documents that seahorses *do not* have differentiated sex chromosomes—that is, every seahorse is chromosomally homogametic. So, paternal gestation in seahorses supports the hypothesis that the gestating sex should be homogametic.[5]

Extending this analysis further into lizards and snakes is currently challenging because there have been both rapid evolution of sexual differentiation mechanisms *and* live bearing, and they are not always known for the same genera and species. Some live-bearing lizards have gestational homogamy (XX/XY in Phrynosomatidae, Liolaemidae, many Scincidae), and other lineages do not (ZZ/ZW in Chamaeleonidae, Lacertidae, and all live-bearing snakes). However, some of these live-bearing reptiles exhibit egg retention with little or no small molecule transfer during gestation, which would not be inconsistent with maternal heterogamy. In some species, sexual differentiation may also occur entirely after birth.[6]

The unexpected relationship between parental homogamy and gestation is a performative hypothesis because it cannot be derived from any analysis of gene-level selection, adaptation of alleles to other genes in the genome, or even theories of maternal-offspring conflict (see appendix 6). Gestational parents will be equally related to all of their offspring regardless of variation in their sexual development mechanisms. Rather, the hypothesis arises directly from a conceptual focus on the role of environmental agencies—the gestational environment—in the development of sexually differentiating individuals.

This performative account of sexual differentiation and placentation highlights another interesting difference in sexual development mechanisms of most XX and XY mammals after gonad differentiation begins. Why do Female and Male mammal bodies differ strikingly in the use of endocrine and paracrine signaling systems during embryonic sexual development? The embryonic reproductive tract and genital development in XY mammals depends extensively upon hormonal signals produced by the testes—including testosterone, which can be converted locally into DHT and anti-Müllerian hormone. In contrast, the embryonic differentiation of these features in XX individuals proceeds *entirely* through local paracrine, *not* endocrine, signaling.

Thus, Müllerian ducts persist and develop into fallopian tubes, uterus, and vagina in the absence of any ovarian steroid hormones, and must otherwise be actively blocked from developing by anti-Müllerian hormone produced by the testes. In contrast, the Wolffian ducts will wither and degenerate in XX individuals, *unless* they are actively *induced* to develop into the vasa deferentia and ejaculatory ducts by testosterone. Similarly, in the absence of external hormonal signals, the genital tubercle, urogenital folds, and genital swellings will develop into a clitoris and labia, while these same structures can develop into the penis and scrotum in response to endocrine DHT. Furthermore, the fusion of the Müllerian ducts to form the upper portions of the vagina and the uterus, and the fusion of the upper Müllerian and lower cloacally derived portions of the vagina are regulated by still under documented paracrine signals. After gonad differentiation in mammalian embryos, Female sexual development proceeds through paracrine signaling alone, while Male development involves both endocrine and paracrine pathways.

From a performative perspective, we can understand how these profound and previously untheorized differences in sexual development mechanisms have coevolved in concerted coordination with the homogametic gestational environment. Heterogametic XY embryos may have evolved to use distinctive hormonal signals in their embryonic sexual development because they constitute a private, or buffered, communication channel that cannot (usually) be confused with cross talk from their homogametic XX gestational environment. However, XX embryos have evolved to rely on localized, endogenous paracrine signals because they must buffer their own sexual differentiation mechanisms from the potentially disruptive endocrine signals from their XX gestational environment, which is similar to their own. This hypothesis is further supported by the observation that only following birth—when gender/sex development can proceed outside of a gestational environment—are XX individuals released from this constraint to employ hormonal mechanisms in their reproductive development, maturation, and behavior.

The discovery of endocrine signaling and the expansive role of hormones in the development of Male reproductive tracts and external genitalia occurred decades before the mechanism of local, cell-to-cell, paracrine signaling was identified in any organism and characterized in mammalian sexual development. This temporal lag between scientific discoveries combined with cultural conceptions of active/assertive Maleness to create the mistaken impression that Female sexual development was a passive default for mammalian embryos,[7] an erroneous view that guided scientific research for decades by establishing Male "sex determination" as synonymous with the entire topic of mammalian sexual development. However, the embryonic sexual development of mammalian Males only gave the appearance of being a more "active" process because it involved hormonal signals that were discovered decades before the structure of DNA was described, and fully seventy-five years before paracrine signals were identified. As we have seen, recent research has now revealed the active role of paracrine signaling in the development and maintenance of ovarian identity and morphology. Inexcusably, our knowledge of Female sexual differentiation is *still* quite patchy. For example, the paracrine signaling pathways involved in Müllerian tube fusion, and in the fusion of the inner, Müllerian-derived, and outer, urogenital-sinus-derived, portions of the composite vagina are still very poorly understood.[8] Investigating these developmental events should be a major future research priority.

Again, I do not know of any previous remark on this striking difference in the use of endocrine versus paracrine signaling pathways in the sexual differentiation of XX and XY mammals, let alone an attempt to explain it. Gene-level selection and parental-offspring conflict simply make no relevant predictions on the matter. Yet, a performative perspective on sexual development both draws our attention to this previously uninvestigated phenomenon, and provides a coherent explanation of it that can guide productive further research in the sexual development of mammals and other live-bearing animals.

Performativity of Illness and Disability

In the opening of *Anna Karenina*, Leo Tolstoy declares, "Happy families are all alike; every unhappy family is unhappy in its own way." Like novelists, physicians also want to tell vitally important stories. From the physician's perspective, a healthy patient, like a happy family, has a pretty boring and trivial story. (Of course, we know this is not actually true, but so did Tolstoy. However, for the purposes of novelists and practicing doctors, it is.) Like an unhappy family, the stories of illness and malady provide more absorbing content, salient details, and diagnostic twists and turns for the physician to work with. Each of us ages differently, with different organs and tissues falling apart at different rates and in response to our distinct genomes, our individually realized material bodies, and our variable life experiences and exposures. Thinking about illness and disease as performative mis-realizations of the self—self-performances gone awry from the individual's own perspective—focuses our attention on the fact that there are indeed a greater *variety* of ways to go wrong than right. This is the truth that Tolstoy understood.

Although it would take a whole additional volume to fully elaborate on the performativity of illness, disease, and disability, I want to point out here how using this framework can improve our understanding of disease and medicine. Thinking about illness and disease as manifestations of the performative phenotype explicitly focuses our attention on the *individuality* of patients—their unique genomes, personal histories, capacities, life exposures, environmental risks, and random life influences. The performative framework encompasses the full variety of challenges to understanding wellness and health, from congenital genetic variations to cultural diversity and economic precarity, from novel pathogens to dietary saturated fats.

The performativity of the phenotype doesn't end at birth, at sexual maturity, or adulthood; it applies to the entire, ongoing arc of the life cycle from conception until death. Thus, it is productive to think of illness, disease, and maladies as aspects of the continuing process of individual becoming. The same tensions found in the biological sciences between reductive intellectual experimentation and the uniqueness of every individual human apply in the understanding of health and medicine as well.

Of course, physicians know that their patients are individuals—unique material realizations with unknowably complex histories full of rich possibilities. But, as applied natural scientists, doctors draw conclusions about patients by grouping them together into diagnostic categories, which are concepts about the causes and potential treatments of ailments. Diagnoses allow doctors to establish whether specific treatments are effective in treating certain ailments. For many illnesses, diagnosis does a great job of guiding appropriate treatment. For others, not so well. Chronic fatigue syndrome, Guillain-Barré syndrome, and now long COVID-19 are examples of complex diagnoses applied to patients with variable symptoms, manifestations, and trajectories that have poorly understood etiologies. Indeed, it is those persistent diseases that resist reductive explanation, in which the individualized, performative nature of the ailment is greatest, that the diagnostic methods of biomedicine perform least well.

The tensions between reductive generalizations about defined diagnostic *classes* of patients sharing a specific defined malady and the lived experiences of individual patients are omnipresent in medicine. As in biological science more broadly, the gap between scientific explanations and lived experiences of disease can be addressed by a shared performative understanding of both. For example, the risks of illness or the effectiveness of medical treatments are often expressed in terms of the incidence of occurrence, or the odds. Physicians say that a specific treatment for a disease is known to be effective in a certain percent of patients, or that individuals of a specific age and body mass index have a certain likelihood of experiencing heart disease or stroke. But we know from the start that these summaries are not really a guide to the actual risks or outcomes for any particular individual. The odds that physicians express are summaries of observations of a predefined population of patients. To draw on Barad's concepts, predefined diagnostic classes of patients are material-discursive apparatuses through which the observed causal relation is *created*. The patients' risks and outcomes result from the performative properties of their specific individual genetic makeup, current physiology, immune status, prior exposures, and material environmental influences interacting with cultural phenomena including race and ethnic categorization, economic status, immigration status, zip code, proximity to a grocery store, access to health insurance and health care, and many other factors. Our performative individuality means that many of the properties that will contribute to our own response to treatment are already highly specified—by the content of our genomes, our histories, environments, and social factors. Our individuality already precludes certain treatment outcomes or responses, and makes certain others possible or even inevitable.

Physicians also express the efficacy of a drug or therapy in terms of the odds. For example, a certain drug may be effective at alleviating certain symptoms in 45 percent of cases. This does not mean that if *you* took this treatment one hundred times, you would expect to be cured forty-five times. Whether the drug actually will work for you or any another specific patient is almost certainly already given before the treatment is attempted; the individual potential response to that treatment is simply unknown to the patient and physician.

Thinking performatively keeps us focused on the challenging gap between the individuality of every patient, and the capacity of biomedical science to make progress by generalizing across patients. Every instance of medical diagnosis consists of an ontological individual—a patient—that has somehow arrived at a state corresponding to the diagnostic criteria for membership in an ontological class—a defined medical condition. Each one of those patient paths is unique and contingent. Like other aspects of the phenotype, these individual paths, or etiologies, are best understood performatively. For some ailments—like a broken arm or appendicitis—the tension between performative individuality and reductive diagnosis may not be great. For other illnesses, the gap can be vast.

As I write, physicians, epidemiologists, and biomedical researchers around the world are struggling to understand the frightening breadth of patient responses and outcomes to infection by the novel coronavirus SARS-CoV-2. The symptoms of COVID-19 may include fever, dry cough, difficulty breathing, muscle aches, lethargy, brain fog, life-threatening pneumonia, blood clotting, stroke, loss of sense of smell or taste, hives or patchy rashes, inflamed "frostbite" toes, kidney failure, diarrhea, intestinal pain, conjunctivitis, a poorly understood multisystem inflammatory syndrome in small children, or frequently no symptoms at all. The possible explanations of this variation in symptoms and responses to a single infective agent are enormous and equally varied—age, prior medical conditions, gender/sex, race, ethnicity, income, profession, blood type, diet, having a specific human DNA sequence of Neanderthal origin, population density and housing, initial viral "dosage," history of multiple exposure, vaccination status, and more. Yet, none of these risk factors is individually decisive. This complexity of outcomes has led to urgent research focused on the diverse and complex responses of the innate and adaptive immune symptom—perhaps the ultimate in bioperformative phenomena (see appendix 2)—to SARS-CoV-2 infection. Given the long-term medical trend toward the increasing frequency of autoimmune diseases as life-threatening infectious diseases have been controlled, the opportunity for performative thinking in biomedicine will only be expanding.[9]

The absence of performative thinking in medicine has also contributed to unproductive and avoidable problems. Drawing on the history of medical genetic research on inbred strains of lab organisms, contemporary medicine became fully invested in the genes-as-causes paradigm. The sequencing of the human genome contributed to a wave of enthusiasm for personal genomic medicine based on the sequence of our individual genomes. Genomic medicine institutes were built, and transformational predictions and promises were made. What happened? The discovery of the "missing heritability" problem (see appendix 3) established that the vast majority of genetic influences on our health function through a complex network of gene-by-gene-by-gene- ... intra-actions enacted by the multiple, hierarchical agencies of our developing and physiologically regulating bodies. As a result, even though the individual risks of these complex diseases may be highly heritable, each instance of these traits is the result of a unique combination of genetic variation, gene intra-actions realized in a distinctive environment.

Genomic medicine failed to realize its initial promise because humans are far too genetically variable. For example, a study of thousands of human genomes has shown that 82 percent of the simplest DNA sequence variations—single nucleotide polymorphisms, or SNPs (pronounced "snips")—occur at frequencies of *less than* once in every fifteen thousand people, or < 0.006 percent.[10] An individual human would only need three or four of these variations to be genetically *unique* among all the world's approximately eight billion people. However, each of our genomes has *thousands* of such variations! It is almost impossible to conceive of how unique nearly every human genome really is. Understanding the map between this variation, disease, and future health outcomes will demand a performative approach that incorporates all the agencies relevant to the enactment of the individual.

Instead of pioneering a future of personalized genomic medical treatments to our complex ailments, genomic medicine has discovered the overwhelming fact of *human individuality*.[11] What's the path forward? Although the newest technological approaches sound similar—proteomics, metabolomics, endobiomics, etcetera—the focus has turned back to physiology and the phenotype itself. This reluctant return to the study of physiology—the primary subject of most medical research *before* genomics came along—comes with many new conceptual and empirical tools developed in the genomic era, but it requires abandoning the strict conception of genes as causes in biology and disease (see below). In short, the future of medicine requires understanding phenotypic performativity.

A performative view of health and well-being may also contribute to a productive framework for thinking about human disability. *Impairment* can be medically defined as an individual functional limitation. In contrast, *disability* refers to the disadvantages that individuals with impairments experience due to social and physical barriers that limit their participation or impede their potential thriving.[12] In parallel with historic analyses of gender, proponents of disability rights have viewed disability as culturally constructed—an interaction between an individual impairment and the structure and normative expectations of the society in which they live. Indeed, as philosopher Ron Amundson and Shari Tresky write, disability can literally be a consequence, of the *physical construction* of the human environment—such as curbs, stairways, and exclusively visual signs.[13]

However, like gender/sex and illness, a performative framework for thinking about disability could support a view of disability as arising from *intra-actions* between individual impairment and norms manifest in the social environment. Like the wave or particle qualities of light, the phenomenon of disability emerges through material intra-actions between the varieties of individual ability and physical-social apparatuses in their environment. As in performative analysis of gender/sex, a performative view of disability could also illustrate pathways to political and cultural changes to those norms and expectations that create disability, providing new tools to work for the thriving of all.

Recalibrating Causality

As if to reassure ourselves about the power and status of our science, biologists have historically packaged their findings in the decisive language of strong linear causation with clear, decisive, often binary outcomes. However, a consideration of the true complexity of organismal development and evolution demonstrates that biological science needs to recalibrate its expectations about causality and scientific explanation. The buffering, or canalization, of development from genetic and environmental variation; the ubiquity of multiple intra-acting, gene regulatory, and physiological pathways; and the complex details of metabolic homeostasis all eliminate the assignment of necessary and sufficient genetic causes in biology. We need to become comfortable with the scientific investigation of natural phenomena that lack causal closure. We need to appreciate mechanisms that are a product of hierarchically nested intra-actions that are so complex, rich, and diverse that a complete reductive explanation of causes will always elude us. We need to become comfortable with some nonreductive descriptions as *scientific* explanations. It is this emergent, iterative, irreducible complexity of the multicellular organism that is captured best by a scientific concept of the performative phenotype.[14]

The traditional perspective of causality in developmental genetics arises in part from the experimental methods themselves, which require the control, that is, elimination, of individual genetic and environmental variation that is unrelated to the specific, mechanistic (micro)hypothesis being tested by a specific intervention into an invariant system. By making claims of *necessity and sufficiency* of genetic causes, biologists confidently conclude that the generalizations drawn from such experiments apply to phenotypes in the absence of these careful controls. These researchers know full well that their own future work will show these conclusions to be incomplete and wrong—that is, these conclusions are contingent upon assumptions of genetic and environmental uniformity that are known to fail in the real world.

Experiments in development can and do produce empirical progress. Many of the biological details discussed in this book were the result of exactly this style of research. But, by itself, this experimental paradigm will always fail to encompass, or be able to conceive of, the actual breadth of the intellectual challenge posed by the many simultaneous agencies involved in genetic, developmental, and evolutionary biology. The incommensurability of experimental genetic methods and the performative individuality of the phenotype is not an intellectual failure of either the experimental method or the concept of performativity. Rather, the disconnect reflects the central and enduring scientific challenge of investigating and understanding the truly individualized, iterative variability of organismal becoming with methods that *begin* by controlling out of existence exactly the individuality that is the heart of the phenomenon of biology.

In biology, the concept of strict causal explanation derived from controlled experiments has become a barrier to scientific progress. Of course, there *are* material causes of the phenotype, but they consist of the entire developmental and physiological process itself enacted by an individual within a specific environment—in other words, individual history itself. We cannot make observations of the world without being *in* the world. As Karen Barad has shown, the scientific "apparatus" and the scientific "subject" together create the scientific "phenomenon" observed or measured. The singular, atomized causes are scientific phenomena created by the experimental methods used to study them, but they are not independent attributes of the world. Controlled genetic experiments do contribute to our understanding of the phenotype, but the strict causal inferences that experimenters draw from them do not simply add up to an understanding of the real biological world. The sum of genetic developmental experiments does not provide a model of the phenotype because each of those isolated causal inferences is made by controlling for the actual cause of variation in the development of the phenotype—variation in the nested hierarchy of intra-acting agencies of molecules, cells, tissues, organs, and body parts.

Although mechanistic reduction is a great scientific tool, it is not synonymous with science itself. Science consists of accounts and models of the world, and ourselves in it, that are robust, consistent, productive, improving, and enduring from multiple points of view. Twentieth-century biologists have been eager to construct a science of biology with the causal clarity and simplicity of Newtonian physics. Consequently, biologists have adopted inappropriate criteria for what counts as a scientific explanation in the discipline.

This conception of science is detrimental to biology because those complex scientific phenomena that resist reductive causal analysis—the origin of life, the subjective experiences and aesthetic mating preferences of animals, the nature and origin of consciousness, free will, etcetera—end up on a scrapheap of intractable scientific problems. Biologists, psychologists, and cognitive scientists need to get comfortable, as quantum physicists have, with the emergence of irreducibly complex phenomena in the *middle* of our disciplines. Quantum mechanics is a great model for how blurring our ideas of causality and what counts as scientific explanation can actually further science itself. Faced with the irreducibility of electron movement to Newtonian mechanics, physicists admitted the limits of prior reductionist tools to resolve their scientific problem. Instead of defining this irreducible scientific subject as outside the bounds of science, physicists successively created the new, probabilistic theory of quantum mechanics to explore these phenomena. The result of abandoning Newtonian causality was a new, remarkably successful, statistically predictive theory with tremendous empirical advantages. Faced with the reality that there are scientific entities that cannot be reduced, physicists changed their concept of what their science could be. Today, the still irreducible quantum-Newtonian rift is simply accepted as an unavoidable complexity of the discipline of physics. The biology of the phenotype needs to be recognized with a similar status.

Geneticists and developmental biologists may understandably feel frustrated upon being told that the repeated application of reductive, controlled experiments cannot, and never will, add up to the complete, functional understanding of the organismal phenotype that the science of biology requires. Nevertheless, scientific efforts to account for the effects of phenotypic performativity will have to be made. The iterative, hierarchical realization of each individual phenotype is fundamental to biology, and there is no doing biology without taking it into account.[15]

With apologies to physicists, particles are vastly simpler than cell-to-cell signaling networks, X chromosome inactivation, patterns of chromatin remodeling, social interactions, or behavior. Biologists should respond to the genuine challenges posed by these irreducible biological issues of phenotypic variety with the same kind of explanatory urgency and priority with which quantum mechanics was pursued in physics. Biology requires a pluralistic patchwork quilt with overlapping but incommensurate edges to explain, in Nancy Cartwright's apt phrase, our *"dappled world."*[16]

The challenge of investigating events with multiple, complex causal inputs is well recognized, if not well resolved. For example, an important historic contribution, proposed by James Mackie, is the recognition of INUS conditions—*insufficient* but *necessary* parts of *unnecessary* but *sufficient* conditions. As Mackie argues, investigation of a house fire may conclude that the fire was caused by an electrical short circuit that ignited flammable materials stored nearby.[17] Obviously, this is not the only possible way that the fire could have started, so the short circuit cannot be described as a necessary condition for the fire. Likewise, the short circuit alone is not a sufficient explanation because the flammable materials could have been stored elsewhere, thereby preventing the fire. The advantage of recognizing INUS conditions is that they acknowledge the contingent complexity of causation in complex cascades of effects, and make explicit the *ceteris paribus* (all other things being equal) assumptions that are implicit in the scientific description of specific genetic causes as necessary and sufficient.

Despite these advantages, however, the application of INUS conditions to research in developmental biology may be limited by the fact that the discipline investigates evolved/evolving organisms whose genetic developmental mechanisms are historically contingent, iterative, agential enactments. Unlike an accidental house fire, the organism is an active and continuous *doing* of itself. Organisms are engaged in an ongoing becoming that involves numerous, hierarchically related intra-actions. The house fire may, or may not, happen. Burning down is not a design function of a house. However, as a consequence of their evolutionary histories, organisms do enact their individual phenotypes. In that sense, INUS conditions may be more appropriately applied in biomedicine when investigating genetic or physiological contributions to disease (see above).

Successful research programs that incorporate a performative model of the phenotype should recognize the broad array of agencies involved in organismal development and evolution. Even while focusing empirically or experimentally on one particular phenomenon—such as mutations to a specific gene, functions of a specific receptor, a tissue-localized electrical potential, or environmental chemicals—researchers should not lose sight of the other genetic and developmental agencies when turning from data gathering and analysis to scientific explanation.

An explicitly performative biology can provide productive new conceptual avenues and tools to address persistent problems in genetics, developmental biology, and evolutionary biology. The challenge facing practitioners of these scientific disciplines is whether to recognize and intellectually embrace the emergent agencies of, and interactions among, clades, species, populations, social groups, organisms, tissues, cells, and genes that their discipline actually demands, or to continue to undermine the success of and future of biology with strictly reductive concepts of genetic causality.

Biology Is Ready to Think Performatively

Some medical practitioners are already recommending the adoption of performative-friendly (if not yet consciously performative) vocabulary to refer to their patients—for example, the proposed substitution of *differences* for *disorders in sexual development* (DSD).[18] So, there is reason to think that the contributions of a performative perspective will be understood in various areas of contemporary biology. But there may be plenty of resistance among biologists and physicians to adopting a performative view of the phenotype. Beyond the immediate connections to Waddington's historic concept of the "epigenetic landscape" (figure 4), however, there are multiple ways in which performative concepts are already arising in biological and biomedical research. The growth of "systems biology" (see appendix 3) has been driven both by the explosive increase in big genetic and metabolomic data sets, and a dissatisfaction with the scope and rate of progress from classic, incremental, controlled genetic micro-experiments.

The tremendous growth of data on mechanisms of animal sexual differentiation in the past decade has also begun to push that literature toward a tentative reconsideration of the concept of "genetic master switches," though it has not yet settled on a new way to think about the alternatives to a strong, genetic, causal concept of "sex determination." For example, Doris Bachtrog and colleagues cite the concept of a single "master-switch" gene as the first and most fundamental myth about organismal sexual development, but they do not yet discard the concept.[19] However, in a recent, prominent review of the evolution of "vertebrate sex determination" in *Nature Reviews Genetics*, cellular and molecular biologist Blanche Capel documents that, contrary to traditional models of "sex determination," vertebrates do not share a "master regulator" to initiate sexual development, or a "common hierarchy of expression in downstream pathways." Rather, Capel observes, the network structure of genes involved in sexual development are "highly permissive" to rapid evolutionary changes in "key regulators." As an alternative to the "master-switch" concept, Capel proposes that "sex determination may be driven by a *parliamentary decision*," defined as a "decision resulting from the contribution of many factors."[20]

Etymologically, the name of these famously deliberative bodies of representative agents comes from the French *parler*—to speak—and *parlement*—a discussion. In this way, Capel's parliamentary metaphor for sexual development focuses explicitly on the *discursive* and *suasive* processes and distributed causality of the molecular development of the sexual body. After all, parliaments cannot function with "master regulators" in charge, and they were invented precisely to overturn the control of central despotic powers. Likewise, real parliaments do not simply rubber stamp or endorse a previously prescribed essence, blueprint, or destiny, but establish the outcome itself through their own deliberative process. Lastly, parliaments do not "determine" the law by choosing among a set of preexisting (e.g., binary) options; parliaments generate law itself. Accordingly, this parliamentary conception of sexual development is inconsistent with the ubiquitous scientific use of the term *sex determination* to describe this process (including by Capel).

Yet, the growing complexity of the data demonstrate that the contingency, complexity, and evolutionary instability of sexual development is completely inconsistent with the traditional scientific representation of master switches with genes as strict causes of the phenotype. Far from being a theoretical leap or culturally based fantasy, a performative view of the phenotype actually reduces the explanatory gap between the material body and scientific representations of it.[21] Mainstream scientific opinion on the mechanisms of sexual development is both consistent with, and actively striving toward, a performative view of cellular and genetic mechanisms of development. And the literature on sexual differentiation is really no different than the broader science of developmental biology. In short, biology is ready for a performative revolution.

Pluralism and the Phenotype

In *Is Water H₂O?*, historian and philosopher of science Hasok Chang explores the fall of the eighteenth-century theory of phlogiston—a theoretical material substance that was released whenever a chemical compound was burned—and the triumph of Antoine Lavoisier's elemental theory of chemical composition.[22] Phlogiston is usually cited as the most universally recognized example of scientific failure. But Chang shows that the rejection of phlogiston-centered scientific practices led to the abandonment of an entire realm of lucid and productive scientific explanations that the elemental theory could not yet account for, especially in electrochemistry. It was not until an understanding of ionic charges developed in the late nineteenth century—more than eighty years later—that the elemental theory of chemistry was able to make accurate predictions and interpretations of certain phenomena that the phlogiston theory had efficiently explained a century earlier. Thus, the rejection of phlogiston had led to a near century-long *degradation* of the capacity of science to explain the world.

The lesson from Chang's historical and philosophical analysis is that science *needs* pluralism—the contemporaneous discussion of multiple theories and models of phenomena—to maintain a healthy and productive intellectual culture.[23] The intellectually healthy path for biology will require adopting a newly pluralistic conception of the organismal phenotype—maintaining a healthy competition between the traditional, flattened population genetic accounts and new, more complex, performative accounts of organisms. As performative research on phenotypic development, physiology, function, and evolution advances, biologists will be able to evaluate the relative benefits and burdens of competing views, and evaluate which *performs* better at the task of explaining the organismal world.

What Is Evolutionary Biology About?

In chapter 1 (in "The Stakes for Evolutionary Biology"), I proposed that a performative understanding of the phenotype is at odds with many of the reductive intellectual commitments of the twentieth-century New Synthesis in evolutionary biology, including the highly influential concept of exclusively gene-level selection. To keep the focus of this book on the development and evolution of the sexual phenotype, I have placed much of the discussion of these issues in appendixes (see appendixes 3–7). But these intellectual conflicts highlight an important question: what is evolutionary biology *really* about?

The historic twentieth-century redefinition of evolution exclusively in terms of ahistorical, population genetic processes—mutation, migration, drift, and selection—like the Ideal Gas Law, contributed to genuine advances in evolutionary understanding. Likewise, the recognition of natural selection acting at the level of alleles of individual genes created important new insights in evolutionary biology. These advances were made possible by reductionist intellectual tools—creating models and concepts that controlled for, and isolated, other higher-level phenomena like development, phylogeny, and higher-level selection. The problems came not from the application of reductive intellectual tools, but from mistaking the goals of these reductive methods for genuine features of the natural world.

For instance, you imagine, for temporary convenience, that one can understand all of inheritance by simply comparing measures of adult offspring to their adult parents. Or you imagine that you can understand natural selection better by thinking of the process purely as competition among alleles at individual genes, and that the organism is simply a vehicle for the propagation of genes. After making some intellectual progress, you then become convinced that these answers are complete, and that development, phylogeny, internal selection (appendix 7), and multiple levels of selection, are actually unimportant, trivial, or even irrelevant to evolutionary biology. This style of reductive flattening becomes an intellectual goal in itself, an admired property of the discipline, and an imagined feature of the world. Soon, this diminished view of evolutionary process becomes our textbook understanding of what evolutionary biology is *actually* about. But this intellectual flattening has made evolutionary biology simpler, narrower, less productive, less powerful, and less interesting.

Ultimately, intellectual flatitudes (see page 31) come to substitute for insights and progress, and contemporary proponents of the New Synthesis organize themselves to defend this paradigm from changes or challenges.[24] One of the primary lines of defense is to try to exert control over what *counts* as a legitimate question, an appropriate perspective, or an intellectual contribution in evolutionary biology.

Any reader who has gotten this far will understand that I am not immune to intellectual enthusiasm! However, my admittedly ambitious and broadly encompassing proposal of a performative biology differs in important ways from the New Synthesis and adaptationism. I do not deny the contributions of reductive population genetic or gene-selectionist tools. Rather, I am simply arguing that they do not, and cannot, provide a complete explanation of what evolutionary biology is about. I aspire to a biology that can explain more about the world—in particular about the iterative material becomings and evolution of diverse, complex phenotypes and novelties. If, as population geneticist Theodosius Dobzhansky wrote, "nothing makes sense in biology except in the light of evolution," then I would argue that little makes sense in evolution except in the light of the performative enactment of individual phenotypes.[25]

CHAPTER NINE

Performance All the Way Up

The twentieth-century composer, conductor, and pianist Andre Previn excelled in the composition and performance of both classical music and jazz. But when asked why he didn't combine the two genres, he responded, "I think they are each doing just fine on their own."[1] Previn thought that the goals and criteria for creating and evaluating in these different artistic fields are incommensurate, or unbridgeable, without mutual damage to both.

Likewise, some readers may believe it would be better to keep the scientific worlds of molecular, developmental, and evolutionary biology separate from the cultural worlds of feminism, queer theory, and gender so as not obscure the richness and variety of their distinct strengths and accomplishments. Of course, another goal of keeping them separate could be to allow the biological world to continue to ignore the cultural world, or even to exacerbate the ongoing tensions and misunderstandings between them. In concluding this book, I want to address why I think creating and maintaining deeper intellectual interconnections between biology and queer feminism would be meaningful to both ways of understanding, and why it will be productive to take these connections seriously, and pursue them further through parallel and overlapping research, publication, and teaching.

I began by proposing that the phenotype is best understood as a performance of the organismal self. In the subsequent chapters, I presented a detailed case for thinking about molecular-developmental biology as a performative realization of the individual—both in terms of a contextual/environmental performance based on antecedent informational resources (the genome), *and* as an expressive material action that both realizes and regulates itself. Making the case that we are *performance all the way down* required getting seriously into the weeds of molecular genetics, developmental biology, and evolution biology. Now, I want to focus on the connections from the detailed biological context of performativity *back* to the original, cultural concept of performativity in gender/sex—that is, I want to trace *performance all the way up*. In the words of Donna Haraway, my goal is to encourage a richer "shared conversation" between feminism, queer studies, and the biology of the material body.[2]

Sexual Reproduction Is an Intra-action

Most fundamentally, sexual reproduction is a cooperative, coordinated confluence of independent organismal agencies to create a new individual with a unique beginning, and its *own* agency to realize itself. In organisms with nucleated cells (called eukaryotes), sexual reproduction involves an alternation of generations that are characterized by a diploid (2N) or haploid (N) number of chromosomes. In animals, the diploid generation dominates, and the haploid generation is reduced to a single cell—an egg or sperm. But in mosses, for example, the haploid generation is multicellular and dominates the life cycle, while the diploid generation is smaller (though still multicellular) and ecologically inconsequential. The great, iterative milestones in this sexual alternation of generations are *meiosis*—the reduction division of a diploid cell of 2N chromosomes to create haploid cells of N chromosomes—and *syngamy* (or fertilization) which combines two haploid cells of N chromosomes to make a diploid cell of 2N chromosomes.

These features of eukaryotic sexual reproduction demonstrate that sexual reproduction itself is an iconic example of a Baradian *intra-action*. Meiosis and syngamy have no biological functions or purposes except in light of each other. In isolation, either would yield genetic reduction or multiplication ad absurdum. Together, they have propelled the eukaryotes to achieve an unimaginable diversity, complexity, and ubiquity that has transformed the planet.[3] The intra-active convergence of agencies in sexual reproduction is fundamentally a creative, indeed *pro*creative, discourse. The more-than-one-billion-year history of sexuality extends the genealogy of sex into deep biological and geological time.

A Posthuman Genealogy of Performative Discourse

It has been suggested that Michel Foucault "divorced 'sexuality' from 'nature' and interpreted it, instead, as a cultural production."[4] Actually, a performative view of biology greatly expands the role of discourse, which Foucault envisioned as acting in the social *creation* of sexuality, into the evolution of sexual reproduction itself.

When you have multiple intra-acting biological agencies, the outcomes are not determined by laws, developmental programs, descriptive "blueprints," or antecedent essences. Rather, the outcomes are contingent upon the structure and details of the hierarchical intra-actions taking place within a specific environment. In biology (as in culture, psychology, and sociology), those intra-actions consist of communicative discourses among agents with the goal of influence, or suasion, of one another. The expanding complexity of biological discourses has a history across the phylogeny of life that comprises a history of living performativity itself. This genealogy of biological discourse connects the material history of life to cultural discourses in the humanities.

Following Foucault, the material performativity of the body has a history—an organic, evolutionary genealogy of emergences, origins, innovations, and the compounding complexities of agencies and their discourses. This is a posthuman genealogy because humanity is not at the organizing center of this history; it exists without us to theorize it. We can now articulate the conditions for the emergence of the components of performative phenotype in biology from the origin of life itself to the origin of humans, and we can trace their contributions to the origin of human cultural performativity as well. This is *performance all the way up*.

In recent decades, research on the origin of life has focused extensively on RNAs—ribonucleic acids—as the first living material. Unlike DNA, RNAs have the capacity both to encode heritable information *and* to catalyze chemical reactions. Thus, RNAs can both replicate themselves, *and* act in the material world. They are considered to be the first adaptively evolving, purposive, biological agents on Earth—single molecules that have aspects of both the genotype *and* phenotype. The origin of RNA replication would have been the origin of *iterative* realization. The subsequent evolutionary origin of genes encoded in DNA, transcribed into RNA, and translated into proteins within membrane-bound organisms created the first possible distinction between the genotype and the phenotype—the antecedent, historically derived, inherited, genetic lexicon versus the material expression of those genes as actions in the material world. With the evolution of cells differentially expressing genes came the emergence of physiological agency around 3.5 billion years ago, and adaptive evolution among genes for diverse and distinct cellular functions.

The subsequent evolutionary origin of contextual, or environmentally responsive, gene regulation—for example, in response to temperature, food availability, diurnal light cycles, age, or life stage—created the opportunity for the emergence of molecular discourses *within* the cell that both achieve and regulate the various states of gene expression—or in Judith Butler's words, "the reiterative power of discourse to produce the phenomena that it regulates and constrains."[5] Transcription factors, intermediate signaling molecules, microRNAs, and their binding domains coevolved and enriched the complexity of organismal *being* and *becoming*. For example, the bacteria *E. coli* express enzymes necessary to metabolize specific sugars—such as lactose—only when they are present in the environment—a capacity regulated by the famous "lac operon." The evolution of such capacities occurred very early in the evolution of cellular life, likely prior to the most recent common ancestor of all extant life forms approximately 3.5 to 4 billion years ago. This capacity for performative genetic response to the environment evolved through natural selection to further the survival of the cell.

Further advances in performative complexity arose with the origin of intercellular discourses *among* single-celled bacteria and archaea. These discourses involve cell-to-cell signaling mechanisms that bacteria and archaea still use to communicate their density and proximity to one another—a coordinated group behavior called quorum sensing. When individual densities and food availability are high enough, single-celled microbes can cooperatively benefit by all releasing digestive enzymes into the environment and absorbing the resulting nutrients by diffusion. Cell-to-cell signaling discourses allow bacteria and archaea to self-organize into a microbial mat, biofilm, or functional community, and to cooperatively extract resources en masse from the environment. Evidence of biofilm structures go back in the fossil record some 3.5 billion years. Some bacteria and archaea can also exchange genetic material through conjugation, creating a simple, early form of sexual recombination.

With the evolution of the eukaryotic cell with a nucleus and symbiotically acquired organelles—mitochondria and, in some lineages, chloroplasts—the intracellular genetic regulatory discourse became even more complicated, and the consequent diversity in cellular morphology and behavior likewise advanced. In the same lineage of life, the evolution of sexual reproduction through meiosis and syngamy (see above) produced new individuals with their own genetic identity and individuality, and constituted the origin of a new, advanced form of intra-active sexual reproduction.[6]

About 1.3 billion years ago, the origins in multiple lineages of multicellular life forms created a new hierarchy of agencies *among* the cells, tissues, organs, and body parts within single biological bodies of genetically identical cells. In a multicellular organism, the independent agencies of single cells must be regulated to serve the emergent needs of a common, shared, multicellular body—a plant, fungus, or animal. Hierarchical, molecular discourses among cells, tissues, organs, signaling centers, developmental fields, and anatomical parts gave rise to complex physiological and developmental possibilities, complex life cycles, anisogamy (i.e., differentiated gametes), and more.

Later, the origins of sexually differentiated bodies producing one gamete type or another led to the entire class of sexual phenotypic performativity explored in this book. Other elements of performativity emerged with the evolutionary origins of neural and sensory systems, cognitive and subjective experience, social behavior, and the varieties of consciousness, each of which involves processes of *becoming*.[7]

All of these biological varieties of performative discourse evolved prior to the origin of human language, human culture, and human gender. The literature on performativity in the humanities and social sciences focuses exclusively only upon the most recent, exclusively human cultural innovations in performative discourse. But these human performativities are both historically contingent upon, and emergent from, the deep, biological performative continuum. This is *performance all the way up*.

Is There Gender in Nature?

One way to think about the relationship between science and culture is to rethink the boundaries or interfaces between them. Thinking productively about humans and biodiversity requires looking horizontally to other extant branches of the tree of life, each of which is *equivalently distant* from the origin of life as ourselves. Thinking about biodiversity is not just about peering into the human past (down from the top of the *Scala Naturae*); rather, it is about regarding other examples of evolutionary and historical complexity.

This genealogy of performativity raises the compelling question of whether any nonhuman animals actually exhibit gender—that is, culturally mediated components of their gender/sex phenotypes or (using Dawkins's idiom) culturally extended sexual phenotypes. The cultural concept of gender includes both categories of gender *and* the culturally and temporally diverse details of how those individuals realize or enact gender categories. Most sexual organisms are thought to lack any culture, including sexual culture, so these species completely lack gender. To be more precise, the cultural influence on their gender/sex phenotypes is zero. But it is also clear that some animals absolutely do have cultural elements to their extended sexual phenotypes—or nonhuman, animal genders.

For example, approximately half the species of birds of the world learn their songs from other conspecifics. There were four phylogenetically independent evolutionary origins of vocal learning in birds, including the hummingbirds (Trochilidae), the parrots (Psittaciformes), the hyperdiverse oscine songbirds (Passeri, including, for example, thrushes, crows, warblers, orioles, and sparrows), and the four species of Neotropical bellbirds (*Procnias*, Cotingidae). Likewise, Male great whales and porpoises sing elaborate vocal songs that are also learned. Some lineages of bats also learn their vocalizations. These bird, whale, and bat songs are *culturally* mediated forms of social *and* sexual communication. Birds and whales learn their songs through social interactions with other singing birds and whales (usually not their parents). The result is cultural descent with modification that is extensively independent of genetic variation and can be very rapidly evolving. For example, the songs of many widely ranging North American bird species sound differently around Chicago, New York, and Boston for the *same* reasons that the people do. And it's not adaptation to the wind in Chicago, the humidity of the Hudson River, or the smell of baked beans in Boston. These geographic varieties in birdsong are true dialects within the broader social-sexual discourses of these species.[8]

Because they function in intraspecific sexual communication, the culturally mediated elements of the social-sexual behavior of individual animals constitute nonhuman forms of gender. Like human gender, the normative expectations of animal gender *change* over time through cultural evolution. Young individuals iteratively conform to adult behavioral models they interact with, but they learn also with error (or, perhaps more accurately, creativity), and sometimes introduce (or invent) cultural changes or innovations. The result can be rapid intra-population change, and geographic, interpopulational divergence in the cultural components of sexual communication—in other words, cultural evolution of animal gender norms.

For example, in the late 1990s, Michael Noad and colleagues documented a rapid cultural change in the Male sexual courtship song dialect of a population of Humpback Whales (*Megaptera novaeangliae*) on the Pacific coast of Australia after the immigration to the Pacific population of two Males singing novel songs from the Indian Ocean dialect. Within two years, 100 percent of the singing individuals in the Pacific population switched over to singing the novel songs of the Indian Ocean dialect, leading to the local extinction of the decades-old Australian Pacific dialect. On the scale of a Humpback Whale's fifty-year lifespan, this constitutes a cultural, gender *revolution*.

Because the preferences of observers/evaluators of gender/sex communication coevolve with those social signals, cultural change in sexual phenotype can also affect the social efficacy of gender/sex communication in animals. For example, Elizabeth Derryberry conducted an experiment on the culturally evolving songs of White-crowned Sparrows (*Zonotrichia leucophrys*) in the Sierra Nevada, California. Male sparrows sing songs that function in the defense of breeding territories from other Males, and in attracting Female mates to reproduce on those territories. Derryberry played recordings to Female and Male sparrows of contemporaneous Male songs, and Male songs recorded twenty-four years earlier at the same locality. She found that the older songs elicited only about half as many social responses—either Male territorial challenges, or Female sexual interest—as the contemporaneous songs did. In other words, temporally rapid cultural change in Male White-crowned Sparrow gender/sex phenotype affects the content, meaning, and social salience of culturally acquired Male sexual signals.[9]

Just as language plays a large role in the realization of human gender, the existence of cultural variation in sexual communication in many birds, whales, and bats demonstrates the parallel origins and ongoing existence of gender/sex in some nonhuman animals. Although few have pursued ornithology or mammalogy from this perspective, these scientific disciplines encompass the same fundamental breadth of complexities and challenges that many have regarded as unique to the humanities and social sciences, including the complexity of gender/sex.

Of course, these instances of gender/sex in birds, whales, and bats are evolutionarily and historically *independent* of gender/sex in humans and other primates. In other words, the most recent common ancestors of humans and songbirds, humans and whales, or humans and bats did *not* have gender—that is, cultural contributions to their individual sexual becoming. Gender/sex has had multiple evolutionary origins, and the genealogy of gender consists of multiple isolated branches on a tree, not a simple line.

There are only a few, quite rare instances of nonsexual culture outside of primates. For example, New Caledonian Crows (*Corvus modeloides*) learn from their parents how to make probing tools out of *Pandanus* leaves. The specific method used to make these tools differs on the two sides of the island of New Caledonia, constituting an avian example of geographic variation in nonsexual, avian, material culture.[10] By comparison, however, it is notable that vocal gender/sex culture has evolved multiple independent times, and is present in *nearly half* of all bird species, and likely all species of whales and dolphins. This provides evidence from multiple lineages of animals that the elaboration of gender/sex is among the very first, most fundamental, and most transformative consequences of culture in the world.

The existence of nonhuman cultures, including nonhuman sexual cultures, in what we commonly consider to be "nature" establishes the need to abandon the nature/culture and sex/gender dichotomies, and search for analytical concepts and frameworks that facilitate research accordingly.

"What Is Sex?" Revisited

Having introduced a working definition of sex in chapter 2, I want to revisit the implications of thinking of sex as an historically enduring cluster of recurring, embodied, intra-acting reproductive homologies, and not as a fixed, essential, or defined attribute of individual genes, zygotes, or organisms.

The sexual phenotype encompasses a host of iteratively embodied features with evolved reproductive functions. The boundary of the sexual phenotype remains necessarily fuzzy—that is, subject to scientific, social, and socio-scientific debate—because many traits have both reproductive and nonreproductive functions, and because various human cultures have co-opted nonreproductive bodily features to function in cultural aspects of gender/sex. (Think of the frequently gendered expectations for the presentation of many sexual monomorphic body features like scalp hair, arm pit hair, eyelashes, and fingernails.)

As we have seen, even though the history of natural and sexual selection on human reproduction has canalized (i.e., constrained) the material, anatomical variation in the sexual phenotype, this selective history cannot create, dictate, or enable an antecedent, essential, binary fact about individual genes, chromosomes, zygotes, or organisms, including sex. Despite its importance to the origin of each and every human life, sexual reproduction cannot overwhelm, eliminate, revoke, or assert control over the performative process by which those individuals are iteratively materialized. A commitment to the existence of a binary fact of individual sex cannot be scientifically justified by reference to genes, genomes, chromosomes, hormone levels, gene-expression profiles, or any other such data, and such a commitment should be considered a cultural idea, not a justifiable scientific judgment.

Attempts to define individual sex scientifically fail, and will continue to fail, because each of the mechanistic inputs to the process of the embodiment of sex is simply one of the multiple, hierarchically networked, biological agencies involved in the realization of an individual's reproductive possibilities. Each source of influence has only an incomplete, indeterminate effect on the outcome, and shares causal parity with other agencies. As the discussions in previous chapters show, the possession of a specific chromosome, gene, hormone, or gonad morphology cannot be used as reliable criterion for defining individual sex because none of these features can individually determine the realized reproductive capacities of the body. Each is only a single discursive participant in a broader process of individual becoming. Sex is the iterative realization of reproductive possibilities, and not reducible to any of the molecular players in that process. As we have seen, sexual reproduction itself involves the intra-action of coevolved reproductive anatomies and physiologies—in mammals, eggs fertilized by sperm and gestated in a womb—but the historical persistence of this structured reproductive mode does not by itself justify an individual sexual binary. As Karen Barad proposes, "intra-action undoes binaries without collapsing them."[11] Thus, the persistence over hundreds of millions of years of the sexual intra-actions in vertebrate reproduction undoes the individual sexual binary without collapsing the breadth of pure sexual difference that makes that sexual reproduction possible.

Because the scientific "fact" of individual sex is so deeply engrained in science and in culture, it is worthwhile to demonstrate again through a thought experiment how such definitions fall apart. One could start by hypothesizing that individual human sex is "determined" by the presence of one Y chromosome, but that hypothesis is falsified by the existence of fertile XX Males with a fully functioning copy of the SRY gene on one X chromosome. You could then hypothesize that sex is "determined" by the presence of the "master-switch" gene SRY, but that hypothesis would be falsified by the existence of SRY mutations that prevent nuclear translocation of the protein, its binding to the DNA, or the induction of a sufficient kink to the DNA after binding to the SOX9 promoter site to enhance SOX9 expression. The SRY hypothesis would also be falsified by the existence of XY individuals with perfectly functional copies of the SRY gene that lack WT1+KTS expression in the developing gonad; the inability to stabilize SRY mRNAs required for SRY translation into a functioning transcription factor protein would lead to the development of ovaries. Even the necessity of the SRY gene for the development of testes and Maleness would be falsified by the existence of fertile Male XX individuals with a mutation to the SOX9 or SOX3 promoters that is constitutively turned on. Proposing any other specific chromosome, gene, or signaling pathways as defining features of individual human sex will raise exactly the same problems.

Alternatively, you could claim that sex is determined by a certain level of testosterone production, but that criterion would be falsified by the existence of XX individuals with congenital adrenal hyperplasia who have a hypospadic penis and no external vulva with functioning ovaries, a uterus, and a vagina that connects internally, and who menstruate out of their urethras. A testosterone criterion would likewise be falsified by the existence of XXY individuals (Klinefelter's syndrome) with generally Male anatomy but low testosterone levels and infertility.

Every attempt to materially define the sex of an individual slips out of grasp, and proves ephemeral, following scientific scrutiny because any feature or mechanism can only affect sexual development through a co-evolved network of other molecular and cellular agencies. Reproductive anatomy and physiology are products of the entire system of organismal becoming, and cannot be defined as the result of a single, or even a few, key players or essential causes. Again, differences in sexual development are not failures or aberrations of a binary sexual development system but evidence that the individual sexual binary does not exist.

Unfortunately, recent trends in biomedical research have begun to expand, not reject, this problematic scientific commitment to an essential, individual fact of biological sex. To facilitate biomedical research on women's health and wellness through the study of human cell lines maintained in laboratory cultures, the idea of binary sex has been expanded to apply to individual human *cells*. Motivated by the desire for greater inclusion of women's health concerns in biomedicine, the director of the US National Institute of Health Office of Research on Women's Health, Janine A. Clayton, states, "Every cell has a sex. Each cell is either Male or Female, and that genetic difference results in different biochemical processes within those cells." This initiative has contributed to broadening the diversity of laboratory cell culture systems that are used in biomedical research, but it has come under criticism from feminist biologists including Stacey Ritz, Anne Fausto-Sterling, Daphna Joel, and Madeleine Pape as unnecessarily reinforcing the binary sex biases that it aims to alleviate.[12]

Of course, individual human cells do carry all the same genomes that human bodies utilize in sexual development and differentiation. However, having a sex is not decreed merely by the possession of certain chromosomes or genes, but by the material intra-actions utilizing those genes during sexual development. Thus, human sex cannot be achieved by an individual human cell. The possession by a human cell of particular chromosomes or particular genes cannot establish its sex, because a sexual phenotype can only be realized, or enacted, by a developing organism, involving intra-actions among many additional players (or agents) that influence the reproductive possibilities that are realized. These include differentiated anatomical parts expressing differentiated paracrine and endocrine receptors, developmental fields, glandular tissues producing hormones, etcetera. The attempt to define the sex of human cells through their genomes further reifies the view of Maleness as the positive, assertive, and affirmative state, with the confirmed presence of a Y chromosome or the gene SRY, and Femaleness as a passive default, indicated by the absence of a Y chromosome or SRY.

Ironically, this reductive concept of genetic, binary sex and binary sexual difference has been advanced by researchers dedicated to *improving* the quality of research into women's health.[13] However, as immunologist and feminist Stacey Ritz explains, it is important to ensure "that scientific projects aimed at redressing gender inequities in health do not inadvertently perpetuate them by relying on essentialist and deterministic foundations."[14] The problem occurs when inclusion of diversity is specifically accomplished by inclusion of a new category of research subjects, not simply inclusion of a greater breadth of human diversity. The alternative is the scientific inclusion of a sufficiently diverse sample of individuals to broadly characterize sexual difference in the variables under study—that is, the breadth of "pure" sexual differences in the absence of a binary sexual partitioning and analysis of the data (see "Sex and Race as Scientific Apparatuses" below).

In response to the questions posed by Judith Butler in *Gender Trouble*, "What is sex, anyway?" and "Does sex have a history?" we can now offer the biological answer that sex *is a history*—a hierarchical history of co-evolved bodily traits with intra-active reproductive functions, *and* the iterated, individual histories of the material realization of those reproductive possibilities. Sex is not a prior or given fact about any body, because the genetic-discursive process of becoming an organism affords no inputs for any such essence, predetermined truth, or antecedent fact about sex. The sex of the individual is a self-realization, the product of a performative becoming. It is "turtles all the way down."[15] Or, as I prefer, turtles *all the way up*.

Toward a Scientific/Cultural Concept of Gender/Sex

Because the sex of the individual body is not a fixed, prior, or given biological fact that is then gendered by cultural action, a unified concept of gender/sex appropriately expresses the continuity between the biological and cultural components of the individual gender/sex phenotype. In this context, the words *sex* and *gender* serve only as a shorthand for the distinctive (but ultimately continuous and interconnected with feedback) biological and cultural components of the gender/sex phenotype. An important goal of identifying the performative continuum in gender/sex—from zygote to gender—is to establish a common language for productive conversations between science and culture studies, between material mechanisms and cultural processes, between biological science and queer feminism.

In chapter 1, I referred to Sarah Richardson's valid criticisms of the "geneticization" of sex.[16] Having explored the development of human sexual bodies in detail, we can now see how a performative understanding of gene action provides a newer, more productive conception of what being *genetic* means. *Genetic* refers to the historically derived (i.e., evolved), constitutive, molecularly networked, material-discursive resources utilized by an individual organism in its development, differentiation, and becoming. From this perspective, feminist objections to "geneticization" are not conflicts with the details of genetics or the developmental biology of organisms per se; rather, they dispute the simplistic, reductive views of genetic causation, and the traditional a priori assumption of a predetermined, essentialist, individual sexual binary that have dominated contemporary biological thought and research on sexual development. Furthermore, the performative framework I have extended to biology expands the role of queer feminist analysis, not merely as a critique of science, but as an actively engaged program of research *in* genetics, developmental biology, *and* evolutionary biology.

The material-cultural continuity of the performative influences on individual gender/sex does not reduce them to any single causal mechanism. Biological performativity does not flatten, eliminate, or explain away any of the complexity or intersectional relevance of psychological, cultural, or sociological influences on gender/sex, which are central to queer and feminist analysis. Rather, it provides a common posthuman framework that can allow us to productively identify similarities, differences, connections, and dissonances among the many hierarchical agencies, influences, and phenomena along the material/cultural performative continuum within and beyond the human.

As I outlined in chapter 6, biomedical research on the genetic mechanisms of human sexual development and differentiation has proceeded through the investigation of queer bodies and lives that present a mismatch between their individual sexual realizations and expected bodily norms. Many researchers have told science stories of these queer lives in ways that presume and reinforce the naturalness and essentialism of binary sexual norms that these individuals themselves defy. Yet, as many early twentieth-century biologists clearly understood and expressed, this narrative mode is a social phenomenon, not a requisite part of scientific explanation. As we have seen, in the mid-twentieth century, both developmental geneticist Conrad Waddington and physiologist Frank Lillie wrote about the fundamental scientific ambiguity of individual sex. Lillie stated adamantly, "There is no such biological entity as sex."[17]

The scientific commitment to sex as a binary fact about individual genomes, cells, and bodies was advanced and expanded during the late twentieth century through development of contemporary molecular biology, and the application of reductive, experimental genetic methods to the investigation of sexual development. So, Richardson is correct that the "geneticization" of biology has aggravated the situation. However, as recent biomedical revisions to the recommended "best practices" in clinical treatment of "*differences* in sexual development" show, the culture of biomedical science may again be changing, and admitting to the natural, empirical spectrum of human sexual possibility.

Performative Perspectives on Transsexual Experience

Without speaking for trans people or attempting to characterize trans experience, I want to introduce the possibility that understanding gender/sex as a performative continuum—spanning the phenomena of the molecular and cellular development of sexual morphology, hormonal physiology, developmental psychology, and individual sociocultural development—can contribute to understandings of trans experience as an individualized extension of the shared, common process of individual sexual becoming. Molecular biologist and trans-activist Julia Serano and other trans people have argued that biology is anti-essentialist, and provides a supportive framework for understanding trans experience. Although performative gender theory, in particular, has been viewed by Jay Prosser and some other authors as antithetical to being transgender, I think there are productive ways to connect these positions.[18]

If the development of sexual anatomy and gender identification lie on a continuum of performative processes, rather than being externalized representations of individual essences, then trans experience and medical responses to it—including hormone replacement therapies and sexual affirmation surgeries—are extensions of the performative process of human gender/sex becoming. In this view, the employment of biomedical and pharmaceutical technologies in trans gender/sex self-realization constitutes another, complex role for culture to influence individual human gender/sex development.

If there is no biological essence of individual sex prior to, or independent of, the developmental enactment of individual gender/sex, then there can be no essential identity to be abandoned, transgressed, or adopted through transgender experience or transition. (As elsewhere in this book, "essence" here refers to a feature that is both inherent and indispensable.) The frequency of sexually transitioning species of animals and plants, and the diversity of their transition trajectories (Female to Male; Male to Female;

or simultaneously hermaphroditic), specifically conflicts with the existence of an essential biological fact of individual sex.

Of course, the absence of an independent or antecedent biological fact of individual sex does not imply that having/experiencing a specific individual gender/sex identity or sexual orientation—whether straight, queer, trans, nonbinary, asexual, or other—is unscientific, irrational, or unjustified. The reality of the gender/sex continuum means that *all* levels of agency—from the molecular to the psychological and cultural—participate in and influence the ongoing becoming, existence, and diversity of individual gender/sex.

Sex and Race as Scientific Apparatuses

Karen Barad's "agential realism" has not yet had much intellectual impact on the study of biology, creating important opportunities to examine additional scientific and cultural implications of this perspective. In this section, I want to explore how the uses of sex and race categories in biological research constitute Baradian scientific apparatuses, and then to ask what ethical obligations these apparatuses demand of scientific, psychological, sociological, and economic researchers that may use them.

Barad proposes that the use of a scientific apparatus involves an intra-action with the material world to produce a scientific measurement, observation, or result. According to quantum physics, the scientist is not independent, or outside, of the material world being investigated, so the scientific distinctions between cause and effect are not an independent feature of the world. To Barad, however, the use of a scientific apparatus induces an "agential cut" that resolves the prior indeterminacy between cause (the object) and effect (the measurement/phenomenon obtained by a scientist). Proper scientific apparatuses are specifically designed to create the conditions through which a cause-and-effect relation can be created, thus resulting in an objective observation or coherent measurement of a phenomenon. Accordingly, Barad writes, "causal relationships cannot be thought of as specific relations between isolated objects; rather, causal relations always entail the specification of the material apparatus that enacts an agential cut."[19] Causal relations are a *result* induced by the scientific apparatus used in intra-action with the world.

Again, Barad's classic example involves optical apparatuses that are designed to create the observable effects of light as a wave or a particle. Prior to the experimental measurement, the particle/wave status of the light is indeterminant. The design and application of an optical apparatus resolves that ontological indeterminacy to produce an effect, or phenomenon.

Barad asserts that the ethical obligations of the scientist also arise from our immersion in the world that we share with all objects of investigation. Because "we are an agential part of the material becoming of the world," the ethics of science is not about an obligation to some "other," somehow outside of us. Rather, our ethical obligations as scientists arise because we are agents in the world; "we" and "they," "us" and the "other" are all actively co-constituted in the world's ongoing becoming.[20] Thus, to Barad, scientific objectivity requires being simultaneously accountable for the specific effects of our scientific agency—the apparatuses we design, invent, and deploy, and the "effects" and phenomena they create in the world. We cannot separate our objectivity from the ethical consequences of the design and use of our scientific tools.

Barad concludes, "Ethical concerns are not simply supplemental to the practice of science, but an integral part of it. . . . Values are integral to the nature of knowing and being. . . . It is not possible to extricate oneself from ethical concerns and correctly discern what science tells us about the world. Realism . . . [is] about the real consequences, interventions, creative possibilities, and responsibilities of intra-acting within and as part of the world."[21] Just as we cannot entirely exclude or control for our own influences on the scientific measurements that we make, we cannot avoid the web of ethical obligations invoked by making those measurements.

Built on Bohr's analysis of quantum physics, Barad's perspective provides productive insights into the biological investigation of sex and race difference. Although Barad focuses on scientific apparatuses as physical objects that intra-act with the material world, the same concept applies to any scientific method that aims to establish causal relations from data sets acquired from measurements made in the world, whether they come from the discipline of astronomy, genetics, animal behavior, psychology, or economics. The conceptualization and structure of the analysis itself *creates* an agential cut that gives rise to the opportunity for a phenomenon and causal inference that the apparatus was designed to investigate, observe, or measure. Thus, the design and deployment of any such statistical model, analysis, or test demands ethical accountability and responsibility for its use. Scientists do not get the license to imagine themselves and their objective results as existing freely, outside of the ethical obligations of the biological, cultural, and social world from which the data—especially human data—is gathered. Therefore, there is no ethical responsible claim to scientific objectivity that is independent of the scrutiny of the design of the scientific tools that have been used in the first place.

Accordingly, the investigation of proposed phenotypic sex differences using data partitioned by binary sex involves the deployment of a conceptual scientific apparatus that has been designed to cause the effect of sex difference, regardless of the effect size—that is, the magnitude of the differences between the imposed data classes compared to the variation in the total sample. As we discussed in chapter 2, the scientific analysis of binary human sex differences was historically initiated to justify white, educated, economically privileged, Anglo-European cultural norms of masculinity and femininity as biologically fundamental, evolutionarily adaptive essences. The historical abuses and ongoing costs of rigid concepts of binary gender/sex categories—including the waves of anti-gay and anti-trans laws proposed and advancing in many American states—bring explicit ethical obligations whenever binary sex categories are deployed in human biology and social science.

Consequently, biologists, geneticists, psychologists and social scientists are obligated to ask themselves, "Under what conditions is it ethically responsible to conduct analyses or statistical tests based on human binary sex categories? Or human racial categories?" I cannot yet answer these questions except to say that the ethical bar should be quite high. For example, we can clearly question the ethical validity of the investigation of sex or race differences that are not overtly discernable in a joint distribution—that is, as obvious bimodal peaks. Because sex and race difference research often seeks causes for effects that cannot be independently argued to even exist, this scientific apparatus is designed to create binary difference where none may actually exist. Although a statistically significant sex or race difference in some unimodally distributed variable can be an objective result of empirical analysis of data from the world, such a result does not mean that the cause of the proposed sex difference is binary sex or race. Rather, the result most likely means the more efficient explanation of the variation is some other variable that is simply differently distributed in its effects between the imposed categories of binary sex or race. These are not the same thing. For any statistically significant sex or race difference that is not diagnosably distinct a priori, a more effective causal explanation is simply waiting to be revealed by another, future scientific apparatus that is more precisely and accurately designed to investigate the phenomenon.

This view is not an argument for science as a cultural construction, or any other relativist view of science. Barad's program is called *agential realism*. The agential cut performed by any scientific apparatus occurs in the material world, or to a data set of real measurements taken from the world. The question is when the apparatuses of sex and race categories are scientifically worthwhile and ethically responsible.

Sex and race difference researchers, and many readers of this literature, are likely to respond to these questions by stating, "But this is simply what an objective analysis of the data tells us." However, as Barad writes, "It is not possible to extricate oneself from ethical concerns and correctly discern what science tells us about the world." It is not merely a question of whether individual researchers are personally free from sexist agendas or discriminatory animus. Regardless of our intentions, all scientists are accountable and ethically responsible for the effects of the scientific apparatuses we deploy. If the scientific apparatus was originally designed to reinforce specific cultural ideals of binary sex difference by recreating them as natural, essential facts, then the choice to use such apparatuses contains this inherent, designed property and effect. This observation is not a politically correct injunction against scientific freedom; all scientists today recognize that there are ethical constraints on the practice of their science. Rather, it is a call to recognize more broadly that the design and application of scientific apparatuses—even statistical analyses—are not free from ethical obligations and complications.

In chapter 2, we also discussed feminist research that documents how the scientific concepts of sex and race differences were historically *co-constituted* in the late nineteenth and early twentieth centuries. In other words, the scientific inquiry into human sexuality, human sexual variation, and the causes of proposed human sex differences was initiated specifically in response to the fear that unregulated sexual desire posed a grave threat to the eugenic purity, survival, and continued dominance of an adaptively superior white race. The scientific concepts of sex difference and sexual pathology were invented *because* it was taken for granted that the actual variability of human biology, psychology, and behavior would lead to an inexorable degradation of white racial purity and cultural superiority. This is not ancient history. Our contemporary biological and cultural concepts of masculinity and femininity remain deeply rooted in their racist origins despite the subsequent repudiation of eugenics by all of mainstream science.

In this regard, the historical ontology I have developed and applied in this book to the definition of sex (see chapter 2) may also provide a productive perspective to understanding human biological and genetic diversity. Like sex, human diversity *has* an explicit history—including genetic evolution, geographic dispersal, geographic isolation among populations, phenotypic diversification, migration, secondary contact, and introgression. Like sex, however, the explicit genetic, geographic, and phenotypic history of human diversity within and among populations *does not* imply that race is an essential scientific fact about individual alleles, genes, genomes, cells, bodies, or human lives. Like human binary sex categories, the scientific investigation of human race, racial categories, and race differences was used to historically establish culturally constituted concepts as scientific truths, leading to pseudoscientific support for racist social policy, injustice, and even genocide.

The application of an historical ontology to sex and race—leading to the abandonment of sex and race difference in favor of a broad focus on sexual difference and total human diversity without race categories—provides an opportunity for *deconstitution* of the eugenic, racist origin of the scientific apparatuses of sex and race.

Accordingly, the scientific analysis of race differences through statistical models, analyses, or tests that explicitly partition subjects on the basis of race categories (even self-identified race) involves the deployment of an essentializing framework onto groups of individuals about whom there is no definable individual scientific fact of race. Because such analyses were designed to explicitly fulfill the racist objective of creating genetic and environmental *causes* of white racial supremacy, these scientific apparatuses retain that fundamental design function.[22] I do not think that one can ethically claim that race is a cultural concept, and then responsibly use these same essentialist racial categories to investigate human population genetic and phenotypic diversity. One cannot use scientific tools in human research that were designed for racist or sexist research purposes—that is, designed to create racist scientific phenomena—and later claim that the results are simply objective facts independent of any ethical obligations invoked by that racist and sexist history. I would not argue that these tools can never be used; rather, I argue that one cannot escape accountability and responsibility for the history and designed function of these scientific tools themselves. Given the myriad of other environmental and sociological variables that are currently co-distributed with race and sex categories in our highly inequitable, stratified, segregated and sexist contemporary societies, I think one would need a very good justification to deploy the historically tainted racist and sexist categories in scientific research.

The explosion of genomic analyses of contemporary human and subfossil human remains in the last twenty years has created a tremendous expansion of detailed scientific investigation of the genetic history of ancient and contemporary human diversity. But the unresolved tension between the explicit history of evolved human diversity and the social construction of race continues to be exacerbated. In *Race to the Finish*, sociologist of science Jenny Reardon chronicles the development of these contemporary research efforts through the history of the Human Genome Diversity Project, and documents the historic and ongoing obfuscation between the cultural vocabulary of race and the genomic research into human genetic diversity.[23] Our ongoing scientific, ethical, and cultural challenge is to create scientific concepts, vocabularies, and analytical apparatuses to investigate the explicit history of human genetic evolution, and phenotypic and cultural diversity that are *not* shaped by the racist origins of human race categories and historically eugenic population genetic concepts.

When is it ethically responsible to use individually defined sex and race categories in a scientific analysis? What scientific apparatuses can be ethically used to study the evolutionary history of human genetic and phenotypic diversity? I cannot provide complete answers here. But I can point out how our framing of these questions flows from a Baradian understanding of the performativity of scientific inquiry.

The ubiquity and ease of use of the scientific apparatus of sex and race categories demonstrate how effective these tools are at fulfilling their designed functions, and how easy it is to conduct sexist or racist science by using them, regardless of our intentions. We each individually act to further the sexist and racist purposes of these crude apparatuses by being satisfied with, or even fascinated by, the shallow, facile answers they provide.

Sex and Race Categories in Biomedical Research

Perhaps the most ethically compelling possible reason to deploy sex or race categories in science would be to advance human health and well-being, or even to save lives. However, prominent efforts to use these scientific apparatuses in biomedicine continue to raise problematic issues.

Cara Tannenbaum, Londa Schiebinger, and colleagues present a boldly optimistic summary of the importance, and potential contributions, of analyses using binary sex categories in biomedical and engineering research.[24] Despite the fact the US National Institutes of Health began requiring equal inclusion of men and women in drug and experimental studies in 1993, and equal representation in tissue culture research in 2014, Tannenbaum and colleagues can only report a few bona fide clinically applicable results of different dosage recommendations in men and women for a handful of drugs. Furthermore, their outline of best practices for analyzing sex and gender difference in scientific and engineering research includes no recommendation to evaluate the statistical effect size of any of the proposed gender/sex differences identified. In other words, all statistical findings of sex difference are assumed to be equally worthy of further investigation at deeper environmental, genetic, or physiological levels regardless of the scale of the inferred sex differences to the overall variation in the data. Unfortunately, if genuinely interesting and biomedically important sex differences are uncommon and difficult to find, then the *vast majority* of positive findings of sex difference will be spurious, false-positive results, creating a *huge* systemic distortion of our biomedical understanding of the biology of sex and sex difference. This would appear to be an accurate description of the current impact of these funding policies on biomedical research. Tannenbaum and colleagues do not express concern that making statistical analysis by binary sex *obligatory* regardless of prior expectations could generate an enormous scientific background noise of false-positive sex difference results. In this they fall far short of the assertion of their essay's title—"Sex and Gender Analysis Improves Science and Engineering."

This exact event occurred recently in the early reports about striking sex differences in morbidity and mortality from COVID-19 in the United States. Early observations that men had higher COVID-19 mortality rates led to wide speculation about the endocrinological and immunological mechanisms that could contribute to the apparent pattern for this newly emerging pathogen. However, subsequent longitudinal analyses have shown that sex differences in COVID-19 outcomes in the United States have not been consistent over all men, all localities, times, or American states.[25] The proposed sex difference in COVID-19 morbidity and mortality are more likely the result of other social factors that are differentially distributed between men and women in different areas.

More appropriately, in a recent report on gender/sex inclusion in scientific research, Londa Schiebinger and Ineke Klinge provided a modified recommendation: "When reporting the results of cross-group comparisons, provide information on the within-group variability and between-group overlap of the distributions. Be cautious not to overemphasise differences between individuals or groups. Ensure that information on both differences and similarities is properly reported in the text, tables and figures, and that sensitivity to nuance is maintained throughout the report. In quantitative research, statistical interactions (and effect-measure modification) should be reported in sufficient detail to enable readers to interpret the effect size and practical significance of the findings."[26] It would be hard to overemphasize the importance of these brief recommendations for ensuring that the obligatory use of binary sex as a scientific variable in biomedical research does not exacerbate the health and social inequities that the recommendations were designed to alleviate. Yet, this issue is mentioned only once (quoted above in full) in an appendix to this 244-page document. Rather than make the search for sex differences obligatory, the biomedical community can do a better job of addressing the biomedical needs of everyone by the inclusion of sufficient sexual difference—total sexual variation—in their studies.

Human genome scientist David Reich maintains that race can be a vital and productive scientific variable in biomedical genetic research, and he worries that the cultural fears of the historical misuse of racial categories will prevent necessary scientific research on the diverse causes of human disease.[27] For example, in research investigating genetic causes of higher rates of prostate cancer in African American men, Reich and colleagues searched for genome regions that were simultaneously associated with West African ancestry and prostate cancer risk in a large sample of African Americans. They identified a genome region that was highly associated with both, and were able to locate a number of specific candidate genes in this region that, in aggregate, statistically explained the higher heritable risk of prostate cancer in their sample of African American men.[28] This research could directly lead to improved health outcomes through individual genetic screening among African American patients for this gene region and enhanced preventive care for individuals at greater risk of prostate cancer.

In this case, Reich has used the scientific apparatus of race difference to structure an investigation of the heritable risk of prostate cancer. However, the same finding—including the same prospect of real, positive health outcomes—could have been made by genetic analysis of any sufficiently broad sample of human diversity in the absence of any race categorization. We know this is true because it has already happened. The association of certain BRCA gene mutations with breast cancer was identified *before* it was realized that these mutations are present in higher frequency among Ashkenazi Jews than other human populations.[29] This finding was made possible because the original sample of patients included a sufficient number Ashkenazi Jews, who had access to medical care and were simply viewed as *legitimate subjects* for inclusion in an investigation of the heritable risks of breast cancer. In the same way, the genetic variations associated with heritable prostate cancer risk in African Americans could have been uncovered in any sufficiently diverse sample of humans. One does not need to impose race categories onto human patients in order to learn how to improve health and save lives. Researchers can, and should, improve health and save lives by including all patients as equally legitimate subjects of health care and biomedical research. Progress in genetic biomedical research can be facilitated by making conscious efforts to base research upon a sufficient sample of human diversity, not by the creation of racial categories of analysis and the recruitment of study subjects to populate them. Although it is laudable to want to apply genetic research to improve the health of African Americans, equitable inclusion of African Americans in the health care system should be sufficient to achieve this. The scientific apparatus of race is not necessary to do so.

In a precise reversal of my argument here, Reich actually argues that the use of racial categories in biomedical genetics can be justified by "learn[ing] from the example of the biological differences that exist between males and females." Reich writes that, "The differences between the sexes are far more profound than those that exist among human populations, reflecting more than 100 million years of evolution and adaptation. Males and females differ by huge tracts of genetic material—a Y chromosome that males have and that females don't, and a second X chromosome that females have and males don't."[30]

Before returning to Reich's suggestion that sex difference research is a positive example to be emulated in genetic studies of race difference, let's first reexamine the "huge" genetic sex differences that Reich describes. As we have seen, the Y chromosome has only approximately seventy-five active genes, compared to the more than eight hundred functioning, protein-coding genes on the homologous human X chromosome, among the approximately twenty-one thousand genes in the entire human nuclear genome. So, approximately 0.357 percent of human protein-coding genes are different between Males and Females, and 100 percent of those differences are genes unique to XY individuals. However, we have also seen that, in all XX mammal embryos, one of the two X chromosomes is randomly inactivated at the gastrula stage of two hundred to three hundred cells. Thus, one set of the alleles from the eight hundred genes on each of the two X chromosomes—or approximately 3.8 percent of the human nuclear genome—will be differentially expressed in two hundred to three hundred different cell lineages within the performative mosaic of every mammalian XX body due to X chromosome inactivation. Cell by cell, then, there is *no difference* in the number of expressed X chromosomes between Female and Male bodies. Rather, the main genetic differences created by variation in X and Y chromosome number are not between Male and Female bodies, but *among* different cells *within* the individual bodies of XX individuals. The scientific apparatus of sex difference specifically fails to actually explain the genuine biological consequences of differences in X and Y chromosome number in mammals.

Are the sex differences Reich cites—0.357 percent of protein-coding genes—either "huge" or "profound"? Of course, the difference of even a single DNA base pair *can* have a profound effect on the phenotype if that difference is, for example, a dominant lethal mutation. But these are not the types of differences that Reich is citing to justify the genetic sex differences as an exemplar for the study of genetic race differences. Whether or not this magnitude of genetic difference has "profound" impacts on the phenotype depends precisely on the way that phenotype is realized in the world through its development and ongoing physiology, and upon the total scale of genetic and phenotypic variation among all humans. These are the very topics that this book suggests should be investigated within a performative framework.

To me, any phenotypic sex or race difference that is not diagnosably distinct—that is, does not generate distinct peaks in a joint frequency distribution—can hardly be described as profound. What is profound is that so many biologists, psychologists, and sociologists continue to describe the differences generated by the scientific apparatus of sex and race difference as fundamental even though it is not possible to reliably identify a human Male or a Female by their body mass, their height, their body strength, their brain scans, or the vast number of other phenotypic features that are subject to sex or race difference research. What is profound is the amount of effort that studies frequently must make in order to find statistically significant sex or race differences within huge data sets, along with ongoing efforts to ignore, obscure, or conceal their effect sizes.

Thus, sex difference research cannot be held up as a positive exemplar of how the genetics of race should be analyzed, as Reich suggests. That genetic variation among human populations is much less than the variation within them is not, in itself, reason to imagine that research constructed to find genetic differences among historic race categories will present, in Reich's own words, "only a modest challenge to accommodate."

I hope that the detailed discussions of genetic and molecular mechanisms in human development in this book will provide new impetus for biologists, psychologists, and others to reevaluate the use of the scientific apparatus of sex and race categories in human genetic, biomedical, psychological, and sociological research. Efforts to broaden research sampling protocols and experimental systems to incorporate the full diversity of human difference and diversity is a more appropriate way to achieve inclusion than consciously applying the scientific apparatuses of sex and race. Equitable inclusion does not require analytical sequestering of populations by sex and race.

Post-disciplinary Material Feminisms

For decades, feminist biologists, historians of science, and "new materialist" feminists have argued for a more detailed engagement between genomic and postgenomic molecular biology (see appendix 1). Here, I want to outline some of the emerging intellectual connections between material feminisms and a performative view of biology, genetics, and development.

In her groundbreaking essay of the same name, Donna Haraway proposes that feminist science involves "situated knowledges" that abandon the search for universal laws in favor of local, "situated" investigations that recognize the agencies of the entities being studied. Haraway argues that science needs to adopt the "view from somewhere." The performative perspective on the material body involves exactly this shift in viewpoint. It acknowledges the agencies of the many molecules, cells, tissues, signaling centers, and developmental fields involved in genetic regulatory intra-actions, abandons lawlike generalizations drawn from strict interpretations of gene causality, and refocuses on mechanistic descriptions that are situated *within* the process of organismal development.[31]

In "Situated Knowledges," Haraway advocates for a science in which "we give up mastery but keep searching for fidelity."[32] Likewise, I argue that biology and biomedicine should refocus their explanatory tools and powers away from the illusory mastery of strict genetic causation toward more complex models and explanations with a greater fidelity to the individualized processes of developmental becoming. This does not mean that we should abandon genetic experiments, but that we should adopt a situated understanding of the contributions that experimental results can make to the genuine explanation of phenotypic variation. The performative view requires abandoning the all-knowing, scientific "voice from nowhere" to embrace a richer description of the individual organisms and their development, functions, and variations.

In encouraging a "shared conversation" between scientific and cultural knowledges, Haraway observes that the scientific search for "translation, convertibility, mobility of meanings, and universality" become reductive "only when one language (guess whose?) must be enforced as the standard for all translations and conversations."[33] Accordingly, I am proposing a nonreductive scientific/cultural framework for understanding gender/sex that is rooted not in the traditional language of science but in the conceptual and analytical language of queer feminist theory. Furthermore, this is not a scientific accommodation to culture, I argue, but an empirical and conceptual improvement to biological science on its own terms.

On another front, biological performativity contributes to Karen Barad's goal of bringing the discursive and the material into closer proximity by analyzing gene regulation and development as the product of material discourses that are outside of language, prior to language, outside of human intention, yet deeply within the human. The performative continuum implies something fundamental about the relationship of biology and culture, and it supports broader science/culture research programs that untangle the complex cultural influences on our biology and health—mediated by class, race, nationality, poverty, economic inequality, ableism, food policy, migration, and more.

As Eve Sedgwick remarks, performativity is anti-essentialist.[34] Once we recognize the power of discourse—whether textual, verbal, gestural, *or molecular*—to make and affect reality, then we realize that such discourses do not, indeed cannot represent, express, or make material a fundamental, prior, or inherent individual essence. Rather, discourses are generative processes in themselves; they are the means of achieving both materialization and meaning. Thus, the performativity of the phenotype undermines all biological essentialisms, including the individual sexual binary. As bodies develop, they are not representing their essential identities. Rather, they are becoming themselves—selves with reproductive capacities and gender/sex possibilities (including nonbinary and asexual possibilities). We are each one of us performances of ourselves as bodies, as minds, and as sexual, psychological, social, and cultural beings.

On yet another front, the performative perspective in biology opens up a new conversation about gene regulation, development, and selection as regulatory and creative forms of posthuman power. Foucault's vision of power as a distributed property of hierarchical social systems, coming "from below," and acting from "innumerable points" is strikingly congruent with the developmental and regulatory mechanisms of the complex, homeostatic, multicellular body. Foucault writes that "there is no power without a series of aims and objectives," but these objectives are not "results from the choice or decision of an individual subject."[35] All of these properties are apt descriptions of the consequences of natural and sexual selection acting on populations of organisms. The evolved anatomical, physiological, ecological, and behavioral "strategies" of organisms are indeed realizations of aims and objectives. But these goals are the immanent, authorless teleology of selection acting on populations, and not the result of conscious, individual decisions.

We can see with greater clarity now that, like the Foucauldian concept of power, selection and development both constrain *and* create possibilities, both eliminate *and* design, acting as both regulatory *and* innovative forces. The complex genetic discourses that constitute the material "apparatuses of bodily production" are a realization of the evolved aims and objectives of organismal becoming, and operate from the bottom up through communicative intra-actions among molecules, genes, and cells. Developmental mechanisms are both constraining and creative, regulatory and innovative. The distributed and decentralized regulation of organismal development is organized not around the human agency of language, but by the evolved discursive molecular agencies of genes, biomolecules, cells, tissues, and bodies. The activities of the heart, liver, lungs, kidneys, brain, etcetera are critical to the whole organism. Yet their development and physiology are not managed by any central, top-down control; they are distributed and locally invested within the cell types and tissues of these organs themselves. This distributed power is inherent to all the complex organs and physiological processes of the body.

Fortunately, progress in biological research has greatly expanded the opportunity for detailed engagement between queer feminist thought and the development and evolution of the material body. The detailed discursive properties of gene action in development has really only been uncovered in recent decades. In the 1980s, Christiane Nüsslein-Volhard and Eric Wieschaus first identified fifteen developmental regulatory genes involved in the early embryonic development of fruit flies, which turned out to be among the first developmental paracrine signaling molecules, receptors, and transcription factors ever identified.[36] Through a mammoth research effort over subsequent decades, biologists have elaborated an enormously detailed picture of the molecular discourse of the developing animal bodies. One cannot emphasize too strongly how novel, extraordinary, and unexpected these discoveries have been. The networked, discursive properties of gene action were neither obvious nor inevitable given prior understanding of genetics and gene action.

In the science of sexual differentiation, these findings have also been transformative. Readers can compare two prominently published reviews of vertebrate sexual differentiation mechanisms written by women who were leaders in the field: Ursula Mittwoch in *Nature* in 1971, and Blanche Capel in *Nature Reviews Genetics* in 2017.[37] The forty-six years of research in between them constituted a staggering advance in detailed knowledge and a wholesale shift away from top-down determination to networked, "parliamentary" (i.e., performative) realization. This avalanche of new, empirical details provides a rich opportunity for cross-disciplinary analysis, engagement, and research.

An Intellectually Queer Space in Science

Recognizing a performative continuum in biology, gender/sex, health, medicine, and disability can help to render all bodies and lives as simultaneously scientifically and culturally legible. Queer bodies and lives are expected outcomes of a common, shared performative process. This posthuman move to expand performativity to molecular, developmental, evolutionary biology, and physiology provides a broader scientific foundation for queer and feminist accounts of individual difference and bodily change over the life span.

In contrast to Foucault's view "positioning science as the enemy of art and sex," queer literary scholar Sam See advocates for a new consideration of the queerness of evolution itself. In his essay "Charles Darwin: Queer Theorist," See proposes that "Charles Darwin is a queer theorist of the material world, who conceptualizes nature as a non-normative, infinitely heterogeneous composite of mutating laws and principles."[38] See further observes that Darwin's theory of sexual selection introduces an aesthetic dimension into evolutionary process—into nature itself—that is potentially independent of adaptation by natural selection. Furthermore, See argued that if, as Darwin proposed, all sexual feelings are aesthetic feelings based on individual subjective responses and desires, then we have cause to acknowledge "how queer the concept of nature was in its Darwinian conception." Bruce Bagemihl's *Biological Exuberance* and Joan Roughgarden's *Evolution's Rainbow* have provided ample evidence of queer sexual behavior in a wide variety of contexts and a great diversity of nonhuman species.[39] Following See, however, I want to argue more broadly for an understanding of evolutionary process as fundamentally queer.

The queerness of evolution is manifest in the central role of variation from the norm in the mechanism of evolutionary change. In the physical sciences, variation is error or uncertainty around some measured value—whether it is the mass of the proton or the angle between the bonds in a molecule of liquid water at 20°C. In contrast, Charles Darwin demonstrated that biology is the science of variation, because variation has a novel *function* in biological science. At every level in biology, heritable variation is the engine of diversification, the source of evolutionary change, and, therefore, ultimately all biological diversity and complexity.

To further connect the biological and cultural functions and consequences of variation, it can help to intellectually connect (but not synonymize!) the concept of cultural norms with the norms (or modes) of a frequency distribution—such as the peak in a normal distribution (or bell curve). Both norms address expectations about the world, against which observations of the world may vary. As David Halperin writes, the queer is "in opposition to the norm," and "by definition *whatever* is at odds with the normal, the legitimate, the dominant. *There is nothing in particular to which it necessarily refers.*"[40] Likewise, in biology, the evolutionary function of heritable variation arises only by reference to the norm of a population. The effect of a new heritable variation and mutation is to displace or disrupt—and thus to *queer*—the distribution of a trait away from its prior historical norm, creating a deviation from prior expectation.

All evolutionary change arises from the differential propagation of variations that transform the norm. Thus, evolution proceeds through the differential success of *incremental queerings* of historic normative expectations. Variations in gender performance function culturally to initially *queer* gender norms, and ultimately to transform them to a new recalibrated expectation that incorporates the rights of all individuals to thrive with freedom, autonomy, and respect. As Karen Barad writes, performativity contributes to "unsettling nature's presumed fixity, and hence an opening up of the possibilities for change."[41] Heritable variations in biological traits contribute to biological evolution in the same manner by unsettling fixity and opening up possibility. From another perspective, the inherent queerness of life can be understood as an unavoidable outcome of genetic and environmental variability that inevitably creates "tails" to the frequency distributions of phenotypic traits. Explicitly recognizing the queerness of evolutionary change contributes to new intellectual connections between biological science and the cultural understanding of sexual and gender diversity.

Of course, my expansion of Sam See's observations on the inherent queerness of Darwinian evolution could be considered as too obvious, shallow, or simplistic to be taken seriously. But I would like to suggest another view. In *Between Men*, Eve Sedgwick explains how "The Gay Closet" functions not only to conceal individual sexual variation from broader public perception and acknowledgment, but also to render much that is publicly obvious invisible, to obscure that which is in plain view to all, and thus make the queer illegible.[42] As Sedgwick asserts, the illegibility function of The Closet can be more culturally critical than its concealing function, which applies only to a sexual minority, because queer illegibility allows the presumed heterosexual majority to maintain its ignorance of, and distance from, the queer content of everyday lives and discourses. I suggest that the illegibility function of The Closet can apply as rigorously to the intellectual activities in the biology department, the laboratory, or the classroom, as it does to the activities in the boardroom, the locker room, or the bedroom. We may be hesitant to recognize Darwinian evolution's explicitly queer content because the implications are too destabilizing to our view of the discipline, our institutions, and ourselves.

Performative biology contributes to a queer science-culture space in the intellectual expanse between Butler, Barad, Waddington, Turing, and Nüsslein-Volhard. This queer intellectual space at the heart of both biology and culture provides an urgent and productive opportunity to connect the genotype to the extended phenotype, materiality to culture, the sciences to the humanities and to lived human experience.

The idea of an essential sexual binary is such a foundational commitment within contemporary biology that rejecting it, or even questioning it, can be quite destabilizing to the field. I do not imagine that changes I am advocating can be simply accommodated by footnotes within current biological frameworks. Rather, structural changes to how biological science is conceived, conducted, communicated, and taught will be required. This book is simply one voice in that enormous effort. We will need to keep thinking deeply and clearly about what we mean when we say Male and Female. (I am certain that I myself have only started to do this.) We will need to reconceive our understanding of how phenotypes relate to genotypes. Disciplines built in various ways upon biology, including medicine, psychology, sociology, and economics will need to reconsider how they respond to such major conceptual changes and to interrogate their own structuring assumptions about the nature of the phenotype. And we will all need to think deeply about the implications of sexual science for the lives of real people.

As many women and queer people have experienced, it can be difficult—and even dangerous—to be asked to fit into a narrowly constraining binary world that assumes its own naturalness, and thereby precludes one's own existence. Biologists must recognize how the culture of science has unnecessarily reified the individual sexual binary, and the distinct genetic causes that have been considered to "determine" it. We should take responsibility for the impact those scientific concepts and assumptions have on the lives of real people, and the culture at large.

The potential impacts of a queer performative biology on biology and culture are broad indeed. If individual organisms are each performances of themselves, then the multispecies, ecological assemblies that comprise food webs, ecological communities, and biomes are ecological co-performances. Returning to Evelyn Hutchinson's analogy of *The Ecological Theater and the Evolutionary Play* and to Hutchinson's own research in community ecology, performative developmental and evolutionary biology connect directly to niche theory, competition, and ecological biogeography, providing new directions and opportunities in queer ecology.

If individual organisms are each performances of themselves, then symbiotic assemblies of organisms are intra-active confluences of these organismic individualities. Symbiosis does not erase or confound these histories—the genome in each of my trillions of mitochondria is still traceable back to their proteobacterial ancestors 1.5 billion years ago. Rather, obligate symbioses, in which neither participant can survive without the other, involve the origin of new, reticulate individualities that retain the separate historical strands of their participants entangled in new and emergent ways. The boundaries of organisms may become queerly blurred, but their historical continuity is enriched by innovating themselves with new intra-active partners.

Pursuing these directions will take new, expanded scientific vision. Scientific training fails to prepare us to think about how we can have, or make, a different science, how we can practice science in ways that help create changes that support more ways of living. Incrementally, these changes can contribute to a biology of the future—a discipline that is more attentive to the full details of genetic, developmental, and ecological process, *and* more connected to the lives and experiences of real people.

Because queerness exists through its distinction from the norm, however, *queering* the culture and content of the biological sciences will only be a transitory state. In my vision for a biology of the future, once the research and teaching of biology become fully conscious of the queer and performative elements of primary topics of the discipline—the genotype-phenotype relation, sexual development, adaptation by natural selection, aesthetic evolution by mate choice, ecological intra-actions, etcetera—these concepts and ideas will cease to be queer. Rather, they will simply become good, productive science, part of a new, more diverse, better functioning, and culturally relevant understanding of biology and ourselves.

This book is only a beginning, a call for a different approach to research. Importantly, the queering of biological science—both intellectually and institutionally—will require the research and intellectual input of queer scientists—research biologists, lab scientists, and teachers whose queer, trans, nonbinary, asexual, and other life experiences inform their views and research in the sciences, leading to new insights, new scientific tools, new approaches to teaching, and new ways of viewing and understanding biology and culture. This performative framework for understanding the biology and culture of gender/sex expands opportunity for any and all who are inspired to contribute to a transformation of biological science and with it the science/culture interface.

Acknowledgments

A unique friendship taught me that that thinking seriously about queer feminist theory and evolutionary biology could be scientifically productive, and thereby contributed greatly to the intellectual trajectory that led to this book. In Spring 2012, I met Sam See, an assistant professor of English at Yale, when he gave a lecture at the Whitney Humanities Center titled "Charles Darwin: Queer Theorist." Completely unexpectedly, I found Sam's consciously queer reading of Darwin's theory of sexual selection to be strikingly in tune with my own scientific research on aesthetic mate choice in birds. I was immediately intrigued by Sam's bold interpretation of the diversity, variety, and queerness of Darwin's intellectual world.[1]

Sam and I followed up with a series of fascinating conversations about evolution, queer theory, science, literature, and birds, which were mutually challenging and inspiring. Unfortunately, our exchanges were cut short by Sam's sudden and untimely death in 2013. However, Sam's bold and vibrant intellectual example persisted as a personal provocation to seek deeper interconnections between biology, evolution, queer theory, and human experience.

In 2000, Jeff Moran, a friend and colleague at the University of Kansas Department of History, invited me to give a guest lecture on the biology of sex to his undergraduate class on the History of Sex. In preparation, I cracked open an old vertebrate anatomy textbook and learned for the first time about the development of mammalian genitalia. I became convinced that familiarity with human genital development could make a meaningful contribution to *everyone's* life, and sought various ways to pursue this goal. In retrospect, this book might never have come to exist had I not had that chance encounter with this extraordinary subject. For that invitation, I am deeply grateful to Jeff. Also at the University of Kansas, my biology colleague Michael Christianson was a good friend, a creative and curious intellect, and an endless inspiration to think more seriously about developmental biology. His intellectual example influenced my work on feather evolution and, ultimately, this book.

I want to thank Richard and Barbara Franke for their thoughtful gift to create the Franke Program in Science and the Humanities at Yale University. I had the pleasure of becoming the first director of the Franke Program, which provided me with an intellectual laboratory to explore interdisciplinary research, and deeply influenced my personal academic goals. I also thank Maria Rosa Menocal and Gary Tomlinson for inviting me to be a fellow of the Whitney Humanities Center, where I was exposed to so much cutting-edge research in the humanities, and for always making me feel welcome there.

At Yale, my colleague Günter Wagner has been an inspiration to think creatively about development and evolution. In summer 2019, I mentioned some of my earliest, inchoate ideas about performativity in biology to Günter. Despite his irritation at some aspects of my proposal, or perhaps because of it, he strongly advised me to get on with it, and make this project a priority. For that kick in the pants, and his many subsequent constructive criticisms and insights on various stages of the manuscript, I am deeply grateful.

I would like to thank the anonymous parent of a daughter with classical congenital adrenal hyperplasia who movingly shared with me the challenging and often agonizing experience of observing and managing her daughter's health care and adolescence.

Deborah Coen and Joanna Radin provided invaluable advice and guidance early on in the project, and on various versions of the text. In addition to introducing me to many important issues and concepts in science studies, Debbie and Joanna both told me (more or less simultaneously), "I think you should be reading Karen Barad." Obviously, their advice had a decisive impact on me and this book.

Many biology colleagues and friends have provided helpful, constructive, and critical comments on the manuscript, in correspondence, or in discussions. These include Joel Abraham, Nicolas Alexandre, Polly Campbell, Titia de Lange, Susan Johnson, Hussein Mohsen, Michael Nachman, Rosalyn Price-Waldman, Daniel Stadtmauer, Stephen Stearns, Liam

Taylor, and Günter Wagner. From the perspective of queer, feminism, science studies, and the humanities, the manuscript greatly benefited from comments by Deborah Goldgaber, Claire Jackson, Bri Matusovsky, Kate McNally, John Durham Peters, Megan Poole, Beans Velocci, and an anonymous undergraduate student at Yale. David Sepkoski provided stimulating discussions of the history of the study of human races in evolutionary biology. Kathryn Lofton provided insights on the history of Erving Goffman's and Margaret Mead's concepts of social performance. The manuscript was greatly improved by formal reviews for the University of Chicago Press from Stacey Ritz and three anonymous referees.

In writing this book, I have greatly benefited from conversations with Liam Taylor, Cody Limber, Haysun Choi, Sam Snow, Jake Musser, Casey Dunn, Erika Edwards, Tom Near, Jack Hitt, Lisa Sanders, and Elizabeth Maddock Dillon.

Tony Wilson from Brooklyn College and Camilla Wittingham from University of Sydney kindly provided their unpublished insights into sea horse (*Hippocampus*) genomics and sexual development. Martin Cohn of the University of Florida kindly gave permission to reproduce his photograph of a mouse embryo with an in situ stain for HOXD13 expression (figure 16).

Rebecca Gelernter created all the illustrations with great care, style, and attention to detail.

Karen Merikangas Darling at the University of Chicago Press provided intellectual insights, needed encouragement, and great editorial suggestions. In her capacity as one of the new academic editors of the science·culture series, Joanna Radin gave invaluable advice, textual suggestions, conceptional insights, and intellectual encouragement that helped me focus on communicating more effectively to as wide an audience as possible. Adrian Johns, the other academic editor of the series, gave me meaningful encouragement and feedback at several critical junctures. I thank Brockman Inc. for representing me. Of course, I remain responsible for all rough edges, errors, and lapses.

My sincere thanks and gratitude to my wife and life partner, Ann Johnson Prum, for her boundless encouragement, support, patience, and input. Our children Gus, Owen, and Liam and their partners Hiann Lee and Lili Dekker shared our pandemic family bubble at various times during the writing of this book, and enriched my understanding of gender/sex through sharing their views and experiences. In response to Ann's question, "Hey, when are you going to work on birds again, Prum?," the answer is *now*.

Appendixes

These appendixes include discussions of topics that are directly related to the primary topic of this book—performative biology—but were somewhat ancillary to the main argument on the development and evolution of the sexual phenotype. Rather than skip the opportunity to make some contributions on these issues, I have gathered them here as additional resources for readers with broader interest in material feminism, acquired immunity, and evolutionary biology. Although less related to the sexual phenotype, the material in the appendixes is written for general audiences, and does not assume specialized knowledge of these topics.

APPENDIX ONE: Material Feminisms

APPENDIX TWO: Acquired Immunity

APPENDIX THREE: Current Models of the Genotype-Phenotype Relationship

APPENDIX FOUR: Modularity

APPENDIX FIVE: Genetic Assimilation

APPENDIX SIX: Why Gene-Level Selection Is Insufficient

APPENDIX SEVEN: Internal Selection

APPENDIX ONE

Material Feminisms

This book builds upon previous work in the broad field of material feminism, and I present a short introduction to this literature for readers with more interest in exploring it further. Originated by feminist scientists, material feminisms have expanded and diversified in recent decades to include the work of researchers and thinkers in Continental philosophy, the history of science, the philosophy of science, and queer and feminist science studies.

A number of the pioneers of material feminism are professional biologists who expanded their research interests into feminist analyses of biological science, including, for example, Donna Haraway, Evelyn Fox Keller, Anne Fausto-Sterling, and Patty Gowaty. Donna Haraway contributed to numerous breakthroughs in analyses of cultural influence in the study of primatology, the proposal of "situated knowledges" as feminist science, the proposal of an interconnected concept of natureculture as the appropriate focus of feminist investigation, and a concern for multispecies interactions/partnerships. Evelyn Fox Keller's biography of Barbara McClintock, *A Feeling for the Organism*, was a groundbreaking portrait of the social and intellectual path of a brilliant woman in twentieth-century genetics, and an analysis of social barriers to women in science. In *The Century of the Gene* and other works, Keller analyzed the nature of genetics at the dawn of the genomic era.[1] In *Sexing the Body* and other works, Anne Fausto-Sterling documented the history of cultural influences on the science of sex, the sexual body, and the medical treatment of intersex conditions. Fausto-Sterling also advocated for applying developmental systems theory to the development of individual gender/sex (see appendix 3).[2] As behavioral ecologist working primarily on the biology of birds (especially ducks and bluebirds), Patty Gowaty critiqued the sexist origins of concepts and models in active use in evolutionary biology.[3] Recently, biologist Malin Ah-King has used population genetic tools to address the definition of sex (see appendix 3).[4]

In parallel, a number of feminists approached sex and embodiment through Continental philosophy itself. Elizabeth Grosz has presented a series of increasingly material and embodied readings of Charles Darwin, Friedrich Nietzsche, Henri Bergson, Luce Irigaray, and Gilles Deleuze, that call for renewed engagement with the material body and the nature of sexual difference (as pure difference, not difference between binary sex categories). Grosz's proposal of the productivity of investigating sexual difference, rather than the differences between sexes, has been particularly productive in this work (see chapter 2).[5] Rooted in post-structuralist critical theory, Vicki Kirby has investigated the interface of language and the materiality of the body. Specifically relevant to this book, Kirby questions whether the concepts of language, text, and discourse at the heart of Continental philosophy have been too narrowly construed, and might properly include a host of material-biological phenomena, including the capacities of bacteria to evolve resistance to antibiotics and the pheromone trails of ants.[6] I have followed Kirby's suggestion extensively in this book.

Other researchers trained in history or the history of science, including Helen Longino, Sandra Harding, Londa Schiebinger, and others, have enriched material feminism in multiple directions. Longino has focused on the investigation of scientific practice and what constitutes feminist science.[7] Schiebinger has focused on the historical critique of concepts of gender, scientific views of the sexual body, and the role of women in shaping scientific knowledge.[8]

With the turn of the century, a new generation of material feminists arrived with training in feminist science studies itself, many of whom would identify as "new materialists." In a series of books, Myra Hird has called for renewed engagement with the materiality of the body, critique of the gender/sex binary, the recognition of the agency of microbes and of matter itself, historical contingency, and the capacity of biological systems for self-organization. In more recent works, Hird presents a posthuman perspective on biological diversity, and explicitly queer feminist analyses of transsexual rights and the contributions of intersex and trans experience to the critique of the gender/sex binary.[9] Elizabeth Wilson has focused on the interfaces of material feminism, medicine, and psychology.[10]

Feminist science studies scholar Sarah Richardson has explored the historic and ongoing role of sex and gender biases in the scientific search for essential sex differences, focusing on research on human chromosomes and genomics. Richardson shows how feminist critiques of the science of the "sex" chromosomes have contributed to the improvement of scientific understanding. In recent works, Richardson has proposed that pluralistic approaches to the definition of sex in biomedical research can resolve certain conflicts between cultural and biological analyses of gender/sex.[11]

In sole-authored books and collaborations that engage seriously with both the scientific literature and the cultural influences on the production of scientific knowledge, Rebecca Jordan-Young and Katrina Karkazis have critically investigated the medical treatment of intersex conditions, the role of hormones in human brain development, and the influence of cultural concepts about testosterone—or T Talk—on the science of testosterone. Their work is notable for their detailed analysis of the slippery intellectual moves through which scientists and social scientists have maintained a traditional narrative about the causal power of testosterone despite the amazingly complicated, often inconsistent, and context-dependent nature of their research methods and data.[12]

In *Undoing Monogamy* and other works, Angela Willey applies a queer feminist vision to investigate the concept of monogamy as a subject of science, and its role in providing scientific support for heterosexual monogamy as a human ideal. Willey pursues a persistently queer critique of science, in which "natureculture" is not divided, through observational anthropological research in a mouse behavioral genetics lab. Willey has special concern for the intellectual "slippage" that occurs when scientific data come to substitute for the body, and biological science comes to stand for biology itself. Willey is also particularly concerned that critiques of the earlier feminist "flight from nature" and calls for feminists to "engage with the data" do not constrain legitimate queer critique of contemporary science.[13] By posing an explicitly queer critique of science from *inside* science, I hope to address such slippage from another direction.

Material feminism continues to be enriched by scholars whose scholarship is informed by their early training and ongoing research in biology. For example, in *Ghost Stories for Darwin*, plant evolutionary geneticist and feminist science studies researcher Banu Subramaniam analyzes the racist and eugenic roots of common concepts in genetic and evolutionary biology, and the persistence of "eugenic scripts" in contemporary research on invasive species ecology.[14] In *Molecular Feminisms* and other recent works, geneticist and neurobiologist Deboleena Roy envisions a feminist practice of laboratory research that is deeply grounded in feminist philosophy and ethics. Roy advocates openness to nonhuman becomings, and encountering study subjects through kinship and recognition of their capacities.[15]

APPENDIX TWO

Acquired Immunity

The acquired immune system of vertebrate animals provides individuals with a mechanism to identify and counterattack novel pathogens and infectious agents that have invaded the body. But how does it work? How can your body make molecules that will selectively recognize and bind to a pathogen that your body (or, in the case of the novel coronavirus that causes COVID-19, all of *Homo sapiens*) has never experienced before? The full details could fill a small library, but even a short outline of the answer establishes that our acquired immune system is one of the most complexly performative components of the human phenotype.

In a nutshell, the adaptive immune system employs a diversity of genetically differentiated cell types, an ingenious, open-ended system for generating millions of novel antigen-binding receptors and networks of intercellular signaling molecules to orchestrate a nearly infinite diversity of possible and enduring immune responses to novel environmental pathogens.

The acquired immune system works by targeting novel molecular features of pathogenic and infectious agents that are not produced by the body itself. These novel molecules are collectively referred to as antigens. The antibody receptor proteins that recognize and bind to novel antigens are produced by B cells from special, hypervariable immunoglobin receptor genes. The three human immunoglobin genes exist in six, thirty, or one hundred slightly variable copies (respectively) in the genome, and they encode proteins that together form the antibody receptor of the B cell. During B cell development early in infancy (and perhaps continuing somewhat later in life), each of the immunoglobin genes in that B cell's genome is edited out, recombined, and rearranged to produce a *single*, virtually unique, antibody receptor protein that includes only *one* of each of the hypervariable versions of these three immunoglobin genes. The resulting antibody receptor is then built with a virtually unique combination of the variable light and heavy protein chains. For example, one antibody receptor protein could be made up of versions 5, 24, and 71 of the three immunoglobin receptor genes, or 2, 18, and 44. Like a complex lottery, the antibody-producing B cells in one human body can produce over *three hundred million different* possible antibodies. These diverse B cell lineages begin to develop in early infancy and then lurk in the blood and body tissues for your entire lifetime, waiting for the opportunity to become useful to your immune system. Of course, which of the three hundred million B cell possibilities are actually produced by the body from its distinctly variable genome is determined by a stochastic and historically contingent process that will lead to different results even among identical twins. And genetic variation *among* individuals in the actual sequence of the many copies of the immunoglobin genes contributes even more variation in the structure of each individual's antibody receptors.[1]

Upon infection with a new pathogen, its novel antigens will be found and bound by one or a few of the *accidentally* biophysically corresponding antibodies from among all the millions of free-floating antibodies that circulate in the bloodstream and body tissues, or that sit upon the surface of living B cells. Alternatively, cells infected by viruses can also signal their infected state to the immune system by presenting chopped-up pieces of those viral proteins on the *outer* surface of their membranes—revealing the internal details of their illness in the public sphere of the body—where they can bind to the corresponding antibodies of specific lineages of B cells. The infected cells will then be molecularly tagged for elimination by other immune-system cells before the infected cell can produce more virus particles, cutting off the viral reproduction process.

All these modes of antigen-antibody binding initiate the rapid production of more B cells with the specific antibody that matches the novel antigen. The result is the speedy deployment of cells to eliminate the pathogen. At the same time, the immune system is signaled to initiate the subsequent production of special lineages of *memory* B cells that encode the newly effective antibody receptor. Memory B cells live a long time and reproduce slowly to provide the body with a nearly permanent record of those antigens that have been previously useful to the organism during its life. These memory B cells can be redeployed rapidly to attack the pathogen should it appear again, thus providing the body with lifelong immunity. Of course, all of these immune-system processes are mediated by complex genetic machinery, cell-to-cell paracrine and endocrine signaling, and multiple additional cells types that I have glossed over.[2]

From a performative perspective, the adaptive immune system functions as an elaborate and distinct molecular discourse involving the creation, and context-dependent deployment, of biophysical messages that enact a successful immune response. Like the random babbling phonemes of a young child, the immunoglobin genes that encode our antibody proteins constitute a vast, combinatorial resource capable of yielding a myriad of initially "nonsense" antibody expressions. These diverse B cells persist, waiting for the moment when their random expressions may become meaningful to the organism. Meaningful how? When an antigen finally *does* bind to a specific antibody from among the millions of different B cell lineages in the body, it communicates to the other immune-system cells that the body is under attack and by what. B cells that encode this newly meaningful, novel antigen-binding receptor are quickly replicated, and these cells fan out around the body looking to fight the pathogen by binding to them and targeting them for attack. After the battle is won, the memory B cells linger forever creating an embodied and enduring *canon* of historically important immune-system expressions. And, like a classic work by Shelley or Brontë, any of these canonical expressions can be pulled off the shelf and deployed again when they are made newly relevant by the return of a same or similar pathogen. The redeployment of antigens from B memory cells involves the citation of an historic, canonical immune response.

A compelling analogy for the acquired immune system was posed by Jorge Luis Borges in a surreal 1941 short story "The Library of Babel." Borges imagines a fantastical library containing all possible books of 410 pages in length including every possible combination of twenty-eight letters and spaces organized on shelves in hexagonal galleries connected by networks of hallways and spiral staircases. For innumerable centuries, generations of "men of the library" have roamed the shelves and galleries in search of a book—or, even a brief snippet—with any intelligible meaning. The aging narrator reports a rumor that somewhere on an unspecified shelf in some far away gallery of the library there is a book that is a complete jumble of letters until on the next-to-last page it suddenly reads, "*Oh Time thy pyramids . . .*"[3]

Arranged like the individual books in the "Library of Babel," the three hundred million possible antibodies that are produced by the adaptive immune system of a single individual would require 428,571 of Borges's hexagonal galleries. Given the hypervariable immunoglobin genes of humans, the immune-system "books" of all of the Earth's 7.5 billion people could require more than 3 quadrillion (3×10^{15}) galleries. (One could name the *borges* as a new unit of information equivalent to the content of one gallery in the Library of Babel, or 10.0762 gigabits.)

Ironically, elements of Borges's surreal thought experiment are a molecular reality for the acquired immune systems of humans. The labyrinth of microscopic, spongy crypts within the bone marrow, where trillions of immune-system cells are biologically "curated," actually bear a fair physical resemblance to the architecture of the galleries in the Library of Babel. Although Borges's narrator has faith that the library "can only be the work of a god," the unfathomable combinatorial diversity of human antibodies is the product of adaptive evolution for the open-ended capacity to discursively fight off *all possible* microbial challenges, and the performative individuality of every human life. Borges imagines that somewhere in the library there must exist a book with the true story of your death and the translations of every possible book into every other language. But somehow Borges never imagines that, as a result of external events intruding on the library, any one of these nonsense volumes could suddenly become vitally *meaningful*, indeed *critical* to the survival of the entire library itself. But that is exactly how the vast living catalog of our antibodies protects us individually from deadly infection. Furthermore, the origins of these novel meanings are performative, not representational, because they involve the tagging of novel invasive infectious agents through a molecular *doing*. The origin of meaning of any antibody involves the arrival of a new opportunity for intra-action—the creation of the sudden opportunity to bind to a novel antigen that has never been presented to the organism before.

The acquired vertebrate immune system demonstrates a novel way to create *anticipatory* intra-actions. The naive immune system produces millions of antibodies in anticipation that some of them will biophysically correspond to novel antigens in the future, and thereby yield novel meanings with the opportunity to prevent life-threatening infections.

Vaccines work by stimulating your body to make B memory cells that encode antibodies that will fight off a possible future exposure to a pathogen without actually giving your body that disease. Furthermore, vaccines *cannot* create appropriate antibodies for novel diseases like COVID-19. They can only *induce* your body to replicate appropriate antibodies from among the hundreds of millions of *preexisting* B cell lineages in your body that encode antibodies that will selectively bind to the novel spike protein on the surface of the coronavirus, creating B memory cells that will give you an aggressive head start in fighting off future SARS-CoV-2 infections. Among the reasons that vaccinations can have variable outcomes in different individuals is that each individual's immune response to the vaccine is hyper-unique—drawn from their *own* individually unique diversity of B cell lineages that are realized through random recombination of their own unique, hypervariable genome by a highly stochastic molecular editing process. Again, the response of identical twins to the same vaccine could be highly distinct. Likewise, the effectiveness of the vaccines can wane over time if the corresponding pathogen can evolve sufficiently to be less efficiently bound by the antibodies deployed in response to previous infections or vaccinations. The pathogen can also evade a vaccine if it has evolved to reproduce in a body *more rapidly* than the time it takes to generate new active B cells from the appropriate lineages of memory B cells created by vaccination. SARS-CoV-2 appears to be doing both of these things.

In summary, the body defends itself from novel infections with a staggeringly complex system of immune responses that are characterized by discursive intra-action, citationality, hierarchical complexity, stochastic contingency, and multiple agencies. It is the environment that determines which of the vastly large number of antigen expressions will become meaningful to that individual organism. The multiple sources of variability that make the system so versatile and responsive to novelty—including variation in the antigen receptor gene sequences, variation in the editing and splicing process during B cell lineage development in infancy, variation in the survival of B cell lineages over the lifetime, random variation in antigen-antibody encounters within the body, variation in pathogen exposure—also ensure that these responses are unique to each individual. Every body invents its own adaptive immune response. The essential contribution to our individual survival, and to the collective evolutionary success of vertebrates, is a further testament to the vital role of performativity in biology.

The advantageous flexibility of immune response, however, can also contribute to performative malfunction (see "Performativity of Illness and Disability" in chapter 8). Autoimmune diseases are a result of the immune system accidentally recognizing a molecular feature of the body itself as a harmful foreign antigen and mounting an assault upon it. Rheumatoid arthritis, type I diabetes, and multiple sclerosis are all examples of autoimmune diseases. The innate immune system can also create other life-threatening conditions. The "cytokine storm" that has killed many victims of respiratory infections, like influenza and COVID-19, is the result of an explosive overexpression of cytokine paracrine signals that recruit other immune cells and stimulate an inflammatory response to fight the virus, but also pathologically compromise lung function in the process.

APPENDIX THREE

Current Models of the Genotype-Phenotype Relationship

For more than one hundred years, the genotype-phenotype relationship has been central to evolutionary biology. Multiple models and analytical tools have been developed to investigate it and account for its complexity. Here, I argue that the queer performative account of the phenotype provides important insights that other contemporary models lack.

As Peter Taylor and Richard Lewontin observe, the traditional population genetic concepts of genotype and phenotype require the recognition of classes of individuals with identical (enough) genes and identical (enough) traits for the experiment at hand. Originally, biologists used phenotypic classes—like fruit flies with red eyes—to infer genotypic classes—fruit flies with a gene for red eyes. Now, researchers frequently have more detailed genetic sequence information, and are generalizing about classes of phenotypes and environments. The creation of experimental inbred strains of model organisms—which are bizarrely less genetically variable than practically any normal organism—further reifies the concept of the genotype as a definable class. The creation of genotype and phenotype classes for genetic analysis involve an abstraction away from the real genetic and morphological/behavioral differences found *among* the individual organisms (especially pure difference—the differences among individuals without regard to imposed phenotypic classes). After much intellectual progress during the twentieth century, Taylor and Lewontin recognized "the need to reintegrate what has been abstracted away" by the traditional concept of genotype and phenotype.[1] The performative phenotype restores a realistic, individual specificity to the genotype-phenotype relationship.

The reaction norm is a traditional analytical method that quantifies the distribution, or spectrum, of phenotypic variation (and fitness consequences) developed by individuals of the same (or similar enough) genotype class over a gradient of abiotic or biotic environmental conditions, like temperature, food availability, salinity, or predator density. The goal is to understand the relative contribution of genetic and environmental variation to phenotypic variation, and ultimately survival and fecundity. A traditional reaction norm is usually produced from an experiment raising replicate, inbred, or clonal individuals in a variety of environments, and observing the resulting phenotypes. Reaction norms can also be generated using genotype classes, which are presumed to have similar enough genomes for the question being studied (such as the classes of human individuals having XY or XX chromosomes). Reaction norms provide insights into the environmentally induced variability, or plasticity, of realized phenotypes, and thus they constitute a traditional method of exploring some aspects of what I refer to as the performativity of the phenotype. For example, in a series of papers, Malin Ah-King and collaborators propose that sex—including behavioral aspects of reproductive behavior and parental investment—should be considered as a reaction norm.[2]

However, the reaction norm concept involves one or both abstractions that Taylor and Lewontin point out, and it is limited in scope and utility to the narrow experimental circumstances necessary to estimate one, such as rearing of multiple individuals of an identical genotype (or genotype class) across a set of variable environments. Like other forms of experimental genetic control, the reaction norm experiment is a Baradian material-discursive apparatus that *creates* the causal phenomenon it measures (see "Sex and Race as Scientific Apparatuses" in chapter 9). An infinite number of reaction norms can be generated to analyze phenotypic variation for any genotype or genetic class (such as individuals with XX or XY chromosomes), and any set of environmental variables (such as temperature, concentrations of dioxane, etc.). Without additional bounds, this open-endedness limits the applicability of the reaction norm as a general concept of individual sex. Furthermore, because reaction norms do not focus on the actual developmental mechanisms, they typically provide no further conceptual insights into *how* the environmentally induced variations of genotypes come to vary. Thus, they can be a useful experimental tool to extract traditional population genetic inferences, but are insufficient to provide a framework for understanding the generation of phenotypic variation, and the broader genotype-phenotype relationship. Lastly, given the many ways and contexts in which the idea of individual sex is referenced and discussed, it is not often scientifically or culturally productive to reframe the concept of sex in terms of the spectrum of variation resulting from developmental experiments on samples of identical genotypes.

More complex experimental methods are used to investigate the variation in environmental plasticity among genotypes, or genotype-by-environment (GxE) interactions—that is, when different genotypes respond differently to environmental variation. In brief, genotype-by-environment interactions can be identified when reaction norms of different genotypes are different in slope or shape from one another, which is clearest when they intersect at some point. However, thinking of the phenotypic variation as the product of an interaction between genes and the environment fails to account for the hierarchy of multiple agencies involved *within* the body—that is, cells, tissues, organs, signaling centers, electrical fields, mechanical forces, individual behavior, etcetera. (See appendix 7.)

With increasing amounts of genomic sequence data, additional methods have been developed to investigate the relationships between genetic, environmental, and phenotypic variation, including quantitative trait loci (QTL) mapping and genome-wide association studies (GWAS). GWAS has become a common method in the study of interactions among the genetic and environmental influences on complex diseases like cancer, heart attack, drug addiction, and other mental illnesses. While powerful, in practice both methods require the same abstractions about phenotypic or environmental classes that Taylor and Lewontin identify.

Widespread application of QTL and GWAS to study complex human diseases has established an empirical conundrum that is now known as the "missing heritability problem." For instance, if you add up the contributions of all the genetic variations that are significantly associated with a particular complex phenotype (such as heart attack, or breast cancer risk), you are able to explain *less than* 10 percent of the *heritable* risk of these diseases. There are a lot of potential explanations for the effect, but the most likely is that the impact of particular genetic variations on the phenotype depends upon many *other* genetic variations, or gene-by-gene interactions (GxG), which biologists call epistasis. In epistasis, the impact of some genetic variations depends upon the specific state of another or, frequently, many other genes. What the missing heritability problem is likely telling us is that, in genetically diverse populations like humans, the impact of a specific genetic variation on the phenotype is dependent upon the specific state of many other genes (GxGxGx . . . , etc.). The impact of gene-by-gene-by-gene effects are common, complex, and extremely specific to individuals and their closest relatives. In other words, there are many, many different ways for genetic variations to contribute to the risk of complex phenotypic traits, and most of them are the result of specific, subtle, and diverse effects compounded across the genome.

These gene-by-gene-by-gene interactions are precisely the kind of effects that are abstracted away by current QTL and GWAS experiments, and unaccounted for in canonical accounts of gene action. Furthermore, QTL and GWAS are Baradian apparatuses that impose on the data a specific conception of gene function, and the empirical discovery of "missing heritability" demonstrates how broadly these tools and concepts fail to explain complex phenotypic features.

However, the contingency of any gene's action on the actions of other genes is the specific focus of a performative view of the phenotype as an individual enactment. These complex gene-by-gene-by-gene interactions vary so specifically because the developmental and physiological mechanisms of organisms involve coevolved *intra-actions*, placing the missing heritability problem squarely in the explanatory space of the phenotype's performativity.

In summary, a performative model of the phenotype addresses specific weaknesses in current population genetic tools for understanding the genotype-phenotype relationship—the abstraction away of the numerous, hierarchical agencies that intervene between the genotype and the phenotype. In contrast, the performative view frames the question more broadly than a simple genotype-by-environment (GxE) interaction to include all of the many molecular, cellular, and anatomical agencies that intra-act in the performance of the complex phenotype. The absence of consideration of—the abstraction away from—these agencies contributes to the failure of GWAS studies to explain the vast majority of heritable contributions to complex phenotypes.

From another perspective, developmental systems theory (DST) presents a powerful critique of traditional methods of population genetics in investigating the genotype-phenotype relationship, and the idea of genes as strict causes of phenotypic development.[3] Advocated by biologists Susan

Oyama and Russell Gray, philosopher of science Paul Griffiths, and others, DST calls for the rejection the nature/nurture and genes/environment dichotomies, and the reframing of our conception and study of organismal development to include: multidirectional flow of information between gene expression, the organism, and the environment; joint causality from multiple inputs; context dependence; extended inheritance (including developmental environment); development as "construction"; distributed control; and evolution as "construction" (i.e., the evolution of organismal-environment relations). Focused extensively (but not exclusively) on the evolution of complex behavioral traits, DST's critique of strict gene-level causation, and advocacy of distributed causation is entirely congruent with a performative account of the phenotype. However, the DST literature has not yet engaged in any detailed analysis of molecular genetic mechanisms of development (as discussed in chapters 4–6). In this sense, I think a performative model of phenotype development productively addresses the many challenges raised by DST, and provides a more ample and productive outline for future research programs in the development of complex phenotypes.

Evolutionary biologist Mary Jane West-Eberhard has also posed a focused and well-documented critique of the traditional population genetic view of the genotype-phenotype relationship. In *Developmental Plasticity and Evolution*, West-Eberhard emphasizes the role of plasticity itself—the capacity of phenotypes to vary independently of genetic variation—in the evolution of adaptations and evolutionary novelty.[4] West-Eberhard champions the role of adaptive phenotypic plasticity—the development of traits that improve individual survival and fecundity under specific environments—in the origin of adaptation (see appendix 5). Although West-Eberhard's thorough and nuanced critique is accurate and productive, her focus on developmental plasticity itself as the cornerstone of a new synthesis of genetics, development, and evolutionary biology is insufficient, I think, to provide a productive general model of the genotype-phenotype relationship. It is only by making theoretical space for, naming, and then accounting for the intra-actions of the hierarchy of multiple agencies involved in the development of complex phenotypes—not merely classifying nongenetic phenotypic variation as plasticity—that we can provide an empirically and theoretically sufficient theory of the phenotype.

Most recently, new perspectives on the genotype-phenotype relationship are emerging in the growing, but conceptually quite amorphous, field of "systems biology." Systems biology encompasses a diversity of highly computational, big-data approaches to genetics, transcriptomics, physiology, development, and evolution of the organismal phenotypes. Growing out of frustrations with the limits of traditional inferences based on controlled, developmental genetic experiments involving incremental micromanipulations, systems biology employs big data—often whole tissue or single transcriptomes, proteomes, or complete metabolic pathways—and computationally intensive mathematical models drawn from engineering and physics to understand the development and physiological function of tissues and whole organisms. Systems biology presents an exciting and productive compliment to traditional, incremental, reductionist experiments alone, and it incorporates some of the hierarchical and computational approaches that will be required to pursue research on phenotypic performativity.

However, systems biology currently lacks a unified conceptual core beyond its dedication to computation and big bio-data. For example, a profile of systems biology research at the US National Institute of Health reports that if you "ask five biomedical researchers to define systems biology, . . . you'll get 10 different answers . . . or maybe more."[5] While this open-endedness may be an intellectual asset given the rapidity of new technology and analytical development, adopting a performative framework in systems biology would explicitly recognize what is missing from previous population genetics accounts—the hierarchy of biological agencies, discursive intra-actions, and environmental influences within the genotype-phenotype relation—and would provide a productive conceptual framework for pursuing systems research in biology and biomedicine. In this way, *queer systems biology* could become more than a commitment to big data and the conscious recognition of the limitations of exclusively gene-level explanations of phenotypic variation.

APPENDIX FOUR

Modularity

Although organisms are highly integrated and functional wholes, complex organisms are composed of multiple functionally, developmentally, or molecularly autonomous components, which we are referred to as modules. In recent decades, the concept of modularity has emerged simultaneously and largely independently in the fields of molecular biology, developmental biology, systems biology, and evolutionary biology, creating a large and diverse literature on the topic.[1]

Modules are parts of the body, or its genetic and physiological systems, that function or develop independently, or semi-autonomously, from other parts of the body. Some modules reoccur multiply in a single organism, such as cells, teeth, hairs, feathers, or leaves. Modules can also be nested hierarchically within other, larger, more inclusive modules. Thus, we can think of tetrapod limbs as pairs of modular body parts composed of multiple hierarchically arranged bones ending in five digits. Although these elements covary among all the limbs of the body, functional and developmental differentiation between forelimbs and hindlimbs in some tetrapods created an evolutionarily new distinction between arms and legs, and likewise with digits between fingers and toes. Thus, we can see that modularity is deeply related to organismal function, development, and evolutionary history, and to organismal complexity.

One of the most interesting aspects of biological modularity is how completely unexpected and unexplained it is from the traditional perspectives of genetics and population genetics. Its profound importance in biology did not arise from investigation of the first principles of gene transcription and translation, or from the study of drift and natural selection acting on genetic variation within populations. Rather, biological modularity was discovered through empirical research on the phenotype itself—its anatomy, physiology, development, and biomechanics. Molecular modules were revealed by reconstructing the networked connections (i.e., intra-actions) and hierarchical relations among developmental and physiological signaling molecules. Anatomical and developmental modules were uncovered through the investigation of functional and developmental covariation among body parts.

In contrast, the concept and importance of biological modularity emerges directly from a performative view of the phenotype. The fundamentally discursive properties of organismal development predicts that the growth of complex morphological difference from simpler, embryonic uniformity will require the individuation of molecular signaling networks among cell types, tissues, anatomical regions, organs, and body parts. You cannot build a complex body using molecular discourses if everyone is "shouting" at once in the same space or forum. Developmental discourses must be individuated and compartmentalized from one another so that different body parts and systems can function autonomously of each other. The ontogenetic and evolutionary individuation of developmental discourses automatically implies a *hierarchy* of relations among the multiple, descendent molecular mechanisms, tissues, and body parts. The origin of these individuated developmental discourses creates the opportunity for selection on their phenotypic functions, giving rise to differentiation among modules. Once evolved, molecular and anatomical modules can then be replicated about the body, co-opted for newly evolved functions, and redeployed to other regions of the body, or other times in development. For example, think of the variety of feathers over the plumage of a bird, and their variety of functions including thermoregulation, water repellency, sexual display, flight, and so on. Thus, the organismal body grows from a simple aggregation of undifferentiated cells to a larger, more complex, hierarchically integrated body by exploiting the inherent potential of discourses to differentiate, individuate, and diversify.

Hierarchical modularity is a direct and inevitable consequence of the performativity of complex phenotypes. This observation reconnects modularity to the core issues in biology—the genotype-phenotype relationship.

APPENDIX FIVE

Genetic Assimilation

In a classic experiment published in 1953, Conrad Waddington demonstrated that it was possible for an initially rare, environmentally induced morphological variation in *phenotype* to become evolutionarily assimilated into a heritable feature of the *genotype*. This unusual sounding phenomenon became an important addition to how we think about the relationship between the genotype and phenotype.

Waddington worked on a particular strain of *Drosophila* fruit fly that failed to develop a particular wing morphology—a posterior cross vein—when raised at 25°C. But he found that a small number of individuals would develop a posterior cross vein when the larvae were exposed to 40°C for four hours at the age of twenty hours. He then selected for an increased frequency in the development of a posterior cross vein in individuals exposed to this developmental "heat shock." Some individuals in these newly evolved lines developed the cross vein when raised at 25°C *without* heat shock. From these individuals, Waddington selected new lines in which most individuals developed the cross vein when raised at 25°C or even 18°C.

Waddington concluded that natural selection can favor not only specific morphological traits, but environmentally stable *mechanisms* to develop those morphological traits. Mary Jane West-Eberhard laments that Waddington's concept of genetic assimilation has been so frequently misunderstood and dismissed, and she argues that this intellectual history has prevented research and recognition of the role of developmental variability in the origin of adaptation.[1] From my perspective, however, genetic assimilation establishes a mechanism by which performative individual variation can influence the subsequent outcome of genetic evolution itself.

Genetic assimilation describes the process by which an environmentally induced *phenotypic* state of an ancestor can evolve to become a fixed feature of the *genotype* of an evolutionary descendent. In other words, the process describes how the existence of variable performative realizations of the phenotype in variable, often stressful, environments can allow organisms to explore new morphological, physiological, and behavioral *possibilities* that can later evolve through selection to become dedicated genetic commitments of the organism. Genetic assimilation captures how the *potential* for variation in development and physiology can facilitate the evolution of subsequent features, and further demonstrates that enactments of the organismal self can play a unique, genuinely *creative* role in the process and course of evolution itself.

One of my favorite examples of genetic assimilation comes from the evolution of cacti (Cactaceae). Cacti are leafless succulent plants that have photosynthetic stems with stomata (gas exchange pores) and a special water-saving version of photosynthesis that allows them to accomplish the light-dependent photosynthetic reactions during the day, and conduct the other, gas-exchange-dependent photosynthetic reactions at night, when less water will be lost to evaporation. Phylogenetically, cacti evolved from within the (now paraphyletic) genus *Pereskia*, which includes a variety of semi-succulent species with both leaves and photosynthetic stems. Comparative physiological and evolutionary investigations by my Yale colleague, evolutionary botanist Erika Edwards, and collaborators show that leafy *Pereskia* species, which are closely related to cacti, perform the same water-saving variety of photosynthesis in their stems when they are under drought stress, but *not* under wetter conditions. Thus, the drought-tolerant anatomical and physiological specializations of cacti evolved from the stress-induced physiological performative states of their semi-succulent *Pereskia* ancestors.[2]

Similarly, in my own work, we found that two novel features of the development of the planar feather vane of birds—a ventral new barb locus and the dorsal fusion of barb ridges to form the rachis—evolved through selection on random developmental variations of the growth of unpolarized down feathers.[3] In other words, it was the performatively variable aspects of individual feather development that facilitated the exploration of morphologies that ultimately contributed to the evolution of the planar vane and bird flight. (I know that I am not providing enough of the developmental and molecular details to really document this example, but I do want to document, as I suggested in chapter 1 [page 34], that one could indeed write a book about the performativity of feathers.)

One of the reasons, I think, that genetic assimilation has not had more impact in evolutionary biology, as West-Eberhard documents, is that the traditional population genetic concept of the genotype-phenotype relationship was *designed* to exclude this possibility, to focus solely on genes, and not development, as the cause of the phenotype. A performative model of the phenotype recognizes the agency of the organism to influence its phenotypic enactment, and therefore evolutionary process itself.

APPENDIX SIX

Why Gene-Level Selection Is Insufficient

In the enormously influential book *The Selfish Gene*, Richard Dawkins proposes a grand unifying view of biology through natural selection acting on selfish genes. According to this account, genes are the true replicators in biology, and organismal phenotypes are simply vehicles for the reproduction of genes. Specifically, the model of gene-level selection works through the competition among alleles (or gene variants) at single loci, or genes.[1]

The Selfish Gene has been cited over thirty-five thousand times, and the idea has become deeply influential in biology, psychology, economics, and elsewhere. But the elimination of the agencies of cells, signaling centers, placodes, and organisms to the selfish agency of genes renders this paradigm completely unable to explain the development, or even the existence, of complex, differentiated, multicellular phenotypes. In short, the evolution of the mechanisms by which groups of genetically *identical* cells self-organize into tissues and organs composed of cell types with distinct gene-expression states cannot be explained as a result of competition among alleles at individual loci—the foundation of gene-level selection.

In *From Darwin to Derrida*, evolutionary biologist and ardent, self-identified gene selectionist David Haig recently attempts to connect this paradigm to a more nuanced view of the phenotype. Haig describes gene expression as a form of molecular discourse that gives rise to biological meaning and value, providing an intellectual bridge between science and the humanities. Haig accurately writes, "Meaning resides in the interpretation, not in the [genetic] information, because the same information can be different things to different interpreters."[2] However, Haig's admirable effort ultimately falls short of creating a genuine intellectual connection between biology and the humanities because his strict gene selectionism prevents the recognition of any genuine agencies beyond those of selfish genes. Since cells in the body all have the same genes, he fails to recognize the very agents that can determine meaning within his framework.

Haig proposes to capture the complex phenomenon of organismal development by reconceiving of all the genes in the genome as providing the "environment" for each particular gene, or locus, within the genome. Accordingly, each gene must adapt to advance its own survival and fecundity in the context of all the other genes in that particular genome. This framework implies a mechanism for the evolution of an orderly physiological and developmental process through gene-level selection. However, because gene-level selection involves only competition among alleles at an individual locus, it cannot explain *why* and *how* the many genetically *identical* cells within a single body enact *different* responses to the same extracellular signals in different contexts, resulting in the development of livers, lungs, limbs, or lymph nodes. Beyond competition among alleles, gene selection is rudderless and has simply nothing to offer to the explanation of complex multicellular phenotypes. The only way to understand the coordinated development of a complex phenotype is to recognize the emergent agency of cells, cell types, tissues, organs, signaling centers, etcetera *despite* their genomic uniformity.

Implicitly recognizing this limitation, Haig proposes that we simply view the investigation of intergenerational transmission of genes (evolution) and the development of complex organismal bodies as different, complimentary, orthogonal research programs. This extension of the traditional move by the twentieth-century New Synthesis to isolate developmental biology from evolution is, of course, a substantial strategic retreat from the grand goal of a unified biological synthesis that the gene-selection paradigm promises. It is ironic that the gene-selection framework prevents us from understanding any of the fundamentals required for the material production of the multicellular "vehicles" upon which natural selection acts. Having gone all in for the ultimate, controlling agency of genes, the paradigm is powerless to explain the other emergent agencies upon which the complex phenotypes depend.

Population geneticists typically respond to such criticisms by stating that they simply don't care, or don't need to care, about the details of the development of the phenotype. They can measure natural selection, and the genetic responses to it, generation by generation, and measure the historical signature of selection on genome sequences. The events that happen in the black box between genomes and phenotypes are simply not considered to be relevant.[3] It is important to recognize, however, that this view is an argument in favor of ignorance. Similar arguments were made in the 1970s and '80s by architects of the New Synthesis, like Ernst Mayr, against the intellectual value of reconstructing organismal phylogenies. Of course, these positions will (or should!) go down in history as egregious intellectual errors. Science progresses by identifying, investigating, and solving inconvenient outstanding problems, not ignoring them.

Gene-level selection has other notable weaknesses as a general framework for evolutionary biology. Gene-level selection posits that the evolutionary advantage of any genetic variation is equivalent to the average of its contributions to survival and fecundity in combination with all other genetic variations in the population. Of course, in real populations of finite size, this is impossible to know. Yet, this little piece of mathematical convenience, first introduced by Ronald Fisher in the early twentieth century, has become the foundation for an ambitious, century-long, and expanding intellectual enterprise to reduce evolutionary explanation to exclusively genetic causes.

Borrowing a brilliant analogy posed by evolutionary geneticist Michael Wade, we can conceptually compare this assumption of pure gene-level selection to the child's card game War.[4] War is the simplest of card games, which children learn before they are old enough to understand the rules of Go Fish. Two players split a deck of cards in half, and holding their pile of cards facedown, play one card faceup at a time. The card with the higher value wins, and the winner takes both cards. If the cards are tied, the players shout "War!," lay three cards facedown, and then play a fourth card faceup. Winner takes all. Repeat until one player wins (a very protracted battle), both players are bored, or the losing player feels humiliated and quits, usually by throwing down their remaining cards in frustrated disgust.

In War, the value of any card is universal—simply its average value against all the other cards in the deck. In a similar way, gene-level selection proposes that the relative contribution of any gene variation to individual survival and fecundity is equivalent to its average contribution against all other alleles in the population.

However, as Mike Wade's decades of theoretical and experimental research has made clear, genetic variations in real, finite populations occur in specific, nonrandom combinations and contexts that give rise to specific genetic and environmental interactions. A more realistic (but computationally less convenient) analogy of evolutionary process would be a game of Poker. Although an ace is of great value generally, what really matters most in a game of Poker is a card's value in interaction with the *specific* combination of other cards in the player's hand. In "Darwin's Poker game," even a lowly deuce, in combination with the right other cards, can contribute to a winning pair, full house, flush, or straight when an otherwise highly valued ace would not.

In Wade's view, the context and interactions that are abstracted away by the gene selectionist's "strategy of averaging" *always matter*. And Wade's decades of theoretical and empirical work demonstrates that paying attention to context and interactions supports a rich theory of multilevel selection acting on the specific combinations of genetic variation partitioned among alleles, genomes, individuals, family groups, clans, social groups, populations, species, and clades. Each context creates emergent interactions and evolutionary phenomena that cannot be completely reduced to, or explained away by, the phenomena at underlying levels below, including the level of alleles at individual genes. Wade's work demonstrates that paying attention to context and interactions—adopting a *situated* perspective in evolutionary process—enriches our understanding of evolution and the natural world. (What remains unknown is the extent to which these gene-by-gene interactions are actually intra-actions—coevolved correspondences among discursive agents.)

Philosopher of biology Eliot Sober writes that, "The strategy of averaging over all contexts is the magic wand of gene level selectionism. It is a universal tool, allowing *all* selection processes, regardless of their causal structure, to be represented at the level of the single gene."[5] As the history of the last fifty years of evolutionary biology documents, this intellectual magic can be powerfully seductive. In response, David Haig candidly admits to being seduced by Fisher's magic wand, viewing it "as a strength rather than a weakness of gene-level selectionism."[6]

Like an optical instrument used to investigate the wave or particulate properties of light, the gene selectionist "strategy of averaging" is an example of a Baradian scientific apparatus that *creates* the phenomenon of individual genetic causality, while excluding from observability any other phenomena that could influence evolutionary process (see "Sex and Race as Scientific Apparatuses" in chapter 9). While some evolutionary questions may be addressed efficiently this way, it is not a sufficient intellectual tool for the breadth of phenomena evolutionary biology raises and requires. In contrast, Wade's hierarchical conception of evolutionary process, which situates analyses hierarchically among multiple contexts, has the power to quantify and analyze gene-level selection and selective impacts at multiple other levels—including the family, colony, group, subpopulation, species, and clade.

Although there are many phenomena for which gene-level selection is entirely appropriate, the deficiencies of gene-level selection to explain development of a complex phenotype, and its power to preclude understanding of other phenomena constitute an intellectually depauperate framework for evolutionary biology.

Despite its admirable breadth and productive theoretical and experimental rigor, Michael Wade's hierarchical framework for evolutionary biology still focuses on evolutionary population genetics, and does not yet incorporate developmental biology, physiology, or biomechanics. This missing piece involves the internal intra-actions among cells, tissues, and body parts that influence the organism's body and function. The intellectual challenge ahead, posed in more detail in appendix 7, is to create hierarchical theories of cell, cell type, and tissue gene expression and discursive regulation that recognize and quantify the impact of these agencies on the phenotype. Only then can we really understand how natural and sexual selection lead to genetic evolution in multicellular organisms.

How do we account for the breadth and depth of the intellectual influence of gene-level selectionism given the stark limits on its actual explanatory reach? The answer appears to require a combination of both philosophy and sociology of science. As philosopher of science Ramus Winther argues, Dawkins's gene-level selectionism suffers from what William James called "vicious abstractionism" via the "philosopher's fallacy."[7] Dawkins's locally productive abstraction of selfish genes as the sole important replicators employs a set of strong assumptions—the strategy of averaging; the absence of gene-by-gene-by-gene interactions; the absence of the evolved, emergent agency of cells, tissues, or organs; and so on—that could be otherwise. But the paradigm still proceeds as if these assumptions were universally true.

Sociologically, the success of gene-level selectionism rests on pure, reductive intellectual chutzpah—the bravura appeal of the bold gesture that feels like real science—and the gaping ignorance of developmental biology that most of evolutionary biology has cultivated for almost a century. But the result is a narrowed, flattened, and impoverished theory of evolution that fails to capture the complexity of genuine organisms.

APPENDIX SEVEN

Internal Selection

In appendix 6, I argued that the insufficiency of gene-level selection as an explanation of phenotypic development makes it a failed general paradigm for biology. I further suggested that multilevel selection, as formalized, for example, by Michael Wade, was a productive alternative.[1] However, Wade's traditional population genetic framework still lacks explicit tools to understand the evolution of development, physiology, and biomechanical function of complex phenotypes. To make a multilevel selection framework fully performative, we need to identify, and conceptually recognize, the level of *internal selection* on the development and function of the individual organism.

Implicit in the writings of late nineteenth- and early twentieth-century developmental biologists Wilhelm Roux, August Weismann, Conrad Waddington, and others, the evolutionary mechanism of internal selection was formally proposed by L. L. Whyte in his 1965 book *Internal Factors in Evolution*. Kurt Schwenk and Günter Wagner provide the most practical, working definition of internal selection as heritable variations in survival and fecundity among genotypes or morphologies that are invariant (or rank-stable) over variations in the external environment.[2] In other words, internal selection arises from differences in survival and fecundity that are a consequence of how individual organisms function *internally* rather than how they function in relation to their environments. Rather than lead to an improved fit between an organism and its environment—that is, traditional adaptation by natural selection—internal selection leads to an improved functional correspondence among the innumerable performative, intra-acting agencies *within* the individual organism. By this definition, internal selection *is* a kind of natural selection, but its importance only arises by recognizing the hierarchical agencies of the complex developing, functioning phenotype that are ignored by the traditional population genetic framework of evolutionary mechanisms.

Internal selection acts upon the intra-active, discursive developmental, physiological, and biomechanical mechanisms of the organismal body. Because intra-active biological agencies must coevolve to maintain their discursive functions, mutation and genetic drift (i.e., random effects in genetic evolution) in any details of these networked systems will require internally adaptive *compensatory* changes to maintain their functions. Numerous examples of this intra-organismal "Red Queen" phenomenon are presented and discussed in "Why Sexual Differentiation Mechanisms Are Generatively Queering" in chapter 7.

Thus, internal selection on the stability and coherence of molecular discourses of bodily becoming are the source of Waddington's developmental "canalization." Internal selection advances the integration of developmental, physiological, and biomechanical parts and functions within the multicellular body. Internal selection is critical to whether any individual phenotype establishes the opportunity to be exposed to natural selection on its performance within an external environment. Likewise, internal selection is the evolutionary force that leads to the control of aging, cancer, and its metastasis.

Although internal selection is obviously required for the evolution of complex phenotypes and is implicitly invoked in traditional genetic descriptions of lethal mutations, many evolutionary geneticists are loath to acknowledge the process formally. For example, James Cheverud has argued that the effects of internal selection can be fully captured by the variance-covariance matrix in quantitative genetic analysis, and therefore microevolutionary population genetics does not need to consider developmental biology to have a complete understanding of evolutionary process. However, this view fails to capture, or even make predictions about, the rich details of molecular-developmental discourse, the modularity of complex phenotypes, the variational potential of developmental systems, or the creativity of internal selection to produce innovations and evolutionary novelties.[3]

Schwenk and Wagner argue that internal selection produces evolutionarily stable configurations whose properties are more strongly shaped by their integrated internal functions—what we know here as intra-actions within the body—than by the environmentally driven, external natural selection. Think of the complex jaws and feeding mechanisms of fishes and reptiles, or the bones, muscles, and feathers that constitute the wings of a flying bird. The parts of these complex anatomies must function appropriately in relation to each other before they can provide a coherent ecological function for the organism in its environment. Thus, we can see that internal selection acts similarly on both the developmental *and* the physiological/biomechanical integration of the phenotype.

A broad conceptual model of internal selection would require the recognition of the many hierarchically nested genetic, cellular, and multicellular agencies that are involved in organismal development, physiology, and biomechanical function. Functional integration among these agencies will coevolve through internal selection for robust, discursive or mechanical intra-actions among them. The internal selection forces that shape the phenotype operate between gene-level selection, and external selection on the organismal phenotype in its environment. Such models should be an outstanding goal in developmental, functional, and evolutionary biology. Just as Michael Wade has expanded, and explicitly modeled, the population genetic concept of heritability to higher-level groups of families, social groups, populations, and species, we should conceptually identify and investigate those phenomena that constitute mechanisms of "internal selection" within the *higher-level* units of selection—such as monogamous pair bonds, colonial organisms, eusocial colonies, and so on. In other words, those properties of families, colonies, social groups, subpopulations, and species that enhance group survival and fecundity among their constituents independent of environmental variation can evolve by internal selection in these units of selection.

Incorporating concepts of internal selection into hierarchical models of multilevel selection should be an important direction for theoretical development in evolutionary biology.

Notes

PROLOGUE

1. For more on manakins, see Prum 2017.
2. Prum 1990, 1992.
3. Lorenz 1971.
4. I have never realized until now, but the word *adaptation* never appears in the text of my 338-page doctoral dissertation, and the phrase *natural selection* is mentioned only once (by comparison to the rate of sexual selection). Yet the research was ultimately featured in several textbooks of ornithology and evolutionary biology.
5. Prum 1999.
6. Xu et al. 2001; Prum and Brush 2002; Harris et al. 2002, 2005.
7. My research program on structural colors has ultimately involved detailed collaborations with a couple of dozen undergraduate and graduate students, physicists, engineers, applied physicists, materials scientists, and mathematicians over nearly thirty years. For example, see Prum et al. 1998; Prum 2006; Dufresne et al. 2009; Noh et al. 2010a–c; Saranathan et al. 2012, 2021.
8. Jones 2002.
9. Dufresne et al. 2009, Prum et al. 2009, Sicher et al. 2021.
10. Barad 2007.
11. Brennan et al. 2007, Brennan, Clark, and Prum 2010; Brennan and Prum 2012; Snow et al. 2019. Patricia Brennan was a postdoc in my lab at Yale University from 2005 to 2010, and is now an associate professor of biology at Mount Holyoke College.
12. Brownmiller 1975, Gowaty 2010.
13. Brennan and Prum 2012.
14. Prum 2017, 156–57.
15. A more complete account of sexual conflict and the evolution of sexual autonomy in waterfowl appears in my book *The Evolution of Beauty* (Prum 2017, 149–81). In subsequent chapters of the book, I investigated a variety of other evolutionary responses to sexual conflict in birds, and explored the role of female mate choice and

sexual autonomy in the evolution of human beings, human sexuality, and human sexual diversity. See also Brennan and Prum 2012 and Prum 2015.

16. In subsequent work, Sam Snow and colleagues (2019) have established the plausibility of this proposed mechanism for the evolution of sexual autonomy with population genetic modeling.

17. Prum 2013, 2020, 2022.

18. Haraway 1988, 592.

19. Crenshaw 1990.

20. Cho, Crenshaw, and McCall 2013, 795. They further state that "our main objective is to illustrate the potential for achieving greater theoretical, methodological, substantive, and political literacy without demanding greater unity across the growing diversity of fields that constitute the study of intersectionality."

21. Haraway 1988, 580.

22. Many thanks to Elizabeth Maddock Dillon, professor of English at Northeastern University, who interjected during one of our academic conversations, "Man, you're an *ornithologist for intersectionality*!" Ever since, I have been eager to explore what that can mean and be.

CHAPTER ONE

1. Green, Benner, and Pear 2018.
2. Green, Benner, and Pear 2018, Simmons-Duffin 2019.
3. See https://www.washingtonpost.com/politics/2022/03/10/texas-trans-kids-abortion-lgbtq-gender-ideology/.
4. The American Civil Liberties Union keeps an up-to-date database of legislation effecting LGBTQ+ rights in the United States. See https://www.aclu.org/legislation-affecting-lgbtq-rights-across-country.
5. Butler 1988.
6. Van Anders 2009, 2015; Fausto-Sterling 2019.
7. For example, intellectual histories of new material feminism are presented by Hird 2004, Willey 2016, Subramaniam 2016, and Roy 2018. In particular, Willey (2016, 17–22) focuses on debates in the history and future directions of "new material feminism" and whether certain critiques of the second-wave feminist "flight from nature" should be considered gestures, interventions, novel contributions, or not appropriately acknowledging earlier feminist antecedents. See also Ahmed 2008 and Davis 2009.
8. Barad 2003, 801.
9. Wilson 2004b, 18; 1998, 14–15; 2004a, 70.
10. Cheng 2020.
11. Longino 1987, 56.
12. Wilson 2004a, 86.
13. Following on its popularity in academic analysis, the word *performative* has become quite common in popular writing, where it has come to mean something like "an overly stylized action done with conscious awareness that one is being observed." This common usage is unrelated to its academic meaning or its usage in this book. See "What Is Performativity?" in chapter 3.

14. Richardson 2013, Fausto-Sterling 2018.

15. As reviewed in Fausto-Sterling 2000 and Richardson 2013, the chromosomal and genomic theories of "sex determination" are only the most recent of a century of binary scientific conceptions of sexual development. The "Decade of the Gonad" (Fausto-Sterling 2000) in the early twentieth century was followed by research focusing on hormones, then chromosomes, then genes, and most recently genomes, genomic modifications (epigenetics), and transcriptomes.

16. Butler 1993, 26.

17. Willey 2016, 558.

18. Sedgwick 1993, 12.

19. Kay 2000, 22. Quote is from Kay's description of Mary Hesse's concept of metaphor in science.

20. Barad 2007, Chang 2012.

21. Bono 1990, 60, cited in Kay 2000, 29. In this passage, I am indebted to Lily Kay's elegant introduction to metaphor in science from *Who Wrote the Book of Life?*

22. Arbib and Hesse 1986, 156 (emphasis mine); cited in Kay 2000, 22.

23. Butler 1993, xv.

24. Richardson 2013, 225.

25. Amundson 2005.

26. Prum 2017, 320.

27. The continued marginal position of development biology in evolution biology is evident from perusing the content of disciplinary journals, and the organization of funding at the US National Science Foundation. The intellectual alienation of developmental biology from the adaptationist New Synthesis is documented and reviewed by Ron Amundson (2005). The intellectual history of the struggle for the incorporation of phylogenetics into evolutionary biology is analyzed by David Hull (1988). Even now, however, the vast majority of evolutionary analysis using phylogenies conceives of phylogenetic history as a statistical problem to be controlled for rather than as the proper focus of scientific research itself. The result is a continued intellectual effort to remove the problem of history from the discipline rather than understand history as the central focus of the discipline of evolutionary biology.

28. I could have simply adopted and applied insights from the concept of performativity to developmental and evolutionary biology laundered of all references to its origins in queer theory (on the down-low, as it were). But this would have been shameless cultural appropriation and simply less effective. Barad (2007, 410–11) points out that Pickering (1993, 1995) adopted a concept of performativity in the philosophy of science without acknowledging its inherently queer political history, or its role in queer feminist studies.

CHAPTER TWO

1. Some of the classics among the many works that have exposed the problematic cultural histories of scientific terms in biology include Beldecos et al. 1988; Martin 1991; Gowaty 1992, 1997a–c.

2. Clayton and Collins 2014. Individual human cells cannot have a sex because the

sex of a vertebrate animal is only realizable through the development of an organismal body. The presence of specific chromosomes or specific genes is not the equivalent to sex because none of these chromosomes or genes is actually a direct cause of sexual development. Chromosomes and the genes they carry can only affect sexual development through a cascade of other developmental players with their own independent variations.

3. The debate over historical ontology in evolutionary biology began with proposals by Ghiselin (1974) and Hull (1978) that species should be understood as ontological individuals. The concept was expanded by Wiley (1980) to include historical entities that include members that are not interacting, but Ghiselin specifically rejected contiguity and ongoing interaction as a necessary requirement of individuality. De Quieroz and Donoghue (1988) propose that the idea of species and higher taxa as individuals was proposed earlier by Hennig (1966). A recent application of the concept to homology is presented by Wagner (2014, 232–40). For a recent review of all of the breadth of concepts of individuality in biology, see Pradeau (2016). In *Everything Flows*, Nicholson and Dupré (2018) edit a volume of essays proposing an alternative process ontology for biology, in which processes are considered the fundamental ontological categories. An alternative ontology of sex could be proposed accordingly, but this mirror-image ontology would produce the same result—sex as an historical entity (process) that cannot be defined as an essence of individual organisms.

4. Hardinson 2012.

5. Pavlicev et al. (2016) connect this historical ontology to the science of emergence by proposing that "emergence is the origin of novel [ontological] individuals."

6. Wong 1976, Koonin and Novozhilov 2009.

7. I should note, however, that my posthuman concept of historical ontology is unrelated to Michel Foucault's (1978) and Ian Hacking's (2002) use of the same term to apply exclusively to the human history of ideas, concepts, and modes of expression. Their concept excludes all natural kinds that are not products of human cultural activity.

8. Togashi et al. 2012.

9. The reproductive bottleneck at the origin of human lives has not been eliminated yet by any reproductive technology, but it seems very likely that this constraint will be overcome during the next century.

10. Barad 2007.

11. Ah-King and Hayward 2013, 1; Ah-King and Nyland 2010; Ah-King and Gowaty 2015.

12. Richardson 2013, 21, 197–99. In contrast to "relational property" (Richardson 2013), the concept of sex as a "relational possibility" (here) recognizes the individual agency that is appropriately associated with sexual reproduction.

13. More recently, Richardson (2021) proposes that we should accept that what is meant by sex in biological and biomedical research varies so widely in different contexts that we should embrace these multiple distinct, pluralistic meanings of sex. Richardson's "sexual contextualism" is a response to the requirement by some biomedical funding agencies to require the use of sex as a variable in biomedical research. Proposed as an alternative to an essentialist biological sexual binary, Richardson suggests that a contextual account of sex will be "ameliorative" by requiring

us to avoid sex essentialism, to be explicit about what sex means in our usage, and to become aware of what we want it to mean. However, I fear that sexual contextualism could contribute to the persistence of essentialist-genetic binary concepts of sex within biological research. But, it could be that broad adoption of Richardson's sexual contextualism would force scientific researchers to explicitly state and defend their genetic essentialism, rather than tacitly assume it without justification, as currently occurs in most scientific literature.

14. Arthur Koestler (1978) proposed a similar ontological concept, which he called the "holon." Thanks to Hasok Chang for pointing out Koestler's book to me. My use of the ontological category of historical individuality is unrelated to the philosophical and biological debate over the status, appropriate boundaries, or existence of "individual" organisms. However, I should make clear that I do not think that the ecological or social interconnectedness of many organisms poses any non-trivial challenges to individuality as a natural kind.

15. Schilling, Watkins, and Watkins 2002.

16. Ingalhalikar et al. 2014, Cahill 2014.

17. Jordan-Young and Karkazis 2019.

18. Godfrey et al. 2020. EIF1A stands for "eukaryotic translation initiation factor 1A."

19. Grosz 1990, 339 (emphasis in original).

20. Johnson 1980, x.

21. Fausto-Sterling 2000, 56–63.

22. Markowitz 2001.

23. Krafft-Ebing 1903. Many American editions of the English translation were printed with confusing dates and different paginations. But this quote can be found in the second paragraph of chapter 3, "Anthropological Facts."

24. Markowitz 2001, 392.

25. Ellis 1905, 165.

26. McWhorter 2009. The association between fear of eugenic racial corruption and masturbation has not disappeared. Several contemporary neo-fascist organizations, such as the Proud Boys in the United States, combine virulent anti-immigrant rhetoric with moralistic bans on masturbation and pornography. See https://gen.medium.com/why-are-the-proud-boys-so-obsessed-with-masturbation-c9932364ebe2.

27. McWhorter 2009, 194.

28. Dawkins 1999. In conceiving the concept of the extended phenotype, Dawkins hoped to unify our understanding of the genetic and cultural influences on the organism by reducing them all to forms of adaptation by natural selection acting exclusively at the level of genes. In contrast, although I recognize the distinct and emergent influences of culture on the phenotype, I still find Dawkins's concept of the extended phenotype to be a productive platform for doing so.

29. Google Scholar search on February 19, 2022. For example, see Chen et al. 2013; Couse et al. 1999; Holleley et al. 2015, 2016; Shao et al. 2014; Tamschick et al. 2016. I know of no previous scientific criticisms of the term *sex reversal*, so these citations are uncritical uses of the concept.

30. Butler 1993, xii.

31. Schlosser 2019.
32. Barandiaran, Di Paolo, and Rohde 2009.
33. Latour 2014, 12.
34. In the book, I will sometimes refer to "entities with agency" as *agents*. But when I refer simultaneously to assemblages of multiple agents within agents within agents, etc.—such as genes, extracellular signaling proteins, cells, multicellular placodes, and bioelectrical fields—I use the word *agencies* in recognition of both the multiple individual agents and the many simultaneous, different kinds and levels of agency.
35. Sedgwick 1985.
36. Ferguson 2004, 2018.
37. Sandilands 2016, Mortimer-Sandilands and Erickson 2010.
38. Jagose 1996, 3.
39. Halperin 1997, 62 (emphasis in original).
40. Halperin 1997, 62.

CHAPTER THREE

1. Mead 1935.
2. Goffman 1956. In looking for information about whether Erving Goffman was influenced by Margaret Mead's work on social performance, I consulted Prof. Kathryn Lofton, who is deeply engaged in a project on his work. She described Goffman as famously thin of citations, only footnoting direct quotations and not concepts. Goffman cited Bateson and Mead's *Balinese Character* (1942) in Goffman's pictorial book *Gender Advertisements* (1976), but there are no other citations of Margaret Mead. Lofton concludes that Goffman was likely influenced by Mead's work, but did not acknowledge it.
3. Austin 1962. As we will see in intra-actions in developmental genetics, Austin's account of performative speech is incomplete without including someone to receive and comprehend it, and some mechanism for change in the material world.
4. Sedgwick 2003, 5.
5. Butler 1988, 1990, 1993.
6. Butler 1988, 519.
7. Butler 1988, 522.
8. Barad 1998, 2003, 2007, 2011. Another concept of performativity has also arisen in science studies, but I will not focus on it extensively here. Andrew Pickering (1993, 1995) developed a sociological analysis of scientific research as a system of performative material devices that are deployed by social networks of scientific researchers to investigate the world. Although Pickering's work evidently influenced Barad's philosophy, my goal here is to focus primarily on the "posthuman" direction of Barad's work to explore the performativity of biology—of organisms, cells, and biomolecules themselves—not the cultural performativity of biologists' investigations of them.

9. This brief summary is my own selection of themes and topics from this diverse literature, and it is not intended as a complete or thorough historical review.

10. Barad 2003, 822–23.

11. Butler 1993, xvii.

12. Butler 1990; Barad 1998, 2007.

13. Foucault 1978.

14. A "transcriptome" is a description of all the genes that are being transcribed at a given moment within a sample of tissue or a cell. They are produced by extracting and sequencing all the mRNAs from a tissue/cell at a given moment.

15. Butler 1993.

16. Butler 2015, 45.

17. The recognition of individual agency does not imply complete individual freedom. Rather performativity implies the individual assertion of gender identity in the face of both facilitating and constraining influences/forces in the social environment. In contrast to performativity, the concept of gender "construction" implies that individuals are free to make up whichever gender they choose or can imagine. This view does not recognize the constitutive material and psychological differences among individuals that we all bring to gender realization. Likewise, the terms gender *expression* and gender *presentation* can be understood to imply that, like Austinian representative statements, gender is an external "expression" or "representation" of an inner essence or a priori truth of individual sex. Both of these views have problematic implications, so I will try to avoid using these terms.

18. Butler 2015.

19. Gutting and Oksala 2019.

20. Posthumanism is often thought of as beyond human, involving artificial intelligence, transhuman cyborgs, or alien species. My scientific posthumanism is rooted in the idea that other contemporary organisms can be agents in cultural and biological processes that were long considered the exclusive realm of humanism. As Barad's and this work show, scientific posthumanism can provide a new perspective on human beings by providing a rich diversity of new biotic phenomena and extending the possibilities of traditional humanism.

21. Prosser 1998.

22. Prosser 1998, 31.

23. Salamon 2004.

24. Prosser 1998.

25. Serano 2013.

26. Fundamental works in scientifically engaged material feminism include those of Donna Haraway (1988, 1989), Elizabeth Grosz (1994, 2011), Anne Fausto-Sterling (2000, 2019), Evelyn Fox Keller (1984, 2000), and Karen Barad (1998, 2007). This research trajectory is also represented by recent edited volumes by Stacy Alaimo and Susan Hekman (2008) and Cyd Cipolla and colleagues (2017), and the recent works by Sarah Richardson (2013), Deboleena Roy (2007, 2018), and Angela Willey (2016, 2018).

27. Butler 1993, xviii. Here, Butler gets very close to articulating performativity

explicitly as a process philosophy. However, reviews of this area have yet to recognize or pursue this connection (e.g., Nicholson and Dupré 2018).

28. Barad 1998, 123, note 18.
29. Thomas 1974.
30. Barad 1998.
31. Barad 2007, 376.

CHAPTER FOUR

1. Taylor and Lewontin 2017, section 5. See also Amundsen 2005 and Walsh 2015.
2. Keller 1997.
3. Butler 1993, xii.
4. Identical twins begin life as a single cell zygote that divides into two distinct but genetically identical zygotes within the first week typically before implantation.
5. Some "somatic" mutations in the nuclear DNA can occur among the dividing cells of the body. These are usually minor, but some of them can lead to cancer. See later discussions of cancer.
6. Sadava et al. 2016, Sender, Fuchs, and Milo 2016.
7. There is some question as to whether the spacer sequences actually facilitate folding. The fugu fish gets by with almost no spacer regions, while some butterflies have ten times more spacer regions than humans do (Titia de Lange, pers. comm.).
8. See also Griffiths and Stotz 2013, Walsh 2015, Wagner 2015.
9. Green 2018.
10. Haraway 1988.
11. Genetic transcription and translation do convert inherited genomic information from the DNA sequence into molecules—proteins—that are accurate representations of that genomic information. But these processes are about the creation of elements or a vocabulary of communication, not communication or discourse itself. Beyond this genetically referential or representative process, the functions of gene products in the world are extensively performative.
12. In her brilliant book *Who Wrote the Book of Life: A History of the Genetic Code*, Lily Kay (2000) reviews the complex history of linguistic metaphors in genetics, the genome, and genomic function. Kay argues that the initially helpful metaphors of genetic "information," "language," "code," "message," and "text" morphed from useful analogies to genuine scientific "ontologies" of the gene, with harmful effects on our understanding of genetics. Kay proposed that the question of *Who Wrote the Book of Life?* can be addressed from three perspectives: the objectivist, constructivist, and deconstructivist. Here, I am proposing a fourth, "posthumanist," perspective on the linguistic analogy of the genome and genome action. The posthumanist view agrees with the objectivist view that the discursive structures being described and analyzed are the products and properties of material organic evolution, but unlike them, the posthumanist view entertains a richer set of biological and genetic agencies. Like the constructivists, my posthumanist view invites scientists to critically examine their processes of representation and their construction of scientific apparatuses and

experimental interventions. Like the deconstructivists, I am specifically interested in the power of discourse to affect and enact biological and cultural reality. But unlike the constructivists or deconstructivists, I see these molecular discourses as existing entirely outside of the human. We humans can study and learn about them through our scientific process, but these biological discourses exist outside of our knowledge of them. Studying nonhuman discourses requires adopting the posthumanist frame.

13. Butler 1993, xii.

14. Starting with the zygote, the cells of the roundworm *C. elegans* are highly polarized, with some molecules in higher concentration at one end of the cell than the other. Each cell division within the body of a developing *C. elegans* creates one "daughter" cell from each half of the parent cell—either left and right, or top and bottom. Cell division partitions different concentrations of these polarized molecules in the cytoplasms of the daughter cells, which strictly determines their gene expression states and identities. This process gives rise to a body of exactly 959 somatic cells that is ruled by this strict lineage hierarchy. Other aspects of *C. elegans* development—notably the morphology of the vulva—are also influenced by paracrine signaling among cells.

15. For example, see Harris et al. 2005.

16. Turing 1952. For a reader-friendly introduction to the application of mathematical models to pattern formation, see Meinhardt 1998.

17. Exceptions to the arbitrariness of words include rare onomatopoetic words like *snap* and *murmur*, which encode their meaning in their sounds, and even rarer words like *sesquipedalian*, which encodes its meaning in its ridiculous etymology. Somewhat more common are a class of broader and quite vague phoneme-meaning associations, such as the high frequency of the use of "sharp-edged sounds" in words for *knife* in many languages independently.

18. For a review of evolutionary cooption of paracrine signaling pathways, see True and Carrol 2002.

19. Wagner 2015, 116.

20. Squier 2017.

21. Gilbert 2000.

22. Waddington's term *epigenetics* has a long, complex history in biology that actually predates the concept of genetics and evolution. In the nineteenth century, *epigenesis* referred to the theory that development did not involve preformation of the organismal body within the gametes. To Waddington, and for six decades after in developmental biology, *epigenetics* referred to the general molecular mechanisms of gene expression regulation within the cell—essentially a core subject of this book. In recent decades, however, Waddington's meaning of the word has been eclipsed. *Epigenetics* has come to refer more narrowly to the heritable modification (often methylation) of the surface of the genome, which can be passed down through gametes to offspring and can affect their phenotypic development without changing the nucleic acid sequence of the DNA. This fascinating and poorly understood process poses intriguing implications for parental influence on the performance of the offspring phenotype. Most recently, developmental biologist Mark Ptashne (2013) argues that

epigenetics refers to when "a transient signal or event triggers a response that is then perpetuated in the absence of the original signal," which is much more specific than Waddington, but entirely excludes heritable methylation. As a result of the historical instability in its meaning, I will generally avoid the term *epigenetics* here.

23. Zeller, Lopez-Rios, and Zuniga 2009, Delgado and Torres 2017, DiFrisco, Love, and Wagner 2020.

24. Haraway 1976.

25. For example, see Musser et al. 2015.

26. For an academic introduction to soft-condensed-matter physics, see Jones 2002. For a review of the role of soft condensed matter in the molecular dynamics within individual cells, see Brangwynne 2011.

27. Prum et al. 2009, Dufresne et al. 2009, Saranathan et al. 2012. The influence of the mechanical forces created by aggregations of cells and tissues has been long appreciated in developmental biology. In 1917, British biologist and mathematician D'Arcy Thompson wrote in *On Growth and Form* that, "Cell and tissue, shell and bone, leaf and flower, are so many portions of matter, and it is in obedience to the laws of physics that their particles have been moved, moulded and conformed.... Their problems of form are in the first instance mathematical problems, their problems of growth are essentially physical problems, and the morphologist is, *ipso facto*, a student of physical science" (Thompson 1917, 7–8). Thompson's material-structural approach to organismal development was sidelined as development biology focused more and more on genetic investigation over the twentieth century.

28. Brunet et al. 2013.

29. Landrein et al. 2015, Trinh et al. 2021.

30. The distinct morphology of the zygodactyl grasping foot of parrots, which evolved in an ancient ancestor of parrots and perching birds over fifty-six million years ago, is realized by recruitment of physical forces of the developing leg muscles during the development of the bones of the fourth toe. Botelho and colleagues (2014) found that two muscles typically insert on the fourth toe of the anisodactyl bird foot, and they develop early in embryogenesis. However, in parrots (Psittaciformes), the *musculus abductor digiti IV* (ABDIV) is strongly developed, but the *musculus extensor brevis digiti IV* (EBDIV) fails to develop. The asymmetrical contraction of ABDIV during parrot development reverses the fourth toe. Apparently, the anisodactyl perching birds (Passeriformes) evolved from a zygodactyl common ancestor shared with parrots. They have also evolutionarily lost both ABDIV *and* EBDIV, so the evolutionary return to anisodactyly did not occur through the restoration of symmetrical muscle action on digit IV but rather through the elimination of a role for muscular forces on the shape of the fourth toe.

31. Likewise, embryonic peristalsis of the intestinal muscles plays a critical role in the elongation and morphogenesis of the vertebrate digestive tract (Khalipina et al. 2020).

32. Harris 2021.

33. Reviewed in Pai et al. 2012.

34. Daane et al. 2021. For an accessible discussion of the role of electrical fields

in development and regeneration, see Matthew Hutson's (2021) portrait of developmental biologist Michael Levin in the *New Yorker*.

35. Kirby 2012, 200.

36. Walsh 2015, 123.

37. "Single-gene diseases" are not diseases caused by genetic variation in a single gene, but diseases that can only be caused by genetic variations of a single gene. However, it is important to realize that many different mutations can eliminate or cause disfunction at any of the genes for a structural protein. Six different types of human sickle-cell anemia mutations have been identified. Over 1,700 different mutations have been identified as causes of cystic fibrosis. Some of them are non-sense or mis-sense mutations in the gene for chloride transmembrane transport channel protein itself, but many others are in other loci that function in the regulation of the production of the protein. See sickle-cell anemia: https://www.nhlbi.nih.gov/health-topics/sickle-cell-disease, and cystic fibrosis: https://www.cff.org/What-is-CF/Genetics/Types-of-CFTR-Mutations/.

38. Barad 2003, 815.

39. Barad (2007, 176) writes that "agential intra-actions are causal enactments." A material-discursive practice enacts an "agential cut" that creates agential separability of the cause (object) and effect (measurement) within the phenomenon. Thus, in gene regulation and development, the signaling molecule or transcription factor (object) and the receptor or binding site (effect) become separable through their mutual enactment.

40. Thanks to my colleague Günter P. Wagner for first making this observation to me.

41. The reification of experimental conditions as reflecting the real world poses more than a simple intellectual gap. It has contributed to genuine intellectual lapses. For example, after the sequencing of the first human genome, the emerging field of genomic medicine was heralded as likely to rapidly contribute revolutionary new cures to most complex human diseases, from heart disease and cancer to mental illness and addiction. Researchers soon found that the most significant genes "explained" less than 10 percent of the *heritable* risks of these complex diseases. Why? Because most complex aspects of the phenotype—whether typical traits or diseases—are the consequence of gene-by-gene interactions among many variable loci. Genetic variation at such sites is overwhelming, and its interactions are complex (e.g., Keinan and Clark 2012). Thus, genomic medicine has identified some new opportunities for traditional pharmaceutical treatments, but it has failed to deliver on its promises. This broad-scale, entirely predictable—indeed, predicted—failure of genomic medicine to contribute to cures for complex diseases is a direct result of the empirical and conceptual gap between controlled experimental genetics and the highly variable, pleiotropic, performative reality of the etiology of complex diseases. See appendix 3.

42. Butler 2015, 45.

43. Barad 2007, 235.

44. Barad 2007, 214.

45. See Michael Wade's (2016) *Adaptation in Metapopulations: How Interaction*

Changes Evolution. Given that many of the interactions that impact evolution are likely to be mediated by discursive communication, the book provides a theoretical underpinning for various performative mechanisms in evolution. See appendixes 6–7.

46. The view of the ontology of natural kinds in biology is unrelated to the (to me) more trivial question of how to define the boundaries of an individual organism (e.g., Quammen 2018). Many biological entities have composite histories that do not undermine their historical individuality. For example, all eukaryotes, including ourselves, are composites of a nucleated cell and mitochondria of bacterial origin. This does not interfere with our capacity to investigate birds, human liver cells, or Abraham Lincoln. Individuality is not about being completely independent of, or isolatable from, other individuals. Few individuals are genuinely ecologically or socially independent of all others. Rather, ontological individuality is about the historical nature of individual existence, not its boundaries. See also chapter 2.

47. Dawkins 2006.

48. Historian of science Ramus Winther (2014) argues that Dawkins's gene-level selectionism suffers from what William James called "vicious abstractionism" via the "philosopher's fallacy." Dawkins's locally productive abstraction of selfish genes as replicators employs a set of questionable assumptions, yet it is universalized and reified to explain all of morphology and behavior. The result is a narrow and impoverished theory that fails to capture the complexity in morphology and behavior that evolutionary biology has the responsibility to explain. See appendix 6.

49. The simplifying assumption that justifies Dawkins's reductionist view of gene-level selection draws upon the views of early twentieth-century population geneticist Ronald Fisher, who proposed that one could assume that the fitness of an allele—that is, a heritable gene variant—within a population would behave like the average of its fitness over all possible combinations with other alleles in the population. In contrast, Sewall Wright emphasized that the fitness of any allele would be determined by which other alleles it is combined with, what population geneticists call epistatic interactions. Although Fisher recognized the importance of epistatic interactions, he created a set of mathematical tools to describe natural selection that were consciously modeled on physical laws in thermodynamics. Both Fisher's and Wright's views have remained core concepts in evolutionary biology, and both have contributed to major advances. Yet Fisher's view, based on a temporary mathematical convenience, ultimately contributed directly to the idea of natural selection on the level of alleles as a strong and deterministic force. For further discussions of Wright and Fisher, see Frank 2012 and Wade 2016. See also "Why Gene-Level Selection is Insufficient," appendix 6.

50. DiFrisco et al. 2020.

51. At the risk of acronym overload, we could extend Wagner, DiFrisco, and Love's ChIMs to become character identity mechanistic performativities—or ChIMPs.

52. Gutting and Oksala 2019.

53. The origin of the complex, integrated, organismal body through natural selection on survival and fecundity supports an hypothesis for the mechanistic origin of

Foucauldian self-regulation in human societies: group selection on the survival and fecundity of social groups.

54. It is worth noting that computers and algorithms do not create well-regulated and stable results by the distributed self-regulation of their hierarchical components. Neither computers nor algorithms get cancer. Because plants have entirely redundant organs (i.e., no liver or heart that the entire organism relies upon) and because their cells migrate very little, plants do not get cancer either. But computer "organs," or subroutines, are not redundant. Although cancer is due to gene mutations, many organisms have evolved regulatory machinery to identify those mutations and fix them, or eliminate cells that have them. Lastly, the fact that cancer cells form a distinct genetic lineage—a historical individuality *within* the individual body—only highlights that there are many, many non-cancerous lineages of cells with somatic (non-germline) mutations throughout the body that do not expand to threaten the life of the individual. These cells remain Foucauldian "docile citizens."

55. Rothman, Stearns, and Shulman 2021.

56. This conception of perception as an *activity* is pursued creatively by philosopher Alva Noë (2007). More broadly, a performative account of the biology of behavior, psychology, and cognition would establish new intellectual connections between organismal biology and the phenomenological tradition of Merleau-Ponty and others.

57. Emlen and Wrege 2004.

58. Prum 2017.

59. The only exceptions to the coevolution of display traits and mating preferences are during their initial origin, or display evolution by sensory bias. See Prum 2012, Rosenthal 2017, and Ryan 2018.

60. Prum 2017.

61. Prum 2013, 2020, 2022.

62. Caro 1986. Alternative explanations imply that stotting is a warning display for other conspecifics.

63. Hutchinson 1965.

CHAPTER FIVE

1. A Google Scholar search on July 12, 2020, for "'sex determination' in humans" recovered 77,200 citations.

2. Richardson 2013; Fausto-Sterling 2000, 2018, 2019; Fausto-Sterling et al. 2012a, 2012b; Hird 2004, 2008; Jordan-Young 2010; Jordan-Young and Karkazis 2019; Karkazis 2008; Wilson 1998, 2004a, 2004b; Gill-Peterson 2018; Willey 2016, 2018.

3. Sadava et al. 2016, 258–59.

4. Mammalian red blood cells are an exception to this rule; they are eukaryotic cells that do not have a nucleus when mature.

5. For example, see Richardson 2013.

6. Smith et al. 2009. DMRT1 stands for "doublesex and mab-3 related transcription factor 1."

7. Richardson 2013.

8. See also Martin 1991.

9. Nüsslein-Volhard and Wieschaus 1980.

10. For example, the gene "fringe" encodes a glycosyltransferase enzyme active in the "notch" signaling pathway; it was named for the distorted wing morphologies produced by its mutation in fruit flies. Later, vertebrate homologs of this gene were named Radical Fringe, Manic Fringe, and Lunatic Fringe. (Get it? Funny, huh?) However, this flippant nomenclature has led to culturally awkward communication problems. For example, the research paper title "Mutation of the LUNATIC FRINGE Gene in Humans Causes Spondylocostal Dysostosis" combines insider lab jokes with a serious human disease (Sparrow et al. 2006).

11. For information about the Human Genome Nomenclature Committee, see https://www.genenames.org/.

12. "Sexual difference-making region on the Y" would be a more accurate basis for the SRY acronym.

13. SOX9 stands for "SRY-related high-mobility group box 9." SRY was the first gene identified in the SOX-gene family, but they decided not to call these other related genes SRY-2, etc. The word "box" found in many gene names comes from the graphical "box" shape drawn around the group of conserved nucleic acids shared among genes aligned together from multiple species. These conserved "boxes" correspond to functionally important, highly conserved areas of the protein that are often involved in protein or DNA binding.

14. NR5A1 stands for "nuclear receptor subfamily 5 group A member 1."

15. FGF9 stands for "fibroblast growth factor 9." PGD2 stands for "prostaglandin D2."

16. https://www.imdb.com/title/tt0031066/. In full disclosure, like the famous non-quote from Humphrey Bogart's Rick Blaine in *Casablanca*, "Play it again, Sam," Mickey Rooney doesn't exactly say, "Hey, let's put on a show!" in *Babes in Arms*. https://www.youtube.com/watch?v=SRZ5400UKSc.

17. Baetens et al. 2019. WT1 stands for "Wilms' tumor protein 1," GATA4 stands for "GATA-binding protein 4," and FOG2 stands for "friend of GATA2."

18. Baetens et al. 2019.

19. Pronounced "Wint Four," WNT4 stands for "Wingless-related integration site protein 4."

20. LEF1 stands for "lymphoid enhancer binding factor 1."

21. FOXL2 stands for "Forkhead box L2."

22. RSPO1 stands for "R-spondin 1."

23. Baetens et al. 2019. DMRT1 stands for "doublesex and mab-3 related transcription factor 1."

24. Wagner 2015.

25. Because of their anatomical association with the embryonic kidney, or mesonephros, the Wolffian and Müllerian ducts can also be referred to as the *mesonephric* and *paramesonephric* ducts, respectively.

26. Shapiro, Huang, and Wu 2002.

27. Seifert et al. 2009.

28. Cohn 2011. HOXD13 stands for "HomeoBox domain D13." The "homeobox" refers to the conserved sequences of the DNA-binding domain shared by all HOX proteins.

29. Seifert et al. 2009.

30. DKK stands for "Dickkopf," which is German for bullhead.

31. For example, see Robboy et al. 2017.

32. Why do the embryonic testes produce a small quantity of testosterone, and amplify its effect locally in genital tissues? It could be that this is an adaptation to reduce the cross-placental impact of hormones produced by the genetically distinct XY embryo on the physiology and development of the gestating XX female. The use of local, tissue-specific conversion of testosterone to the more powerful gene regulator DHT may have evolved to provide males with a "private channel" for genital development that does not disrupt maternal hormonal homeostasis. However, in chapter 6, we will see that this reliance on the 5α-reductase enzyme to convert testosterone in genital development creates the opportunity for further performative variation in the human sexual phenotype.

33. Seifert et al. 2009, 2010; Cohn 2011; Gredler 2015.

34. The mature testes are surrounded by two layers of peritoneal membrane, called the tunica vaginalis and the tunica albuginea, with the peritoneal space in between.

35. Fausto-Sterling 2012, 2019; Fausto-Sterling et al. 2012a, 2012b.

36. Jordan-Young and Karkazis 2019.

37. Pankhurst 2017.

CHAPTER SIX

1. Foucault 1980, vii.

2. Laqueur 1992.

3. Laqueur 1992; Lillie 1939, 3, cited in Fausto-Sterling 2000, 178; Waddington 1940, 91. Following Laqueur (1992, 16), I too am interested in the "space between [embodied sex] and its cultural representation," but I will be approaching this science-culture interface through the molecular genetic development of the sexual body.

4. Kessler 1998, Fausto-Sterling 2000, Hird 2004, Karkazis 2008, Davis 2015, Gill-Peterson 2018, Velocci 2021.

5. Fausto-Sterling 2000, Hird 2004, Karkazis 2008, Davis 2015, Gill-Peterson 2018.

6. The term *intersex* has been adopted as a culturally powerful personal identity by many individuals with differences in sexual development, and as a political identity organizing the recognition of their concerns by the biomedical community. In this book, however, I will avoid the term *intersex* further because I think that it unnecessarily implies that differences in reproductive anatomy and physiology consist of a mixture of, an intermediacy between, or the copresence of elements of two distinct things—maleness and femaleness. To me, the term is not an accurate way to charac-

terize most differences in sexual development. Furthermore, I do not think that there are two distinct individual sexes to be mixed, copresent, or intermediate between. Rather, all bodies are the individualized enactments of a general performative process.

7. Karkazis 2008, Davis 2015.
8. Pinker 1994.
9. Gastaud et al. 2006.
10. Fausto-Sterling 2000, 57–58. Fichtner et al. 1995.
11. For analysis of women's perceptions of labia size and labiaplasty, see Clerico et al. 2017 and Widschwendter et al. 2020. For a general description of female genital anatomy, see Puppo 2013. For an educational presentation of human labial diversity, see the Labia Library presented by Women's Health Victoria—www.labialibrary.org.au/. I know of no prior attempt to investigate the genetic covariance of inner labial and penis size.
12. Young 1937, 139–42.
13. Gill-Peterson 2018, 76. Gill-Peterson presents extensive evidence of the "racialization of plasticity" in the Johns Hopkins University medical school group of pediatric endocrinologists, urologists, and surgeons who pioneered the treatment of differences in sexual development. According to evidence in various case files, Black trans girls were more likely to be pathologized and criminalized as sexual deviants than were similarly aged white trans girls.
14. Hughes 2008.
15. For recent biomedical classifications and discussions of the varieties of differences in sexual development, see Hughes 2008, Ono and Harley 2016, and Baetens et al. 2019.
16. Kessler 1998, Fausto-Sterling 2000, Hird 2004, Karkazis 2008, Davis 2015, Gill-Peterson 2018.
17. Gartler, Waxman, and Giblett 1962, Dewald et al. 1980, Berger-Zaslav et al. 2009. Another quite common mechanism of genetic chimerism occurs following pregnancy when fetal blood stem cells can cross the placenta, and take up residence in the bone marrow of the gestating mother. These offspring stem cells may persist for the rest of the mother's life.
18. Baetens et al. 2019, table 1.
19. Baetens et al. 2017.
20. Google Scholar search 20 April 2023.
21. Kashimada and Koopman 2010.
22. Kashimada and Koopman 2010, Chen et al. 2013.
23. Chen et al. 2013. The absence of a phosphorylation site on SRY in nonprimate mammals documents the ongoing nature of coevolution in transcription factor structure with other interacting features of the genetic regulatory cascades. Such evidence of evolutionary performativity is discussed further in chapter 7.
24. Sedgwick 1990.
25. Kusz et al. 1999, Gunes et al. 2013.
26. Migeon 2014.
27. Young (1937, 159–71) describes such a patient, but the genetic details are unknown.

28. Baetens et al. 2019.

29. Imperato-McGinley et al. 1974, 1979. Prostate development is also mediated by DHT, so individuals with 5α-reductase deficiency have small prostates. This observation contributed to the discovery that prostate enlargement and cancers can be treated with 5α-reductase inhibitors.

30. The term "androgen receptor" erroneously reinforces the concept that testosterone and DHT are actually "sex" hormones. The androgen receptor (AR) gene can also be called NR3C4, which stands for "nuclear receptor subfamily 3, group C, member 4."

31. Because the androgen receptor gene is located on the X chromosome and causes sterility in XY individuals, CAIS cannot naturally occur in XX individuals. However, it is likely that some XX heterozygous carriers of a nonfunctional androgen receptor gene (i.e., mothers of XY CAIS individuals) could have fertility problems as a result of X chromosome inactivation.

32. ESR2 stands for "estrogen receptor 2." Couse et al. 1999, Baetens et al. 2018.

33. Liu et al. 2015. EMX2 stands for "empty spiracles homeobox 2 gene."

34. Heinonen 1984.

35. Baskin, Himes, and Colburn 2001.

36. Baetens et al. 2019, Baskin et al. 2001.

37. Speiser and White 2003.

38. With the sensitivity of mammalian sperm to the high body temperatures of the warm-blooded mammals, the evolutionary solution of moving the testes into the scrotum is usually described as an obvious necessity. However, birds—the other warm-blooded organisms—have retained internal testes despite having much higher body temperature than any mammals, around 43°C, or 109°F. Obviously, birds have evolved a different physiological solution to the challenge of sperm temperature sensitivity.

39. Baetens et al. 2019.

40. The quotation is from Baetens et al. 2019; all the examples are from Baetens et al. 2017 and the references therein.

41. Baetens et al. 2017.

42. Baetens et al. 2017.

43. DAX1 stands for "dosage-sensitive sex reversal, adrenal hypoplasia critical region, on chromosome X, gene 1."

44. SOX3 stands for "SRY-related HMG Box gene 3."

45. Álvarez-Nava et al. 2011.

46. Blanchard 1997, Bogaert et al. 2018. Bogaert's (2006) analysis includes a sample of men raised in blended or adopted families and supports the impact of biological birth order, rather than position in the social family, on the development of male homosexuality. In previous work, I have proposed that human same-sex attraction evolved because it furthers Female sexual autonomy in ongoing sexual conflict with Males over reproduction (Prum 2017). The fact that every human develops within a womb provides Females with a unique, in utero, developmental opportunity to influence the outcome of sexual conflict. If sexual conflict is involved

in the evolution of human sexual preferences, we would predict that in utero maternal influences on sexual development would have stronger effects on Male sexual preferences than on Female because sexual conflict exists between the sexes and not within. Birth-order effects are best explained by in utero maternal influence on offspring sexual development because birth order is independent of a child's genetic makeup. Maternal mechanisms to produce Male birth effects in sexual preference development might evolve if Female sexual autonomy is advanced by *diversity* in male sexual preferences because some fraction of Males will always have had older brothers. Additional theory and data are required to investigate these possibilities. The presence of neuroligins on the Y chromosome poses other interesting questions. Given the apparently common development of maternal immune antibodies to these proteins, these Y chromosome–exclusive genes and their functions would have evolved in a context where some substantial portion of all XY offspring would have had an older XY sibling.

47. Richardson 2013.

48. It would be interesting to investigate the variation in susceptibility to *in utero* sibling effects on sexual differentiation in closely related mammalian species that have multiple births (i.e., litters) versus single births. Species with frequent multiple births must evolve mechanisms to buffer the development of females from the influence of male litter mates. This could result in selection for the development of non-shared placentas, or selection for lowering sensitivity to hormonal signals. This is an example of a performative developmental biology research question that has, apparently, never been asked before.

49. Bütikofer et al. 2019.

50. Ah-King and Hayward 2013, Diamanti-Kandarakis et al. 2009.

51. Paulozzi, Erickson, and Jackson 1997.

52. Gilbert and Barresi 2016.

53. Tamschick et al. 2016.

54. The question remains whether identifying genetic influences on nonnormative sexual development can help the individuals themselves. Some individuals may be genuinely interested in achieving fertility or understanding their sexual differences, but many aspects of this research program do not have obvious benefits for the individuals being investigated.

55. Imperato-McGinley et al. 1979.

56. Imperato-McGinley et al. 1979, 1234, 1236, 1235.

57. Kuhn 1970.

58. Kessler 1998, Fausto-Sterling 2000, Karkazis 2008, Davis 2015, Gill-Peterson 2018.

59. Gastaud et al. 2006. InterACT: Advocates for Intersex Youth is an advocacy organization that champions the legal rights of children with differences in sexual development to delay surgeries until the age of informed consent, and foster the wellness and thriving of variations in sexual development.

60. Griffiths 2021.

61. Laqueur 1992, 11.

CHAPTER SEVEN

1. Gould 1989, Losos 2017.
2. Butler 1988, 1990.
3. Butler 1993, 5.
4. Quammen (2018) discusses reticular deviations from treelike bifurcations in the history of life created by horizontal gene transfer and endosymbiosis. However common these processes may be, they do not disrupt our capacity to reconstruct phylogeny. This observation is amply demonstrated by the evolutionary history of photosynthesis in eukaryotes, which is characterized by multiple endosymbiotic and secondary endosymbiosis events, none of which obscures our capacity to infer the phylogenetic relationships of the endosymbiotic hosts and the donor cyanobacterial lineages.
5. For information about sexual recombination in bacteria, see Vos 2009. The evolutionary advantage of sexual reproduction is a classic question in evolutionary biology. For example, see Stearns 1987.
6. Togashi et al. 2012.
7. Having bodies that make one type of gamete or another is referred to as dioecy or gonochory. Like anisogamy itself, the multiple independent evolutionary origins of dioecy in plants and animals means that "female" and "male" are not always homologous. Although we refer to male and female parts of flowers, and to the sperm and ova of plants, anisogamy and dioecy in plants and animals are convergently evolved.
8. The analogous innovation of the hard-coated seed evolved in seed plants provided them with the same release from reproductive dependence on water, and greatly contributed to their evolutionary success and diversity.
9. In the 1990s and 2000s, American television marketing genius Ron Popeil created the memorable pitch line "*Set it and forget it!*" in the televised infomercials for his Showtime Rotisserie grill.
10. Mittwoch 1971, 432.
11. Ioannides et al. 2021.
12. Agate et al. 2003, DaCosta, Spellman, and Klicka 2007, Zhao et al. 2010.
13. Weintraub 2019.
14. Grützner et al. 2004. The double-branched penis and unusual sperm clusters of the male echidna are described by Johnston et al. (2007).
15. Bachtrog et al. 2014, Nagabhushana and Mishra 2016.
16. Tanner et al. 2019.
17. How could the sea turtle sexual development system evolve to avert extinction of males, and therefore the species? Most realistically, the range of temperatures that switch the molecular expression toward testis or ovary identity could evolve.
18. Bull 1983, Bull and Charnov 1985.
19. Gamble et al. 2015.
20. Pen et al. 2010. Recently, Kwon et al. (2022) show that the mechanisms of genetic sexual development initiation have evolved rapidly through artificial selection in domesticated, ornamental strains of the Siamese fighting fish, *Betta spledens*, from a yet unidentified ancestral mechanism to a dosage-dependent system using DMRT1.

21. Capel 2017; Holleley et al. 2015, 2016. Rapid evolutionary transitions in sexual development initiation mechanisms are facilitated by the "cold-blooded" (i.e., poikilothermic) physiology of lizards, which means that body temperature is a particularly salient feature of their physiology.

22. Shao et al. 2014.

23. Why have placental mammals remained so consistently invested in our XX/XY sexual differentiation system? Or birds in their ZZ/ZW system? The re-evolution of temperature-dependent initiation of sexual differentiation in birds and mammals may have been prevented by the evolution of warm-bloodedness, which led to the evolution of either internal gestation (mammals) or incubation (birds), both of which reduced the potential for environmental variation in temperature on embryonic development.

24. Reviewed recently by Bachtrog et al. 2014.

25. Jaccarini et al. 1983, Bachtrog et al. 2014.

26. Kaur et al. 2021. Various *Wolbachia* have evolved to manipulate the sex of offspring of infected hosts by cytoplasmic incompatibility (i.e., sperm that kills zygotes that develop from ova without *Wolbachia* infection); male killing (genes lethal to development of all male offspring); parthenogenesis (production of only female offspring without incorporation of sperm genome); and developmental feminization (insertion of a *Wolbachia* gene onto one Z chromosome to produce ZZ females).

27. Bachtrog et al. 2014.

28. This scenario for sex chromosome evolution is based on Bachtrog (2013).

29. Katsura et al. 2018.

30. Mammalian X chromosomes do not become laden with detrimental mutations because these variations can be eliminated during crossover with another X during gametogenesis in females.

31. Bachtrog 2013.

32. Bachtrog 2013.

33. Kashimada and Koopman 2010.

34. Kashimada and Koopman 2010, 3927.

35. Polanco and Koopman 2007, 13.

36. Chen et al. (2013, emphasis mine) hypothesize that the highly threshold-dependent effects of SRY in human and mouse cells is a product of balancing multilevel selection on sexual development and the social behavioral effects of fetal testosterone.

37. Fausto-Sterling (2000, 199) and Richardson (2013) have shown how the precarious instability of mammalian male sexual development and Y chromosome function have been interpreted and described by some scientists as a threat to manhood that must be overcome through struggle, self-assertion, and support. However, it is just a historically contingent impact of the performative variability inherent to all sexual differentiation mechanisms.

38. Kashimada and Koopman 2010, fig. 3.

39. Carroll 1920.

40. Kashimada and Koopman 2010, fig. 3.

41. Jiménez, Barrionuevo, and Burgos 2012; Kuroiwa et al. 2010. Recently, Terao et al. (2022) show that Male development in Xo/Xo *Tokudaia* spiny rats is initiated by an alternative, upregulated SOX9 allele on an autosomal chromosome.

42. Couger et al. 2020.

43. The controversy over the inevitability of Y chromosome loss has been reviewed by Richardson (2013).

44. Bachtrog et al. 2014; Vicoso, Kaiser, and Bachtrog 2013.

45. Vicoso, Kaiser, and Bachtrog 2013.

46. Stearns et al. 2012. Framing this evolutionary phenomenon as "intra-locus sexual conflict" is deeply problematic because the term erroneously misconstrues the maladaptive and suboptimal evolution in allele frequency under sexual dimorphism as "sexual conflict"—differences in selection between sexes in relation to sexual reproduction—including mate identity, mating rate, parental care, and reproductive investment. Rooted in a selfish-gene selection paradigm, "intra-locus sexual conflict" requires conceiving of the interests of an allele at a gene locus as being in conflict with the interest of the same allele in a different sexual phenotype. By design, the gene-level selection paradigm does not do an elegant job of analyzing sexual dimorphic phenotypes, given that it is premised on the notion that the phenotype is simply a vehicle for true genetic replicators. However, sexual conflict is about genuine social conflict among individual organisms—such as rape—which has absolutely nothing to do with the inefficiency of adaptive evolution to optimize allele frequencies under sexual dimorphism. Similarly, many genetic variations that foster health and fecundity in youth also contribute to reduced function, infirmity, and increasing debility in old age. This is why senescence is essentially inevitable in the absence of asexual reproduction. However, we don't refer to senescence as "age antagonism" or "intergenerational conflict." Of course, we could, but that would be ridiculous. The biggest problem with the current terminology is that it implies a false equivalence to, or intellectual commonality with, the evolutionary processes that contribute to genuine examples of sexual conflict and sexual antagonistic selection through sexual coercion and sexual violence. This is an example of a faux-synthesis in evolutionary biology—a common intellectual phenomenon in which a reductive simplification of complex, unrelated phenomena is achieved by the conceptual flattening that requires the elimination of agencies beyond the selfish gene. The result is then touted as a triumphant evolutionary "synthesis," which actually functions by the elimination, or flattening, of biological complexity (Prum 2017). Many evolutionary biologists are loath to admit the fact that constraints on the efficiency of natural selection to establish optimal outcomes are ubiquitous. So, rather than put adaptation's constraints at the center of intellectual attention, they create other, erroneous ways of talking about this fundamental challenge to the efficiency of adaptation as a dynamic adaptive "conflict" between sexes. Evolutionary biology already has an accurate term for directionally opposing selection on a trait or genetic variation—disruptive selection. So, the best term for this process is *sexually disruptive selection*.

47. Traditionally, the high incidence of breast and prostate cancers has been causally related to the high rates of cell divisions in these tissues, which can contribute

to increased frequency of mutations contributing to cancer. However, blood and intestinal cells also undergo high rates of cell division. Although they are prominent sources of life-threatening cancers, neither tissue rivals breast and prostate cancer in incidence. Of course, incidence of lung cancer is greatly influenced by environmental factors like smoking tobacco and air pollution. https://www.cancer.org/content/dam/cancer-org/research/cancer-facts-and-statistics/annual-cancer-facts-and-figures/2022/2022-cancer-facts-and-figures.pdf.

48. This observation is congruent with Wagner's (2014) assertion that character identity networks (ChINs) and mechanisms (ChIMs) are more evolutionarily conserved than the spatial or temporal initiation mechanisms.

49. IRF9 stands for "interferon regulatory factor 9."

50. Capel 2017. Recall that, even though SOX3 usually plays no role in sexual differentiation in humans, the duplication of the SOX3 gene, or genetic variations in the SOX3 promoter sequences can result in the development of testes and male full fertility in XX humans. Much remains to be learned about the evolutionary history of such changes, but we can already see that performative variability observed *within* a species reflects important aspects of the evolvable possibilities *among* species.

51. Alternatively, in the most recent mammalian common ancestor, SOX9 could have been induced in testis development by SOX3. In this case, the novel gene SRY may have coopted the ancestral SOX3 binding site for SOX9, and simply diverged in binding specificity to differentiate its function in sexual development from SOX3's functions elsewhere in the body.

52. Herpin and Schartl (2015) have suggested that there is more variation in genetic initiators of sexual development than in later stages of sexual development, describing it as a relative conservation at the bottom and an apparent "diversity at the top." Clinging to the "master switch" concept despite this extensive variation in initiating gene identity and network connections, they propose a multiply problematic "*masters change, slaves remain*" paradigm to explain these data. However, they recognize that this concept cannot explain the still extensive variation in the details of downstream signaling pathways in gonad differentiation.

53. Herpin and Schartl 2015, 1269.

54. Jarne and Auld 2006, Avise and Mank 2009, Bachtrog et al. 2014. Young (1937, chapter 7) provides detailed anatomical descriptions of nine cases of humans with some combination of ovary, testis, or ovotestes.

55. For a discussion of developmental constraints as a creative as well as a limiting force in evolutionary biology, see Gould 2002, chap. 10.

56. For a detailed discussion of the role of development in the evolution of innovations, see Wagner 2015.

57. Gredler et al. 2015.

58. Cohn 2011.

59. Chavan, Bhullar, and Wagner 2016; Chavan, Griffith, and Wagner 2017; Erkenbrack et al. 2018.

60. IL1 stands for "interleukin 1 alpha." IL9 stands for "interleukin gene 9."

TNR stands for "tenascin R." PTGS2 stands for "prostaglandin-endoperoxidase synthase 2." PGE2 stands for "prostaglandin E2." At birth, marsupials do not yet have hindlimb buds, but they have well-developed forelimbs that they use to crawl upward and into the pouch. This developmental decoupling of fore- and hindlimbs further facilitated the evolution of highly unusual adult morphologies in which the forelimbs and hindlimbs are very different in size and morphology, such as in kangaroos.

61. Live birth in other lineages of animals, including various lizards and fishes, is usually accomplished by egg retention or simply the apposition of maternal and embryonic tissues. Interpenetration of maternal and fetal tissues is unique to the mammalian placenta. The mammal clade including horses, cows, and their relatives, has secondarily evolved a derived, noninvasive placenta in which the maternal tissue mechanically resists the invasion of the fetal tissue.

62. TNF stands for "tumor necrosis factor." PGF2A stands for "prostaglandin F2α."

63. Chavan et al. 2016, 2017; Erkenbrack et al. 2018.

64. A few fishes and reptiles and many more invertebrate species have evaded this constraint through the evolution of parthenogenesis—the capacity of females to produce offspring that are genetically identical (or very close) to themselves. Parthenogenesis eliminates the genetic "costs" of reproduction, because all of a female's offspring are entirely related to the single parent.

CHAPTER EIGHT

1. This wise insight comes from my friend and colleague Günter Wagner. I call it Günter's Rule, and it is always a good check on whether any idea is worth pursuing.

2. Gilbert and Barresi 2016.

3. Maddock and Schwartz 1996, Traut and Winkling 2001, Chapman et al. 2007.

4. Hamlett 1989, Pollux et al. 2006.

5. Personal communication from Profs. Tony Wilson, Brooklyn College, and Camilla Wittingham, University of Sydney.

6. Shine and Bull 1979; Stewart, Heulin, and Surget-Groba 2004; Pokorná and Kratochvíl 2009; Rovatsos et al. 2014, 2015; Telemeco 2015; Alam et al. 2018. Although I arrived at this novel (to me) hypothesis by thinking performatively about the broader environmental context in which XX and XY sexual differentiation occurs, I am not the first to identify this relationship. In a series of papers in the 1970s, Ursula Mittwoch (1971, 1975) mentioned a similar idea, despite the rather limited understanding at the time of endocrine and paracrine signaling pathways in sexual differentiation.

7. Fausto-Sterling 2000, Richardson 2013.

8. The epithelia of the Müllerian-derived, upper portions of the developing vagina express PAX2 (PairedBox 2), whereas the urogenital sinus epithelia express HOXA1 (Robboy et al. 2017), but their roles in tissue differentiation, the organization of the

vaginal plate, and the fusion of the Müllerian to the urogenital tissues to form the vagina are unknown.

9. An evolutionary hypothesis for the diversity of symptoms of COVID-19 is that the disease has not yet evolved to specialize in the infection of humans. Diseases like measles and polio are thousands of years old, and have been transmitting from person to person every few weeks at most for those thousands of years. That is tens of thousands of cycles of transmission in a single species of host, providing enough time to specialize on a particular mode of infection, which limits the diversity of pathogenesis. However, SARS-CoV2 first infected humans in December 2019, and each of the billions of persons it has infected so far are all less than one hundred transmission events away from its original nonhuman host. Because the virus has only recently begun to evolve to infect humans, it is uncovering extensive genetic variation in human response and is only beginning to specialize in a particular mode of infection.

10. Keinan and Clark 2012. The primary reason why humans have so many rare genomic variations is the explosively rapid expansion of human population sizes over the last thirty thousand years. Keinan and Clark describe this condition as an "excess" of rare genetic variants, but these variants are only excessive in relation to the assumption of stable or equilibrium population conditions, which are irrelevant to the history of contemporary humans.

11. Complex human traits have many, many distinct and unique genetic influences. This is why the once great promise of personal genomic medicine is still a distant dream. Progress in genomic medicine is possible in those rare cases in which there is significant frequency of a specific genetic variant within a particular ethnic group. But not for most of the genetic variants that contribute to heritability of the complex diseases of most individuals. In theory, scientific research could investigate and unravel the specific nature of the genetic factors underlying one person's risk of heart attack, cancer, and so on, but these results would not be generalizable to many other individuals because they are influenced by unique combinations of your specific rare genetic variations. The genetic influences on the vast majority of our individual heart attacks, cancer risks, *or* sexual preferences are virtually unique. Unfortunately, even though these facts about human genetic variation are undisputed and clear, many researchers in medical and behavioral genetics continue to misrepresent the results of their studies to the public. For example, a recent study on the genetic basis of schizophrenia documents that a genetic variation in C4 MHC gene increases that risk of this disorder by 25 percent (Sekar et al. 2016). Even though this finding is highly statistically significant, this result actually documents that the vast majority of individuals with this genetic variation do not develop schizophrenia, and that the vast majority of cases of schizophrenia are completely unrelated to variation in this gene.

12. Amundson and Tresky 2007.

13. Amundson and Tresky 2007, 545.

14. Pavlicev et al. (2016) define the origin of emergent, irreducible scientific mechanisms to the origin of novel ontological individuals. "We hypothesize that any

major transition in the evolution of life is accomplished through the origin of novel kinds of individuals. Individual in this context refers to a distinct entity, which has an origin, a continuation in time over generations or species even, and potentially an end.... Individuals... are not limited to a particular organizational level; rather, they can be morphological or behavioral traits, cell types, as well as higher-level entities involving multiple organisms."

15. With thankful acknowledgments and apologies to Felsenstein (1985).
16. Cartwright 1999.
17. Mackie 1965.
18. Baetens et al. 2019.
19. Bachtrog et al. 2014; http://treeofsex.org/.
20. Capel 2017.
21. Laqueur 1992, 16.
22. Chang 2012.
23. For example, Chang (2012) describes how the same material object can be referred to as a proton or a hydrogen ion, depending on the intellectual context. The former participates in atomic theory and the latter in the theory of electrochemistry. This is not confusing to chemists, but it points out that we *could* build chemical theories on either foundation. Instead of having a linear periodic table where elements are marked by the incremental acquisitions of protons and neutrons, we could develop an atomic system in which elemental nuclei are conceived as built up from the nested addition of hydrogen, deuterium, and helium ions. Science could be otherwise, and function just as well. Perhaps in some ways, even better.
24. The nearly hysterical response by more than one hundred gene-selectionist biologists (Abbot et al. 2011) to a paper by E. O. Wilson and colleagues (Nowak, Tarnita, and Wilson 2010) to the proposal of a role for group selection in the evolution of eusociality is a telling example of reflexive panic and intellectual retrenchment.
25. Dobzhansky 1973.

CHAPTER NINE

1. Sadly, I cannot yet find and confirm this quote exactly. This report is from my own personal recollections of a radio interview from the 1980s or 1990s on NPR, or perhaps on Robert J. Lurtsema's radio show on WGBH Boston. I find lots of documentation of Previn's opinion (see, for example, https://www.nytimes.com/1991/07/07/nyregion/andre-previn-joyful-about-playing-jazz-for-the-neighbors.html), but no record of this exact quote.
2. Haraway 1988, 584.
3. There is some speculation that meiosis evolved as a mechanism to reduce 4N nuclei that arose through mitotic mistakes (Titia de Lange, pers. comm.). However, this reduction division would still be an intra-action.
4. Halperin 1997, 7.
5. Butler 1993, xii.

6. The complex molecular signaling interactions that occur between gametes during fertilization are extremely well studied in mammals, and a few other model organisms. They constitute an incredibly rich opportunity for the development of a discursive molecular biology of sexual reproduction. For a readable account of the biology of fertilization, see also Gilbert and Pinto-Correia 2017.

7. For example, see Godfrey-Smith 2016.

8. For evidence of the distribution and evolution of vocal learning in birds see Jarvis 2004, Jarvis et al. 2014, and Saranathan et al. 2007. For vocal learning in whales, see Garland et al. 2017.

9. Noad et al. 2000, Derryberry 2007.

10. Hunt and Gray 2003.

11. Quotation from K. Barad in the Second Terry Lecture on November 3, 2022, at Yale University, New Haven, Connecticut. Lectures to be published by Yale University Press.

12. Quote from Dr. J. A. Clayton in Rabin 2014. Ritz 2017, Fausto-Sterling and Joel 2014, Pape 2021.

13. Clayton and Collins 2014. I thank Sarah Richardson for bringing this issue to my attention.

14. Ritz 2017, 324.

15. Herzig (2004, 137) described Barad's (1998) performative use of Donna Haraway's concept of "apparatuses of bodily construction" as "echo[ing] the proverbial 'turtles all the way down.'"

16. Richardson 2013, 225.

17. Lillie 1939, 3, cited in Fausto-Sterling 2000, 178; Waddington 1940, 91.

18. Serano 2013, Prosser 1998.

19. Barad 2007, 175–76.

20. Barad 2007, 178–79.

21. Barad 2007, 37.

22. In Barad's (2007) analyses of scientific apparatuses in optical physics, the design of the apparatus incorporates the a priori intention of the scientific researcher to make a specific agential cut that will structure the kind of phenomenon produced—that is, a diffraction pattern or scattering pattern. Although we usually think of statistical methods of analysis as objective and free of intention or bias toward a specific phenomenon, that view does not fully describe the real use of these "apparatuses" in research, which would include the choice of data measured and the nature of the categorical variables invoked to analyze it. It is also important to remember that many of the pioneering architects of modern frequentist statistics—including notably Ronald A. Fisher—had explicitly eugenic research goals and purposes. Many statistical tools were invented *through* the pursuit of eugenic research. A more detailed analysis of the sources of designed purpose would be productive.

23. Reardon 2005.

24. Tannenbaum et al. 2019.

25. Danielsen et al. 2022.

26. Schiebinger and Klinge 2020, 186–87. For a pdf, see http://gendered

innovations.stanford.edu/GI%20%202%20How%20Inclusive%20Analysis%20 Contributes%20to%20R&I.pdf.

27. Reich 2018a, 2018b.
28. Freedman et al. 2006, Mancuso et al. 2016.
29. Hall et al. (1990) and Miki et al. (1994) document the identification of the association between BRCA1 gene mutations and breast cancer risk. The higher frequency of BRCA mutations in Ashkenazi Jewish people was reported by Tonin et al. (1995) and Struewing et al. (1997). BRCA stands for "breast cancer" gene.
30. Reich 2018a.
31. Haraway 1988, 592.
32. Haraway 1988, 593–94.
33. Haraway 1988, 580.
34. Sedgwick 2003.
35. Foucault 1978, 94–95.
36. Nüsslein-Volhard and Wieschaus 1980.
37. Mittwoch 1971, Capel 2017.
38. See 2020, 11–49.
39. See 2020, Bagemihl 1999, Roughgarden 2004.
40. Halperin 1997, 62 (emphasis in original).
41. Barad 2007, 64.
42. Sedgwick 1985.

ACKNOWLEDGMENTS

1. Sam See's essay was published posthumously in See 2020.

APPENDIX ONE

1. Keller 1984, 2000.
2. Fausto-Sterling 2000, 2018, 2019; Fausto-Sterling et al. 2012a, 2012b.
3. Gowaty 1992, Waage and Gowaty 1997.
4. Ah-King and Hayward 2013, Ah-King and Nyland 2010, Ah-King and Gowaty 2015.
5. Grosz 1990, 1994, 2011.
6. Kirby 2011.
7. Longino 1987, Longino and Doell 1983.
8. Schiebinger 1993, 1999.
9. Hird 2000, 2004, 2008, 2009; Giffney and Hird 2008.
10. Wilson 1998, 2004a, 2004b.
11. Richardson 2013, 2021.
12. Karkazis 2008, Jordan-Young 2010, Jordan-Young and Karkazis 2019.
13. Willey 2016, 2018.
14. Subramaniam 2016.
15. Roy 2007, 2018; Roy and Subramaniam 2016.

APPENDIX TWO

1. For an introductory description of innate and adaptive immune systems, see Sadava et al. 2016, 867–90. For a fuller presentation, see Murphy and Weaver 2017.

2. Acquired immunity can fail to protect the individual if the pathogen evolves rapidly enough so that when it reinfects the same individual, the antibodies that controlled the previous infection no longer bind selectively to the new pathogen. This is the issue with influenza viruses, which evolve so rapidly that acquired immunity will often not protect an individual one or more years later. Measles, on the other hand, is much more infectious than influenza, but it is also very slowly evolving. So, the measles vaccine developed in the early 1960s still works and provides individuals with lifelong immunity from infection.

3. Borges 1962. Thanks to Jack Hitt for the inspiration to pursue this Borgesian analogy.

APPENDIX THREE

1. Taylor and Lewontin 2017, section 5.
2. Ah-King and Hayward 2013, Ah-King and Nyland 2010, Ah-King and Gowaty 2015.
3. Oyama 1985, 2001; Griffiths and Gray 1994; Griffiths and Tabery 2013.
4. West-Eberhard 2003.
5. Wanjek 2011.

APPENDIX FOUR

1. Wagner et al. 2007.

APPENDIX FIVE

1. Waddington 1953, West-Eberhard 2003.
2. Ogburn and Edwards 2009, Edwards and Donoghue 2006.
3. Harris et al. 2002.

APPENDIX SIX

1. Dawkins 2006.
2. Haig 2020, 242.
3. I have heard this argument firsthand from several well-known professional biologists that will remain anonymous.
4. Wade 2016.
5. Sober 1984, 11 (emphasis mine).
6. Haig 220, 95.
7. Winther 2014.

APPENDIX SEVEN

1. Wade 2016.

2. Whyte 1965, Wagner and Schwenk 2000, Schwenk and Wagner 2000, Wade 2016. The history of the concept of internal selection is reviewed by Whyte (1965) and Wagner and Schwenk (2000).

3. Cheverud 1985.

References

Abbot, P., J. Abe, J. Alcock, S. Alizon, J. A. C. Alpedrinha, M. Andersson, J.-B. Andre F. Balloux, S. Balshine, N. Barton, et al. 2011. Inclusive fitness theory and eusociality. Nature 471:E1–E4.

Agate Robert, J., W. Grisham, J. Wade, S. Mann, J. Wingfield, C. Schanen, A. Palotie, and P. Arnold Arthur. 2003. Neural, not gonadal, origin of brain sex differences in a gynandromorphic finch. Proceedings of the National Academy of Sciences 100:4873–78.

Ah-King, M., and P. A. Gowaty. 2015. Reaction norms of sex and adaptive individual flexibility in reproductive decisions. In: T. Hoquet, editor. Current Perspectives on Sexual Selection. New York: Springer, 211–34.

Ah-King, M., and E. Hayward. 2013. Toxic sexes: Perverting pollution and queering hormone disruption. O-Zone: A Journal of Object-Oriented Studies 1:1–12.

Ah-King, M., and S. Nylin. 2010. Sex in an evolutionary perspective: Just another reaction norm. Evolutionary Biology 37:234–46.

Ahmed, S. 2008. Open forum imaginary prohibitions: Some preliminary remarks on the founding gestures of the "new materialism." European Journal of Women's Studies 15:23–39.

Alaimo, S., and S. Hekman, editors. 2008. Material feminisms. Bloomington: Indiana University Press.

Alam, S. M. I., S. D. Sarre, D. Gleeson, A. Georges, and T. Ezaz. 2018. Did lizards follow unique pathways in sex chromosome evolution? Genes & Dev. 9:239.

Álvarez-Nava, F., M. Soto, M. Temponi, R. Lanes, and Z. Alvarez. 2011. Female pseudohermaphroditism with phallic urethra in the offspring of a mother with an adrenal tumor. Journal of Pediatric Endorinology & Metabolism 17:1571–74.

Amundson, R. 2005. The changing role of the embryo in evolutionary thought: Roots of evo-devo. Cambridge: Cambridge University Press.

Amundson, R., and S. Tresky. 2007. On a bioethical challenge to disability rights. Journal of Medicine and Philosophy 32:541–61.

Arbib, M. A., and M. Hesse. 1986. The contruction of reality. Cambridge: Cambridge University Press.

Austin, J. L. 1962. How to do things with words. Oxford: Oxford University Press.

Avise, J. C., and J. E. Mank. 2009. Evolutionary perspectives on hermaphroditism in fishes. Sexual Development. 3:152–63.

Bachtrog, D. 2013. Y-chromosome evolution: Emerging insights into processes of Y-chromosome degeneration. Nature Reviews Genetics 14:113–24.

Bachtrog, D., J. E. Mank, C. Peichel, M. Kirkpatrick, S. P. Otto, T-L. Ashman, M. W. Hahn, J. Kitano, I. Mayrose, R. Ming, et al. 2014. Sex determination: Why so many ways of doing it? PLoS Biol 12:e1001899.

Baetens, D., H. Verdin, E. De Baere, and M. Cools. 2017. Non-coding variation in disorders of sex development. Clinical Genetics 91:163–72.

Baetens, D., T. Güran, B. B. Mendonca, N. L. Gomes, L. De Cauwer, F. Peelman, H. Verdin, M. Vuylsteke, M. Van der Linden, E. S. Group, et al. 2018. Biallelic and monoallelic ESR2 variants associated with 46,XY disorders of sex development. Genetics in Medicine 20:717–27.

Baetens, D., H. Verdin, E. De Baere, and M. Cools. 2019. Update on the genetics of differences in sexual development. Best Practice and Research Clinical Endocronology and Metabolism 33:101271.

Bagemihl, B. 1999. Biological exuberance: Animal homosexuality and natural diversity. New York: St. Martin's Press.

Barad, K. 1998. Getting real: Technoscientific practices and the materialization of reality. Differences 10:87–128.

———. 2003. Posthumanist performativity: Toward an understanding of how matter comes to matter. Signs 28:801–31.

———. 2007. Meeting the universe halfway. Durham, NC: Duke University Press.

———. 2011. Nature's queer performativity. Qui Parle 19:121–58.

Barandiaran, X. E., E. Di Paolo, and M. Rohde. 2009. Defining agency: Individuality, normativity, asymmetry, and spatio-temporality in action. Adaptive Behavior 17:367–86.

Baskin, L. S., K. Himes, and T. Colburn. 2001. Hypospadias and endocrine disruption: Is there a connection? Environmental Health Perspectives 109:1175–83.

Beldecos, A., S. Bailey, S. Gilbert, K. Hicks, L. Kenschaft, N. Niemczyk, R. Rosenberg, S. Schaertel, and A. Wedel. 1988. The importance of feminist critique for contemporary cell biology. Hypatia 3:61–76.

Berger-Zaslav, A.-L., L. Mehta, J. Jacob, T. Mercado, I. Gadi, J. H. Tepperberg, and L. S. Palmer. 2009. Ovotesticular disorder of sexual development (true hermaphroditism). Urology 73:293–96.

Blanchard, R. 1997. Birth order and sibling sex ratio in homosexual versus heterosexual males and females Annual Review of Sex Research 8:27–61.

Bogaert, A. F. 2006. Biological versus nonbiological older brothers and men's sexual orientation. Proceedings of the National Academy of Sciences USA 103:10771–74.

Bogaert, A. F., M. N. Skorska, C. Wang, J. Gabrie, A. J. MacNeil, M. R. Hoffarth, D. P. VanderLaan, K. J. Zucker, and R. Blanchard. 2018. Male homosexuality and maternal immune responsivity to the Y-linked protein NLGN4Y. Proceedings of the National Academy of Sciences 115:302–6.

Bono, J. J. 1990. Science, discourse, and literature: The role/rule of metaphor in science. In: S. Peterfreund, editor. Literature and Science: Theory and Practice. Boston: Northeastern University Press, 59–90.

Borges, J. L. 1962. The library of Babel. Labyrinths: Selected Stories and Other Writings. New York: New Directions, 51–58.

Botelho, J. F., D. Smith-Paredes, D. Nuñez-Leon, S. Soto-Acuña, and A. O. Vargas. 2014. The developmental origin of zygodactyl feet and its possible loss in the evoution of Passeriformes. Proceedings of the Royal Society of London B 281 (1788): 2014.0765.

Brangwynne, C. P. 2011. Soft active aggregates: Mechanics, dynamics and self-assembly of liquid-like intracellular protein bodies. Soft Matter 7:3052.

Brennan, P. L. R., C. J. Clark, and R. O. Prum. 2010. Explosive eversion and functional morphology of the duck penis supports sexual conflict in waterfowl genitalia. Proceedings of the Royal Society London B 277:1309–14

Brennan, P. L. R., and R. O. Prum. 2012. The limits of sexual conflict in the narrow sense: new insights from waterfowl biology. Philosophical Transactions of the Royal Society of London B 367:2324–38.

Brennan, P. L. R., R. O. Prum, K. G. McCracken, M. D. Sorenson, R. E. Wilson, and T. R. Birkhead. 2007. Coevolution of male and female genital morphology in waterfowl. PLoS One 2:e418.

Brownmiller, S. 1975. Against our will: Men, women, and rape. New York: Simon & Schuster.

Brunet, T., A. Bouclet, P. Ahmadi, D. Mitrossilis, B. Driquez, A.-C. Brunet, L. Henry, F. Serman, G. Béalle, C. Ménager, et al. 2013. Evolutionary conservation of early mesoderm specification by mechanotransduction in Bilateria. Nature Communications 4:2821.

Bull, J. J. 1983. Evolution of sex determining mechanisms. Menlo Park, CA: Benjamin/Cummings.

Bull, J. J., and E. L. Charnov. 1985. On irreversible evolution. Evolution 39:1149–55.

Bütikofer, A., D. N. Figlio, K. Karbownik, C. W. Kuzawa, and K. G. Salvanes. 2019. Evidence that prenatal testosterone transfer from male twins reduces the fertility and socioeconomic success of their female co-twins. Proceedings of the National Academy of Science USA 116:6749–53.

Butler, J. 1988. Performative acts and gender constitution: An essay in phenomenology and feminist theory. Theatre Journal 40:519–31.

———. 1990. Gender trouble: Feminism and the subversion of identity. New York: Routledge.

———. 1993. Bodies that matter. New York: Routledge.

———. 2015. Notes toward a performative theory of assembly. Cambridge, MA: Harvard University Press.

Cahill, L. 2014. Fundamental sex difference in human brain architecture. Proceedings of the National Academy of Sciences 111:577–78.

Capel, B. 2017. Vertebrate sex determination: Evolutionary plasticity of a fundamental switch. Nature Reviews Genetics 18:675–89.

Caro, T. 1986. The functions of stotting: A review of the hypotheses. Animal Behaviour 34 (3):649–62.

Carroll, L. 1920. Alice's adventures in Wonderland. New York: Macmillan.

Cartwright, N. 1999. The dappled world: A study of the boundaries of science. Cambridge: Cambridge University Press.

Chang, H. 2012. Is water H_2O? Evidence, realism and pluralism. Heidelberg: Springer.

Chapman, D. D., M. S. Shivji, E. Louis, J. Sommer, H. Fletcher, and P. A. Prodo. 2007. Virgin birth in a hammerhead shark. Biological Letters 3:425–27.

Chavan, A. R., B.-A. S. Bhullar, and G. P. Wagner. 2016. What was the ancestral function of decidual stromal cells? A model for the evolution of eutherian pregnancy. Placenta 40:40e51.

Chavan, A. R., O. W. Griffith, and G. P. Wagner. 2017. The inflammation paradox in the evolution of mammalian pregnancy: Turning a foe into a friend. Current Opinion in Genetics & Development 47:24–32.

Chen, Y.-S., J. D. Racca, N. B. Phillips, and M. A. Weiss. 2013. Inherited human sex reversal due to impaired nucleocytoplasmic trafficking of SRY defines a male transcriptional threshold. Proceedings of the National Academy of Sciences 110:E3567–E3576.

Cheng, F. 2020. X + Y: A mathematician's manifesto for rethinking gender. New York: Basic Books.

Cheverud, J. 1985. Quantitative genetics and developmental constraints on evolution by selection. Journal of Theoretical Biology 110:155–71.

Cho, S., K. W. Crenshaw, and L. McCall. 2013. Toward a field of intersectionality studies: Theory, applications, and praxis. Signs 38:786–810.

Cipolla, C., K. Gupta, D. A. Rubin, and A. Willey, editors. 2017. Queer feminist science studies: A reader. Seattle: University of Washington Press.

Clayton, J. A., and F. S. Collins. 2014. NIH to balance sex in cell and animal studies. Nature 509:282–83.

Clerico, C., A. Lari, A. Mojallal, and F. Boucher. 2017. Anatomy and aesthetics of the labia minora: The ideal vulva? Aesthetic Plastic Surgery 41:714–19.

Cohn, M. J. 2011. Development of the external genitalia: Conserved and divergent mechanisms of appendage patterning. Developmental Dynamics 240:1108–15.

Couger, M. B., S. W. Roy, N. Anderson, L. Gozashti, S. Pirro, L. S. Millward, M. Kim, D. Kilburn, K. J. Liu, T. M. Wilson, et al. 2020. Sex chromosome transformation and the origin of a male-specific X chromosome in the creeping vole. Science 372:592–600.

Couse, J. F., S. C. Hewitt, D. O. Bunch, M. Sar, V. R. Walker, B. J. Davis, and K. S. Korach. 1999. Postnatal sex reversal of the ovaries in mice lacking estrogen receptors alpha and beta. Science 286:2328–31.

Crenshaw, K. W. 1990. Mapping the margins: Intersectionality, identity politics, and violence against women of color. Stanford Law Review 43:1241–99.

Daane, J. M., N. Blum, J. Lanni, H. Boldt, M. K. Iovine, C. W. Higdon, S. L. Johnson, N. R. Lovejoy, and M. K. Harris. 2021. Modulation of bioelectric cues in the evolution of flying fishes. Current Biology 31:5052–61.e5058.

DaCosta, J. M., G. M. Spellman, and J. Klicka. 2007. Bilateral gynandromorph in a White-ruffed Manakin (*Corapipo altera*). Wilson Journal of Ornithology 119:289–91.

Danielsen, A. C., K. M. N. Lee, M. Boulicault, T. Rushovich, A. Gompers, A. Tarrant, M. Reiches, H. Shattuck-Heidorn, L. W. Miratrix, and S. S. Richardson. 2022. Sex disparities in COVID-19 outcomes in the United States: Quantifying and contextualizing variation. Social Science & Medicine 294:114716.

Davis, G. 2015. Contesting intersex: The Dubious diagnosis. New York: New York University Press.

Davis, N. 2009. New materialism and feminism's anti-biologism: A response to Sara Ahmed. European Journal of Women's Studies 16:67–80.

Dawkins, R. 1999. The extended phenotype. Oxford: Oxford University Press.

———. 2006. The selfish gene. 30th anniversary edition. New York: Oxford University Press.

de Queiroz, K., and M. J. Donoghue. 1988. Phylogenetic systematics and the species problem. Cladistics 4:317–38.

Delgado, I., and M. Torres. 2017. Coordination of limb development by crosstalk among axial patterning pathways. Developmental Biology 429:382–86.

Derryberry, E. P. 2007. Evolution of bird song affects signal efficacy: An experimental test using historical and current signals. Evolution 61:1938–45.

Dewald, G., M. W. Haymond, J. L. Spurbeck, and S. B. Moore. 1980. Origin of chi46,XX/46,XY chimerism in a human true hermaphrodite. Science 207:321–23.

Diamanti-Kandarakis, E., J.-P. Bourguignon, L. C. Giudice, R. Hauser, G. S. Prins, A. M. Soto, R. T. Zoeller, and A. C. Gore. 2009. Endocrine-disrupting chemicals: An Endocrine Society scientific statement. Endocrine Reviews 30:293–342.

DiFrisco, J., A. C. Love, and G. P. Wagner. 2020. Character identity mechanisms: A conceptual model for comparative-mechanistic biology. Biology and Philosophy 35:44 (32 pages).

Dobzhansky, T. 1973. Nothing in biology makes sense except in the light of evolution. American Biology Teacher 35:125–29.

Dufresne, E. R., H. Noh, V. Saranathan, S. G. J. Mochrie, H. Cao, and R. O. Prum. 2009. Self-assembly of amorphous biophotonic nanostructures by phase separation. Soft Matter 5:1792–95.

Edwards, E. J., and M. J. Donoghue. 2006. *Pereskia* and the origin of the cactus lifeform. American Naturalist 167:777–93.

Ellis, H. 1905. Studies in the psychology of sex. Vol. 1, Pt. 3. Sexual selection in man. New York: Random House.

Emlen, S. T., and P. H. Wrege. 2004. Size dimorphism, intrasexual competition, and sexual selection in Wattled Jacana (*Jacana jacana*), a sex-role-reversed shorebird in Panama. Auk 121:391–403.

Erkenbrack, E. M., J. D. Maziarz, O. W. Griffith, C. Liang, A. R. Chavan, M. C.

Nnamani, and G. P. Wagner. 2018. The mammalian decidual cell evolved from a cellular stress response. PLoS Biol 16:e2005594.

Fausto-Sterling, A. 2000. Sexing the body: Gender politics and the construction of sexuality. New York: Basic Books.

———. 2018 Oct. 25. Why sex is not binary. New York Times.

———. 2019. Gender/sex, sexual orientation, and identity are in the body: How did they get there? Journal of Sex Research 56:529–55.

Fausto-Sterling, A., C. Garcia Coll, and M. Lamarre. 2012a. Sexing the baby: Part 1 What do we really know about sex differentiation in the first three years of life? Social Science & Medicine 74:1684–92.

———. 2012b. Sexing the baby: Part 2 applying dynamic systems theory to the emergences of sex-related differences in infants and toddlers. Social Science & Medicine 74:1693–702.

Fausto-Sterling, A., and D. Joel. 2017 Dec 6. The science of difference: Let's do it right! HuffPost.

Felsenstein, J. 1985. Phylogenies and the comparative method. American Naturalist 125:1–15.

Ferguson, R. A. 2004. Abberations in Black: Toward a queer of color critique. Minneapolis: University of Minnesota.

———. 2018. Queer of color critique. Oxford Research Encyclopedia of Literature. https://doi.org/10.1093/acrefore/9780190201098.013.33.

Fichtner, J., D. Filipas, A. M. Mottrie, G. E. Voges, and R. Hohenfellner. 1995. Analysis of meatal location in 500 men: Wide variation questions need for meatal advancement in all pediatric anterior hypospadias cases. Journal of Urology 154:833–34.

Foucault, M. 1978. The history of sexuality: An introduction. New York: Vintage.

———. 1980. Herculine Barbin: Being the recently discovered memoire of a nineteenth-century French hermaphrodite. New York: Vintage.

Frank, S. A. 2012. Wright's adaptive landscape versus Fisher's fundamental theorem. In: R. Svensson and R. Calsbeek, editors. The Adaptive Landscape in Evolutionary Biology. Oxford: Oxford University Press.

Freedman, M., L., C. A. Haiman, N. Patterson, G. J. McDonald, A. Tandon, A. Waliszewska, K. Penney, R. G. Steen, K. Ardlie, E. M. John, et al. 2006. Admixture mapping identifies 8q24 as a prostate cancer risk locus in African-American men. Proceedings of the National Academy of Sciences 103:14068–73.

Gamble, T., J. Coryell, T. Ezaz, J. Lynch, D. P. Scantlebury, and D. Zarkower. 2015. Restriction site-associated dna sequencing (rad-seq) reveals an extraordinary number of transitions among gecko sex-determining systems. Molecular Biology and Evolution 32:1296–309.

Garland, E. C., L. Rendell, L. Lamoni, M. M. Poole, M. J. Noad. 2017. Song hybridization events during revolutionary song change provide insights into cultural transmission in humpback whales. Proceedings of the National Academy of Science USA. 114:7822–29.

Gartler, S. M., S. H. Waxman, and E. Giblett. 1962. An XX/XY human hermaphrodite resulting from double fertilization. Genetics 48:332–35.

Gastaud, F., C. Bouvattier, L. Duranteau, R. Brauner, E. Thibaud, F. Kutten, and P. Bougnères. 2006. Impaired sexual and reproductive outcomes in women with classical forms of congenital adrenal hyperplasia. Journal of Clinical Endocrinology & Metabolism 92:1391–96.

Ghiselin, M. T. 1978. A radical solution to the species problem. Systematic Biology 23:536–44.

Giffney, N., and M. Y. Hird, editors. 2008. Queering the non/human. Burlington, VT: Ashgate.

Gilbert, S. F. 2000. Diachronic biology meets evo-devo: C. H. Waddington's approach to evolutionary developmental biology. American Zoologist 40:729–37.

Gilbert, S. F., and M. J. F. Barresi. 2016. Developmental biology. 11th ed. Sunderland, MA: Sinauer.

Gilbert, S. F., and C. Pinto-Correia. 2017. Fear, wonder, and science in the new age of reproductive technology. New York: Columbia University Press.

Gill-Peterson, J. 2018. Histories of the transgender child. Minneapolis: University of Minnesota Press.

Godfrey, A. K., S. Naqvi, L. Chmátal, J. M. Chick, R. N. Mitchell, S. P. Gygi, H. Skaletsky, and D. C. Page. 2020. Quantitative analysis of Y-chromosome gene expression across 36 human tissues. Genome Research 30:860–73.

Godfrey-Smith, P. 2016. Other minds: The octopus, the sea, and the deep origins of consciousness. New York: Farrar, Straus and Giroux.

Goffman, E. 1956. The presentation of self in everyday life. Edinburgh: University of Edinburgh Social Sciences Research Centre.

Gould, S. J. 1989. Wonderful life. New York: W. W. Norton & Co.

———. 2002. The structure of evolutionary theory. Cambridge, MA: Harvard University Press.

Gowaty, P. A. 1992. Evolutionary biology and feminism. Human Nature 3:217–49.

———. 1997a. Introduction: Darwinian feminists and feminist evolutionists. In: P. A. Gowaty, editor. Feminism and Evolutionary Biology. New York: Chapman & Hall.

———. 1997b. Principles of females' perspectives in avian behavioral ecology. Journal of Avian Biology 28:2–9.

———. 1997c. Sexual dialectics, sexual selection, and variation in reproductive behavior. In: P. A. Gowaty, editor. Feminism and Evolutionary Biology. New York: Chapman & Hall, 351–84.

———. 2010. Forced or aggressively coerced copulation. In: M. D. Breed and J. Moore, editors. Encyclopedia of Animal Behavior. London: Academic Press, 759–63.

Gredler, M. L., C. E. Larkins, F. Leal, A. K. Lewis, A. M. Herrera, C. L. Perriton, T. J. Sanger, and M. J. Cohn. 2015. Evolution of external genitalia: Insights from reptilian development. Sexual Development 8:311–26.

Green, D. R. 2018. Cell death: Apoptosis and other means to an end. 2nd ed. Cold Spring Harbor, NY: Cold Spring Harbor Laboratory Press.

Green, E. L., K. Benner, and R. Pear. 2018 Oct 21. "Transgender" could be defined out of existence under Trump administration. New York Times.

Griffiths, D. A. 2021. Queering the moment of hypospadias "repair." GLQ 27:499–523.

Griffiths, P., and K. Stotz. 2013. Genetics and philosophy: An introduction. Cambridge: Cambridge University Press.

Griffiths, P. E., and R. D. Gray. 1994. Developmental systems and evolutionary explanation. Journal of Philosophy 91:277–304.

Griffiths, P. E., and J. Tabery. 2013. Developmental system theory: What does it explain, and how does it explain it? Advances in Child Development and Behavior 44:65–94.

Grosz, E. 1990. Conclusion: A note on essentialism and difference. In: S. Gunew, editor. Feminist Knowledge: Critique and Construct. New York: Routledge, 332–44.

———. 1994. Volatile bodies: Toward a corporeal feminism. Bloomington: Indiana University Press.

———. 2011. Becoming undone: Darwinian reflections on life, politics, and art. Durham, NC: Duke University Press.

Grützner, F., W. Rens, E. Tsend-Ayush, N. El-Mogharbel, P. C. M. O'Brien, R. C. Jones, M. Ferguson-Smith, A., and J. A. Marshall Graves. 2004. In the platypus a meiotic chain of ten sex chromosomes shares genes with the bird Z and mammal X chromosomes. Nature 432:913–17.

Gunes, S., R. Asci, G. Okten, F. Atac, O. E. Onat, G. Ogur, O. Aydin, O. T., and H. Bagci. 2013. Two males with SRY-positive 46,XX testicular disorder of sex development. Systems Biology in Reproductive Medicine 59:42–47.

Gutting, G., and J. Oksala. 2019. Michel Foucault. In: E. N. Zalta, editor. Stanford encyclopedia of philosophy. https://plato.stanford.edu/entries/foucault/.

Hacking, I. 2002. Historical ontology. Cambridge, MA: Harvard University Press.

Haig, D. 2020. From Darwin to Derrida: Selfish genes, social selves, and the meanings of life. Cambridge, MA: MIT Press.

Hall, J. M., M. K. Lee, B. Newmen, J. E. Morrow, L. A. Anderson, B. Huey, and M.-C. King. 1990. Linkage of early-onset familial breast cancer to chromosome 17q21. Science 250:1684–89.

Halperin, D. M. 1997. Saint Foucault: Towards a gay hagiography. Oxford: Oxford University Press.

Hamlett, W. C. 1989. Evolution and morphogenesis of the placenta in sharks. Journal of Experimental Zoology Supplement 2:35–53.

Haraway, D. 1976. Crystals, fabrics, and fields: Metaphors that shape embryos. New Haven, CT: Yale University Press.

———. 1988. Situated knowledges: The science question in feminism and the privilege of partial perspective. Feminist Studies 14:575–99.

———. 1989. Primate visions: Gender, race, and nature in the world of modern science. New York: Routledge.

Hardinson, R. C. 2012. Evolution of hemoglobin and its genes. Cold Spring Harbor Perspectives in Medicine 2012 (2): a011627.

Harris, M. K., J. F. Fallon, and R. O. Prum. 2002. Shh-Bmp2 signaling module and the evolutionary origin and diversification of feathers. Journal of Experimental Zoology (Molecular and Developmental Evolution) 294:160–76.

Harris, M. K., S. Williamson, J. F. Fallon, H. Meinhardt, and R. O. Prum. 2005. Molecular evidence for activator-inhibitor mechanism in development of embryonic feather branching. Proceedings of the National Academy of Sciences USA 102:11734–39.

Harris, M. P. 2021. Bioelectric signaling as a unique regulator of development and regeneration. Development 148:180794.

Heinonen, P. K. 1984. Uterus didelphys: A report of 26 cases. European Journal of Obstetrics and Gynecology 17:345–50.

Hennig, W. 1966. Phylogenetic systematics. Urbana: University of Illinois Press.

Herpin, A., and M. Schartl. 2015. Plasticity of gene-regulatory networks controlling sex determination: Of masters, slaves, usual suspects, newcomers, and usurpators. EMBO Reports 16:1260–74.

Herzig, R. 2004. On performance, productivity, and vocabularies of motive in recent studies of science. Feminist Studies 5:127–47.

Hird, M. Y. 2000. Gender's nature: Intersexuality, transsexualism and the "sex"/"gender" binary. Feminist Theory 1:347–64.

———. 2004. Sex, gender, and science. New York: Palgrave Macmillan.

———. 2008. Animal Trans. In: N. Giffney and M. Y. Hird, editors. Queering the Non/Human. Burlington, VT: Ashgate, 227–47.

———. 2009. The origins of sociable life: Evolution after science studies. New York: Palgrave Macmillan.

Holleley, C. E., D. O'Meally, S. D. Sarre, J. A. Marshall Graves, T. Ezaz, K. Matsubara, B. Azad, X. Zhang, and A. Georges. 2015. Sex reversal triggers the rapid transition from genetic to temperature-dependent sex. Nature 523:79–82.

Holleley, C. E., S. D. Sarre, D. O'Meally, and A. Georges. 2016. Sex reversal in reptiles: Reproductive oddity or powerful driver of evolutionary change? Sexual Development 10:279–87.

Hughes, I. 2008. Disorders of sex development: A new definition and classification. Best Practice & Research Clinical Endocrinology & Metabolism 22:119–34.

Hull, D. L. 1988. Science as a process. Chicago: University of Chicago Press.

Hunt, G. R. and R. D. Gray. 2003. Diversification and cumulative evolution in New Caledonian Crow tool manufacture. Proceedings of the Royal Society of London B 270:867–74.

Hutchinson, G. E. 1965. The ecological theater and the evolutionary play. New Haven, CT: Yale University Press.

Hutson, M. 2021 May 10. Persuading the body to regenerate its limbs. New Yorker.

Imperato-McGinley, J., L. Guerrero, T. Gautier, and R. Peterson. 1974. Steroid 5alpha-reductase deficiency in man: An inherited form of male pseudohermaphroditism. Science 186:1213–15.

Imperato-McGinley, J., R. E. Peterson, F. Gautier, and E. Sturla. 1979. Androgens and the evolution of male-gender identity among male pseudohermaphrodites with 5alpha-reductase deficiency. New England Journal of Medicine 300:1233–37.

Ingalhalikar, M., A. Smith, D. Parker, T. D. Satterthwaite, M. A. Elliott, K. Ruparel,

H. Hakonarson, R. E. Gur, R. C. Gur, and R. Verma 2014. Sex differences in the structural connectome of the human brain. Proceedings of the National Academy of Sciences 111:823–28.

Ioannidis, J., G. Taylor, D. Zhao, L. Liu, A. Idoko-Akoh, D. Gong, R. Lovell-Badge, S. Guioli, J. McGrew Mike, and M. Clinton. 2021. Primary sex determination in birds depends on DMRT1 dosage, but gonadal sex does not determine adult secondary sex characteristics. Proceedings of the National Academy of Sciences 118:e2020909118.

Jaccarini, V., L. Agius, P. J. Schembri, and M. Rizzo. 1983. Sex determination and larval sexual interaction in *Bonellia viridis* Rolando (Echiura: Bonelliidae). Journal of Experimental Marine Biology and Ecology 66:25–40.

Jagose, A., editor. 1996. Queer theory: An introduction. New York: New York University Press.

Jarne, P., and J. R. Auld. 2006. Animals mix it up too: the distribution of self-fertilization among hermaphroditic animals. Evolution 60:1816–24.

Jarvis, E. D. 2004. Learned birdsong and the neurobiology of human language. Annals of the New York Academy of Sciences 1016:749–77.

Jarvis, E. D., S. Mirarab, A. J. Aberer, B. Li, P. Houde, C. Li, S. Y. W. Ho, B. C. Faircloth, B. Nabholz, J. T. Howard, et al. 2014. Whole-genome analyses resolve early branches in the tree of life of modern birds. Science. 346:1320–31.

Jiménez, R., F. J. Barrionuevo, and M. Burgos. 2012. Natural exceptions to normal gonad development in mammals. Sexual Development 7:147–62.

Johnson, B. 1980. The critical difference: Essays in the contemporary rhetoric of reading. Baltimore: Johns Hopkins University Press.

Johnston, S. D., B. Smith, M. Pyne, D. Stenzel, and W. V. Holt. 2007. One-sided ejaculation of echidna sperm bundles. American Naturalist 170:E162.

Jones, R. A. L. 2002. Soft condensed matter. Oxford: Oxford University Press.

Jordan-Young, R. M. 2010. Brain storm: The flaws in the science of sex differences. Cambridge, MA: Harvard University Press.

Jordan-Young, R. M., and K. Karkazis. 2019. Testoterone: An unauthorized biography. Cambridge, MA: Harvard University Press.

Karkazis, K. 2008. Fixing sex: Intersex, medical authority, and lived experience. Durham, NC: Duke University Press.

Kashimada, K., and P. Koopman. 2010. SRY: The master switch in mammalian sex determination. Development 137:3921–30.

Katsura, Y., H. Kondo, J. Ryan, V. Harley, Y. Satta. 2018. The evolutionary process of mammalian sex determination genes focusing on marsupial SRYs. BMC Evolutionary Biology. 18:3.

Kaur, R., J. D. Shropshire, K. L. Cross, B. Leigh, A. J. Mansueto, V. Stewart, and S. R. Bordenstein. 2021. Living in the endosymbiotic world of *Wolbachia*: A centennial review. Cell Host & Microbe 29: 879–93.

Kay, L. E. 2000. Who wrote the book of life? A history of the genetic code. Stanford, CA: Stanford University Press.

Keinan, A., and A. G. Clark. 2012. Recent explosive human population growth has resulted in an excess of rare genetic variants. Science 336:740–43.

Keller, E. F. 1984. A feeling for the organism: The life and workd of Barbara McClintock. New York: Macmillan.

———. 1997. Developmental biology as a feminist cause? Osiris 12:16–28.

———. 2000. The century of the gene. Cambridge, MA: Harvard University Press.

Kessler, S. J. 1998. Lessons from the intersexed. New Brunswick, NJ: Rutgers University Press.

Khalipina, D., Y. Kaga, N. Dacher, and N. R. Chevalier. 2020. Smooth muscle contractility causes the gut to grow anisotropically. Journal of the Royal Society Interface 16:20190484.

Kirby, V. 2011. Quantum anthropologies: Life at large. Durham, NC: Duke University Press.

———. 2012. Initial conditions. Differences 23:197–205.

Koestler, A. 1978. Janus: A summing up. New York: Random House.

Koonin, E. V., and A. S. Novozhilov. 2009. Origin and evolution of the genetic code: The universal enigma. IUBMB Life 62:99–111.

Krafft-Ebing, R. von. 1903. Psychopathia sexualis with special reference to the antipathic sexual instinct: A medico forensic study. New York: Medical Arts Agency.

Kuhn, T. S. 1970. The structure of scientific revolutions. 2nd ed. Chicago: University of Chicago Press.

Kuroiwa, A., Y. Ishiguchi, F. Yamada, A. Shintaro, and Y. Matsuda. 2010. The process of a Y-loss event in an XO/XO mammal, the Ryukyu spiny rat. Chromosoma 119:519–26.

Kusz, K., M. Kotecki, A. Wojda, M. Szarras-Czapnik, A. Latos-Bielenska, A. Warenik-Szymankiewicz, A. Ruszczynska-Wolska, and J. Jaruzelska. 1999. Incomplete masculinisation of XX subjects carrying the SRY gene on an inactive X chromosome. Journal of Medical Genetics 36:452–56.

Kwon, Y. M., N. Vranken, C. Hoge, M. R. Lichak, A. L. Norovich, K. X. Francis, J. Camacho-Garcia, I. Bista, J. Wood, S. McCarthy, et al. 2022. Genomic consequences of domestication of the Siamese fighting fish. Science Advances 8:eabm4950.

Landrein, B., A. Kiss, M. Sassi, A. Chauvet, P. Das, M. Cortizo, P. Laufs, S. Takeda, M. Aida, J. Traas, et al. 2015. Mechanical stress contributes to the expression of the STM homeobox gene in *Arabidopsis* shoot meristems. eLife 4:e07811–e07811.

Laqueur, T. 1992. Making sex: Body and gender from the Greeks to Freud. Cambridge, MA: Harvard University Press.

Latour, B. 2014. Agency at the time of the Anthropocene. New Literary History 45:1–18.

Lillie, F. R. 1939. General biological introduction. In: E. Allen, editor. Sex and internal secretions. Baltimore: Williams and Wilkins, 3–14.

Liu, S., X. Gao, Y. Qin, W. Liu, T. Huang, J. Ma, J. L. Simpson, and Z.-J. Chen. 2015. Nonsense mutation of EMX2 is potential causative for uterus didelphysis: First molecular explanation for isolated incomplete Müllerian fusion. Fertility and Sterility 103:769–74.

Longino, H. 1987. Can there be a feminist science? Hypatia 2:51–64.

Longino, H., and R. Doell. 1983. Body, bias, and behavior: a comparative analysis of reasoning in two areas of biological science. Signs 9:206–27.

Lorenz, K. 1971. Comparative studies of the motor patterns of Anatidae (1941). In: K. Lorenz, editor. Studies of Animal and Human Behaviour. Cambridge, MA: Harvard University Press, 14–114.

Losos, J. A. 2017. Improbable destinies: Fate, chance, and the future of evolution. New York: Riverhead Books.

Mackie, J. L. 1965. Causes and condition. American Philosophical Quarterly 2:245–64.

Maddock, M. B., and F. J. Schwartz. 1996. Elasmobranch cytogenetics: Methods and sex chromosomes. Bulletin of Marine Science 58:147–55.

Mancuso, N., N. Rohland, K. A. Rand, A. Tandon, A. Allen, D. Quinque, S. Mallick, H. Li, A. Stram, X. Sheng, et al. 2016. The contribution of rare variation to prostate cancer heritability. Nature Genetics 48:30–35.

Markowitz, S. 2001. Pelvic politics: Sexual dimorphism and racial difference. Signs 26:389–414.

Martin, E. 1991. The egg and the sperm: How science has constructed a romance based on stereotypical male-female roles. Signs 16:485–501.

McWhorter, L. 2009. Racism and sexual oppression in Anglo-America: A genealogy. Bloomington: Indiana University Press.

Mead, M. 1935. Sex and temperament in three primitive societies. New York: Morrow.

Meinhardt, H. 1998. The algorithmic beauty of sea shells. 2nd ed. Berlin: Springer.

Migeon, B. R. 2014. Females are mosaics: X inactivation and sex differences in disease. Oxford: Oxford University Press.

Miki, Y., J. Swensen, D. Shattuck-Eidens, P. A. Futreal, K. Harshman, S. Tavtigian, Q. Liu, C. Cochran, L. M. Bennett, W. Ding, et al. 1994. A strong candidate for the breast and ovarian cancer susceptibility gene BRCA1. Science 266:66–71.

Mittwoch, U. 1971. Sex determination in birds and mammals. Nature 231:432–34.

———. 1975. Chromosomes and sex differentiation. In: R. Reinoth, editor. Intersexuality in the Animal Kingdom. New York: Academic, 438–46.

Mortimer-Sandilands, C., and B. Erickson. 2010. Queer ecologies: Sex, nature, politics, desire. Bloomington: Indiana University Press.

Murphy, K., and C. Weaver. 2017. Janeway's immunobiology. New York: Garland Science, Taylor and Francis.

Musser, J. M., G. P. Wagner, and R. O. Prum. 2015. Nuclear β-catenin expression supports homology of feathers, avian scutate scales, and alligator scales in early development. Evolution & Development 17:185–94.

Nagabhushana, A., and R. K. Mishra. 2016. Finding clues to the riddle of sex determination in zebrafish. Journal of Biosciences 41:145–55.

Nicholson, D. J., and J. Dupré. 2018. Everything flows: Towards a processual philosophy of biology. Oxford: Oxford University Press.

Noad, M. J., D. H. Cato, M. M. Bryden, M.-N. Jenner, and K. C. S. Jenner. 2000. Cultural revolution in whale songs. Nature 408:537.

Noë, A. 2007. Action in perception. Cambridge, MA: MIT Press.

Noh, H., S. F. Liew, V. Saranathan, S. G. J. Mochrie, R. O. Prum, E. R. Dufresne, and H. Cao. 2010a. How noniridescent colors are generated by quasi-ordered structures of bird feathers. Advanced Materials 22:2871–80.

———. 2010b. Contribution of double scattering to structural coloration in quasi-ordered nanostructures of bird feathers. Physical Review E 81:051923.

———. 2010c. Double scattering of light from biophotonic nanostructures with short-range order. Optics Express 18:11942–48.

Nowak, M. A., C. E. Tarnita, and E. O. Wilson. 2010. The evolution of eusociality. Nature 466:1057–62.

Nüsslein-Volhard, C., and E. Wieschaus. 1980. Mutations affecting segment number and polarity in *Drosophila*. Nature 287:795–801.

Ogburn, R. M., and E. J. Edwards. 2009. Anatomical variation in Cactaceae and relatives: Trait lability and evolutionary innovation. American Journal of Botany 96:391–408.

Ono, M., and V. R. Harley. 2013. Disorders of sex development: New genes, new concepts. Nature Reviews Endocrinology 9:79–91.

Oyama, S. 1985. The ontogeny of information. Cambridge: Cambridge University Press.

———. 2001. What is developmental systems theory? In: S. Oyama, P. E. Griffiths, and R. D. Gray, editors. Cycles of Contingency: Developmental Systems and Evolution. Cambridge, MA: MIT Press, 1–12.

Pai, V. P., S. Aw, T. Shomrat, J. M. Lemire, and M. Levin. 2012. Transmembrane voltage potential controls embryonic eye patterning in *Xenopus laevis*. Development 139:313–23.

Pankhurst, M. W. 2017. A putative role for anti-Müllerian hormone (AMH) in optimising ovarian reserve expenditure. Journal of Endocrinology 233:R1–R13.

Pape, M. 2021. Co-production, multiplied: Enactments of sex as a biological variable in US biomedicine. Social Studies of Science 51.

Paulozzi, L. J., J. D. Erickson, and R. J. Jackson. 1997. Hypospadias trends in two US surveillance systems. Pediatrics 100:831–34.

Pavlicev, M., G. Tomlinson, R. O. Prum, and G. P. Wagner. 2016. Systems emergence—the origin of individuals in biological and biocultural evolution. In: N. Eldredge, T. Pievani, E. Serrelli, and I. Tëmkin, editors. Evolutionary Theory: A Hierarchical Perspective. Chicago: University of Chicago Press.

Pen, I., T. Uller, B. Feldmeyer, A. Harts, G. M. While, and E. Wapstra. 2010. Climate-driven population divergence in sex-determining systems. Nature 468:436–38.

Pickering, A. 1993. The mangle of practice: Agency and emergence in the sociology of science. American Journal of Sociology 99:559–89.

———. 1995. The Mangle of Practice: Time, Agency, and Science. Chicago: University of Chicago Press.

Pinker, S. 1994 April 25. The game of the name. New York Times, Section A.

Pokorná, M., and L. Kratochvíl. 2009. Phylogeny of sex-determining mechanisms in squamate reptiles: Are sex chromosomes an evolutionary trap? Zoological Journal of the Linnean Society 156:168–83.

Polanco, J. C., and P. Koopman. 2007. SRY and the hesitant beginnings of male development. Developmental Biology 302:13–24.

Pollux, B. J. A., M. N. Pires, A. I. Banet, and D. N. Reznick. 2009. Evolution of placentas in the fish family poeciliidae: An empirical study of macroevolution. Annual Review of Ecology and Systematics 40:271–89.

Pradeau, T. 2016. The many faces of biological individuality. Biology and Philosophy 31:761–73.

Prosser, J. 1998. Second skins. New York: Columbia University Press.

Prum, R. O. 1990. Phylogenetic analysis of the evolution of display behavior in the neotropical manakins (Aves: Pipridae). Ethology 84:202–31.

———. 1992. Syringeal morphology, phylogeny, and evolution of the Neotropical manakins (Aves: Pipridae). American Museum Novitates 3043:65 pp.

———. 1999. Development and evolutionary origin of feathers. Journal of Experimental Zoology (Molecular and Developmental Evolution) 285:291–306.

———. 2006. Anatomy, physics, and evolution of avian structural colors. In: G. E. Hill and K. J. McGraw, editors. Bird Coloration, Vol. 1: Mechanisms and Measurements. Cambridge, MA: Harvard University Press, 295–353.

———. 2012. Aesthetic evolution by mate choice: Darwin's *really* dangerous idea. Philosophical transactions of the Royal Society of London B 367:2253–65.

———. 2013. Coevolutionary aesthetics in human and biotic artworlds. Biology and Philosophy 28:811–32.

———. 2015. The role of sexual autonomy in evolution by mate choice. In: T. Hoquet, editor. Current Perspectives in Sexual Selection. New York: Springer, 237–62.

———. 2017. The evolution of beauty: How Darwin's forgotten theory of mate choice shapes the animal world—and us. New York: Doubleday.

———. 2020. Ornament in human and biotic artworlds. In: G. He and K. Bloomer, editors. Natures of Ornament. New Haven, CT: Yale University Press, 48–67.

———. 2022. The ontology of artworlds: a posthuman, coevolutionary framework for aesthetics, art history, and art criticism. In: G. Levine, editor. The Question of Aesthetics. Oxford: Oxford University Press.

Prum, R. O., and A. H. Brush. 2002. The evolutionary origin and diversification of feathers. Quarterly Review of Biology 77:261–95.

Prum, R. O., E. R. Dufresne, T. Quinn, and K. Waters. 2009. Development of the color producing beta-keratin nanostructures in avian feather barbs. Journal of the Royal Society Interface 6S:253–65.

Prum, R. O., R. H. Torres, S. Williamson, and J. Dyck. 1998. Coherent light scattering by blue feather barbs. Nature 396:28–29.

Ptashne, M. 2013. Epigenetics: Core misconcept. Proceedings of the National Academy of Sciences 110:7101–3.

Puppo, V. 2013. Anatomy and physiology of the clitoris, vestibular bulbs, and labia minora with a review of the female orgasm and the prevention of female sexual dysfunction. Clinical Anatomy 26:134–52.

Quammen, D. 2018. The tangled tree: A radical new history of life. New York: Simon & Schuster.

Rabin, R. C. 2014 May 14. Labs are told to start including a neglected variable: Females. New York Times.

Reardon, J. 2005. Race to the finish: Identity and governance in an age of genomics. Princeton, NJ: Princeton University Press.

Reich, D. 2018a March 23. How genetics is changing our understanding of "race." New York Times.

———. 2018b. Who we are and how we got here. New York: Pantheon.

Richardson, S. S. 2013. Sex itself: The search for male and female in the human genome. Chicago: University of Chicago Press.

———. 2021. Sex contextualism. Philosophy, Theory, and Practice in Biology 14:17.

Ritz, S. 2017. Complexities of addressing sex in cell culture research. Signs 42:307–27.

Robboy, S. J., T. Kurita, L. S. Baskin, and G. R. Cunha. 2017. New insights into human female reproductive tract development. Differentiation 201:9–22.

Rosenthal, G. G. 2017. Mate choice: The evolution sexual decision making from microbes to humans. Princeton, NJ: Princeton University Press.

Rothman, D. L., S. C. Stearns, and R. G. Shulman. 2021. Gene expression regulates metabolite homeostasis during the Crabtree effect: Implications for the adaptation and evolution of metabolism. Proceedings of the National Academy of Sciences USA 118:e2014013118.

Roughgarden, J. 2009. Evolution's rainbow: Diversity, gender, and sexuality in nature and people. Berkeley: University of California Press.

Rovatsos, M., M. Pokorná, M. Altmanová, and L. Kratochvíl. 2014. Cretaceous park of sex determination: Sex chromosomes are conserved across iguanas. Biology Letters 10:20131093.

Rovatsos, M., J. Vukić, P. Lymberakis, and L. Kratochvíl. 2015. Evolutionary stability of sex chromosomes in snakes. Proceedings of the Royal Society of London B 282:20151992.

Roy, D. 2007. Somatic matters: Becoming molecular in molecular biology. Rhizomes 14(Summer).

———. 2018. Molecular feminisms: Biology, becomings, and life in the lab. Seattle: University of Washington Press.

Roy, D., and B. Subramaniam. 2016. Matter in the shadows: Feminist new materialism and the practices of colonialism. In: V. Pitts-Taylor, editor. Matering: Feminism, Science, and Materialism. New York: New York University Press.

Ryan, M. J. 2018. A taste for the beautiful: The evolution of attraction. Princeton, NJ: Princeton University Press.

Sadava, D., D. M. Hillis, H. C. Heller, and S. D. Hacker. 2016. Life: The science of biology. Sunderland, MA: Sinauer.

Salamon, G. 2004. The bodily ego and the contested domain of the material. differences 15:95–122.

Sandilands, C. 2016. Queer Ecology. In: J. Amadson, W. A. Gleason, and D. N.

Pellow, editors. Keywords for Environmental Studies. New York: New York University Press.

Saranathan, V., J. D. Forster, H. Noh, S. F. Liew, S. G. J. Mochrie, H. Cao, E. R. Dufresne, and R. O. Prum. 2012. Structure and optical function of amorphous photonic nanostructures from avian feather barbs: A comparative small angle X-ray scattering (SAXS) analysis of 229 bird species. Journal of the Royal Society Interface 9:2563–80.

Saranathan, V., S. Narayanan, A. Sandy, E. R. Dufresne, and R. O. Prum. 2021. Evolution of single gyroid photonic crystals in bird feathers. Proceedings of the National Academy of Sciences 118:e2101357118.

Schiebinger, L. 1993. Nature's body: Gender in the making of modern science. Boston: Beacon Press.

———. 1999. Has feminism changed science? Cambridge, MA: Harvard University Press.

Schiebinger, L., and I. Klinge. 2020. Gendered innovations 2: How inclusive analysis contributes to research and innovation. Publications Office of the European Union, Luxembourg.

Schilling, M. F., A. E. Watkins, and W. Watkins. 2002. Is human height bimodal? The American Statiscian 56:223–29.

Schlosser, M. 2019. Agency. In: E. N. Zalta, editor. Stanford encyclopedia of philosophy. https://plato.stanford.edu/entries/agency/.

Schwenk, K., and G. P. Wagner. 2001. Function and the evolution of phenotypic stability: Connecting pattern to process. American Zoologist 41:552–63.

Sedgwick, E. K. 1985. Between men: English literature and male homosocial desire. New York: Columbia University Press.

———. 1990. The epistemology of the closet. Berekely: University of California Press.

———. 1993. Tendencies. Durham, NC: Duke University Press.

———. 2003. Touching feeling: Affect, pedagogy, performativity. Durham, NC: Duke University Press.

See, S. 2020. Charles Darwin: Queer theorist. In: C. Looby and M. North, editors. Queer Natures, Queer Mythologies. New York: Fordham University Press, 11–49.

Seifert, A. W., C. M. Bouldin, K.-S. Choi, B. D. Harfe, and M. J. Cohn. 2009. Multiphasic and tissue-specific roles of sonic hedgehog in cloacal septation and external genitalia development. Development 135:3949–57.

Seifert, A. W., Z. Zheng, B. K. Ormerod, and M. J. Cohn. 2010. Sonic hedgehog controls growth of external genitalia by regulating cell cycle kinetics. Nature Communications 1:23.

Sekar, A., A. R. Bialas, H. de Rivera, A. Davis, T. R. Hammond, N. Kamitaki, K. Tooley, J. Presumey, M. Baum, V. Van Doren, et al. 2016. Schizophrenia risk from complex variation of complement component 4. Nature 530:177–83.

Sender, R., S. Fuchs, and R. Milo. 2016. Revised estimates for the number of human and bacteria cells in the body. PLoS Biol 14:e1002533.

Serano, J. 2013. Excluded: Making feminist and queer movements more inclusive. Berkeley, CA: Seal Press.

Shao, C., Q. Li, S. Chen, P. Zhang, L. Lian, Q. Hu, B. Sun, L. Jin, S. Liu, Z. Wang, et al. 2014. Epigenetic modification and inheritance in sexual reversal of fish. Genome Research 24:604–15.

Shapiro, E., H. Huang, and X.-R. Wu. 2002. New concepts on the developmnet of the vagina. In: L. S. Baskin, editor. Hypospadias and Genital Development. Boston: Springer, 173–85.

Shine, R., and E. L. Bull. 1979. The evolution of live-bearing in lizards and snakes. American Naturalist 113:905–23.

Sicher, A., R. Ganz, A. Menzel, D. Messmer, G. Panzarasa, M. Feofilova, R. O. Prum, R. W. Style, V. Saranathan, R. M. Rossi, and E. R. Dufresne. 2021. Structural color from solid-state polymerization-induced phase separation. Soft Matter 17:5772–79.

Simmons-Duffin, S. 2019 May 24. Trump administration's proposed HHS rule would redefine what "sex" means. All Things Considered. National Public Radio.

Smith, C. A., K. N. Roeszler, T. Ohnesorg, D. M. Cummins, P. G. Farlie, T. J. Doran, and A. H. Sinclair. 2009. The avian Z-linked gene DMRT1 is required for male sex determination in the chicken. Nature 461:267–71.

Snow, S. S., S. H. Alonzo, M. R. Servedio, and R. O. Prum. 2019. Female resistance to sexual coercion can evolve to preserve the indirect benefits of mate choice. Journal of Evolutionary Biology 32:545–58.

Sober, E. 1984. The nature of selection. Cambridge, MA: MIT Press.

Sparrow, D. B., G. Chapman, M. A. Wouters, N. V. Whittock, S. Ellard, D. Fatkin, P. D. Turnpenny, K. Kusumi, D. Sillence, and S. L. Dunwoodie. 2006. Mutation of the LUNATIC FRINGE gene in humans causes spondylocostal dysostosis with a severe vertebral phenotype. American Journal of Human Genetics 78:28–37.

Speiser, P. W., and P. C. White. 2003. Congenital adrenal hyperplasia. New England Journal of Medicine 349:776–88.

Squier, S. M. 2017. Epigenetic landscapes: Drawings as metaphor. Durham, NC: Duke University Press.

Stearns, S. C., editor. 1987. The evolution of sex and its consequences. Basel: Birkhauser.

Stearns, S. C., D. R. Govindaraju, D. Ewbank, and S. G. Byars. 2012. Constraints on the co-evolution of contemporary human males and females. Proceedings of the Royal Society of London B 279:4836–44.

Stewart, J. R., B. Heulin, and Y. Surget-Groba. 2004. Extraembryonic membrane development in a reproductively bimodal lizard, *Lacerta (Zootoca) vivipara*. Zoology 107:289–314.

Struewing, J. P., P. Hartge, S. Wacholder, S. M. Baker, M. Berlin, M. McAdams, M. M. Timmerman, L. C. Brody, and M. A. Tucker. 1997. The risk of cancer associated with specific mutations of BRCA1 and BRCA2 among Ashkenazi Jews. New England Journal of Medicine 336:1401–8.

Subramaniam, B. 2016. Ghost stories for Darwin: The science of variation and the politics of diversity. Urbana: University of Illinois Press.

Tamschick, S., B. Rozenblut-Kościsty, M. Ogielska, L. A., P. Lymberakis, F. Hoffmann, I. Lutz, W. Kloas, and M. Stöck. 2016. Sex reversal assessments reveal different vulnerability to endocrine disruption between deeply diverged anuran lineages. Scientific Reports 6:23825.

Tannenbaum, C., R. P. Ellis, F. Eyssel, J. Zou, and L. Schiebinger. 2019. Sex and gender analysis improves science and engineering. Nature 575:137–46.

Tanner, C. E., A. Marco, S. Martins, E. Abella-Perez, and L. A. Hawkes. 2019. Highly feminised sex-ratio estimations for the world's third-largest nesting aggregation of loggerhead sea turtles. Marine Ecology Progress Series 621:209–19.

Taylor, P., and R. Lewontin. 2017. The genotype/phenotype distinction. In: E. N. Zalta, editor. Stanford encyclopedia of philosophy. https://plato.stanford.edu/archives/sum2017/entries/genotype-phenotype/.

Telemeco, R. S. 2015. Sex determination in southern alligator lizards (*Elgaria multicarinata*; Anguidae). Herpetologica 71:8–11.

Terao, M., Y. Ogawa, S. Takada, R. Kajitani, M. Okuno, Y. Mochimaru, K. Matsuoka, T. Itoh, A. Toyoda, T. Kono, et al. 2022. Turnover of mammal sex chromosomes in the *Sry*-deficient Amami spiny rat is due to male-specific upregulation of *Sox9*. Proceedings of the National Academy of Sciences USA 119:e2211574119.

Thomas, L. 1974. The lives of the cell: Notes of a biology watcher. New York: Penguin.

Thompson, D. A. W. 1917. On growth and form. Cambridge: Cambridge University Press.

Togashi, T., J. L. Bartelt, J. Yoshimura, K. Tainaka, and P. A. Cox. 2012. Evolutionary trajectories explain the diversified evolution of isogamy and anisogamy in marine green algae. Proceedings of the National Academy of Sciences 109:13692–97.

Tonin, P., O. Serova, G. Lenoir, H. Lynch, F. Durocher, J. Simard, K. Morgan, and S. Narod. 1995. BRCA I mutations in Ashkenazi Jewish women. American Journal of Human Genetics 57:189.

Traut, W., and H. Winking. 2001. Meiotic chromosomes and stages of sex chromosome evoution in fish: Zebrafish, platyfish and guppy. Chromosome Research 9:659–72.

Trinh, D.-C., J. Alonso-Serra, M. Asaoka, L. Colin, M. Cortes, A. Malivert, S. Takatani, F. Zhao, J. Traas, C. Trehin, and O. Hamant. 2021. How mechanical forces shape plant organs. Current Biology 31:R143–R159.

Turing, A. M. 1952. The chemical basis of morphogenesis. Philosophical Transactions of the Royal Society of London 237:37–72.

van Anders, S. M. 2009. Chewing gum has large effects on salivary testosterone, estradiol, and secretory immunoglobulin A assays in women and men. Psychoneuroendocrinology 35:305–9.

———. 2015. Beyond sexual orientation: Integrating gender/sex and diverse sexualities via sexual configurations theory. Archive of Sexual Behaviour 44:1177–1213.

Velocci, B. 2021. Binary logic: Race, expertise, and the persistence of uncertainty in American sex research. PhD diss., Yale University. https://elischolar.library.yale.edu/gsas_dissertations/436.

Vicoso, B., V. B. Kaiser, and D. Bachtrog. 2013. Sex-biased gene expression at homomorphic sex chromosomes in emus and its implication for sex chromosome evolution. Proceedings of the National Academy of Sciences 110:6453–58.

Vos, M. Why do bacteria engage in homologous recombination? Trends in Microbiology 16:226–32.

Waage, J. K., and P. A. Gowaty. 1997. Myths of genetic determinism. In: P. A. Gowaty, editor. Feminism and Evolutionary Biology. New York: Chapman & Hall, 585–613.

Waddington, C. H. 1940. Organizers and genes. Cambridge: Cambridge University Press.

———. 1953. Genetic assimilation of an acquired character. Evolution 7:118–26.

Wade, M. J. 2016. Adaptation in metapopulations: How interaction changes evolution. Chicago: University of Chicago Press.

Wagner, G. P. 2015. Homology, genes, and evolutionary innovation. Princeton, NJ: Princeton University Press.

Wagner, G. P., M. Pavlicev, and J. Cheverud. 2007. The road to modularity. Nature Reviews Genetics 8:921–31.

Wagner, G. P., and K. Schwenk. 2000. Evolutionarily stable configurations: Functional integration and the evolution of phenotypic stability. In: M. K. Hecht, R. J. MacIntyre, and M. T. Clegg, editors. Evolutionary Biology. New York: Kluwer/Plenum, 155–217.

Walsh, D. M. 2015. Organisms, agency, and evolution. Cambridge: Cambridge University Press.

Wanjek, C. 2011. Systems biology as defined by NIH: An intellectual resource for integrative biology. NIH Catalyst 19.

Weintraub, K. 2019 Feb 9. A rare bird indeed: A cardinal that's half male, half female. New York Times.

West-Eberhard, M. J. 2003. Developmental plasticity and evolution. Oxford: Oxford University Press.

Whyte, L. L. 1965. Internal factors in evolution. New York: George Braziller.

Widschwendter, A., D. Riedl, K. Freidhager, S. A. Azim, S. Jerabek-Klestil, E. D'Costa, S. Fessler, A. Ciresa-König, C. Marth, and B. Böttcher. 2020. Perception of labial size and objective measurements—Is there a correlation? A cross-sectional study in a cohort not seeking labiaplasty. Journal of Sexual Medicine 17:461–69.

Wiley, E. O. 1980. Is the evolutionary species fiction? A consideration of classes, individuals and historical entities. Systematic Biology 29:76–80.

Willey, A. 2016. Biopossibility: A queer feminist materialist science studies manifesto, with special reference to the question of monogamous behavior. Signs 41:553–77.

———. 2018. Undoing monogamy: The politics of science and the possibilities of biology. Durham, NC: Duke University Press.

Wilson, E. A. 1998. Neural geographies: Feminism and the microstructure of cognition. Durham, NC: Duke University Press.

———. 2004a. Gut feminism. Differences 15:66–94.

———. 2004b. Psychosomatic: Feminism and the neurological body. Durham, NC: Duke University Press.

Winther, R. G. 2014. James and Dewey on abstraction. The Pluralist 9:1–28.

Wong, J. T. 1976. The evolution of a universal genetic code. Proceedings of the National Academy of Science 73:2236–2340.

Xu, X., Z. Zhou, and R. O. Prum. 2001. Branched integumental structures in *Sinornithosaurus* and the origin of feathers. Nature 410:200–204.

Young, H. H. 1937. Genital abnormalities, hermaphroditism, and related renal diseases. Baltimore: Williams & Wilkins.

Zeller, R., J. Lopez-Rios, and A. Zuniga. 2009. Vertebrate limb bud development: Moving towards integrative analysis of organogenesis. Nature Reviews Genetics 10:845–58.

Zhao, D., D. McBride, S. Nandi, H. A. McQueen, M. J. McGrew, P. M. Hocking, P. D. Lewis, H. M. Sang, and M. Clinton. 2010. Somatic sex identity is cell autonomous in the chicken. Nature 464:237–42.

Index

Page numbers in italics refer to figures.

"abstractionism, vicious," 309, 326n48
acetylation, 93–95, 98, 136, 168, 203–4, 210
acquired immunity, 36, 120, 281, 287–92, 342n2 (app. 2)
adaptation: biological, 52; and developmental variability, 301; and evolution, 31, 52, 193, 267, 297; and feather evolution, 5–6, 217; and gender/sex roles, 52; gene-level, 113, 229; and history of life, 190; by natural selection, 4, 8–9, 31–32, 190, 315n4, 319n28; and New Synthesis, 31–32, 317n27; and selection, 6; and selfish gene, 31–32; and sexual coercion, 10–13, 123, 335n46; and sexual ideals, 52
Adaptation in Metapopulations (Wade), 325n45
adaptationism: and feather evolution, 6; gene-level, 113; and natural selection, 9, 31, 190
adrenal glands, 151, 176, 178–79, 227, 254, 278
aesthetic philosophy, 11
aesthetics, 60; of animals, 12, 238; and art, 124; and beauty, 8–9; and birds, evolution of, 8–9, 12–13; and evolutionary process, 8–9; and queer theory, 12
African Americans, 164, 266–67, 330n13
Against Our Will (Brownmiller), 10
agency: aesthetic, 9; biological, 9, 59–60, 75, 78, 111–14, 118, 227, 247, 253, 298, 312; defined, 59–60, 320n34; in developmental biology, 111–14; and discourse, 68; elimination of, 10, 12; emergent, 102, 113, 240, 306, 309; Female, 10; and immunity, 291; individual, 66, 321n17; living, 59; of organisms, 12; and performativity, 66, 71–72, 112; and selection, 112; of selfish genes, 112–13, 305–6, 309; subjective, 9–12
agential realism, 26, 67, 75–76, 259, 262
Ah-King, Malin, 46, 284, 294
Alaimo, Stacy, 321n26
algae, green, 43, 193
alleles (gene variants), 32, 113, 171, 208, 229, 243, 263, 268, 305–8, 326n49. *See also* genetic variation

AMH. *See* anti-Müllerian hormone (AMH)
amniotes, 192, 194, 215, 218–19
amphibians, 145, 181, 218–19. *See also* reptiles
Amundson, Ron, 236, 317n27
analogies, 85, 96, 98, 210, 275, 289–90, 307–8, 322n12, 342n3 (app. 2). *See also* metaphor
analogs, 44
Anatidae. *See* ducks
anatomy: and behavior, 4, 44, 197; and development, 161, 300; and neurobiology, 121; and physiology, 18, 44–45, 51, 54, 118, 121, 128, 130, 158, 161, 167, 181, 207, 253–54, 302, 329n6; reproductive, 18, 43–44, 128, 158, 161, 166, 197, 218, 253–54, 329n6; sexual, 48, 50–51, 54, 130, 158, 160–63, 172, 181–82, 195, 214, 258; and sexual difference, 50; vertebrate, 277
androgen receptor, 173, 331nn30–31
animal behavior, 3, 260
anisodactyly, 104–5, 324n30. *See also* zygodactyly
anisogamy, 43, 192–93, 249, 333n7
Anna Karenina (Tolstoy), 231–32
anthropology, 34, 52, 59, 106, 319n23; and behavioral genetics, 285; and evolution, 189
antibiotics, bacteria's resistance to, 106, 284
antibodies, 120, 179, 287–91, 332n46, 342n2 (app. 2)
antigens, 120, 287–92
anti-Müllerian hormone (AMH), *135*, 140, 142, 145, 153, 172–73, 178, 180, 229–30
ants: eusocial, 112; haplodiploid sexual development, 199–200; pheromone trails of, 284
apical ectodermal ridge, 101
Arbib, Michael, 27

assimilation, genetic, 36, 301–3
astronomy, 260; and evolution, 189; and geology, 42; as historical science, 42–43; and ontology, 42
atomic theory, 339n23
Austin, J. L., 65–66, 68, 80, 84–85, 107, 119–20, 320n3, 321n17
autoimmune diseases, 234, 291–92
autonomy, sexual. *See* sexual autonomy
avian: beauty, as subjective, 9; biology, 1–2, 8–9, 284; cells, 196; chromosomes, 131, 196–97, 334n23; diversification, 2; evolution, 1–2, 284; flight, 5, 302; geographic variation, 252; penis, 9; phylogeny, 2–4, 12, 104; sexual communication and display, 8; somatic tissues, 196; structural coloration, 6. *See also* birds; ornithology

Babes in Arms (film), 138, 328n16
bacteria: cyano-, 333n4; and resistance to antibiotics, 106, 284; sexual recombination in, 333n5
Baetens, Dorien, 161, 331n40
Bagemihl, Bruce, 273
Balinese Character (Bateson and Mead), 320n2
Barad, Karen, 9, 20–22, 26, 35, 45, 66–69, 71–76, 78, 83, 85, 101–2, 106, 109–11, 125, 233, 237, 246, 253, 259–60, 262, 264, 270, 274, 278, 294, 296, 308–9, 317n28, 320n8, 321n20, 321n26, 325n39, 340n11, 340n15, 340n22
Barandiaran, Xabier, 59
Barbin, Herculine, 157
Bateson, Gregory, 320n2
bats, 192–93, 250–52
β-catenin, *135*, 140, 211, 214
beauty, defined, 8–9
Beauvoir, Simone de, 66, 158
bees: eusocial, 112; haplodiploid sexual development, 199–200

Beethoven, Ludwig van, 63–64
behavior: and adaptive agency of genes, 112; and anatomy, 4, 44, 197; and cellular morphology, 249; and developmental biology, 93; display, 3–4, 15, 123; and gender presentation, 22; and gender/sex realizations, 68; and genetics, 338n11; and homologs, 115; and masculinity, 152–53; and morphology, 249, 326n48; and neurobiology, 121; and pathology, 262; and performativity, 24, 64, 126; and phenotype, 152–53, 302; and physiology, 153–54; reproductive, 294; and selection, 271; and sex, 19; and sexual development, 153–54; and sexual difference, 50, 207; and sexual pathology, 262; and social environment, 214
Beldecos, A., 317n1
Bergson, Henri, 284
Between Men (Sedgwick), 60, 274
biases, 48, 132, 152, 206, 217–18, 255, 285, 327n59, 340n22
Biden, Joe, 17
binary bottleneck, 44, 185, 224–25, 318n9
biodiversity, 273, 284–85; conservation, 16, 34; evolution of, 32; and gene-level selectionism, 32; and history, 190; history of, 3; and performativity of phenotype, 190; and posthumanism, 75; and self-understanding, 32, 191, 249–50
bioelectrical fields, 105–6, 113, 320n34
biological development: and differentiation, 54–55; and evolution, 114; and gender performativity, 114; and genetics, 127; and physics, 102–3; and soft condensed matter physics, 102–3
biological essentialism, 21, 271
Biological Exuberance (Bagemihl), 273

biological performativity. *See* performative biology
biological self, enactment of, 77–126
biology: and agency, 9, 111–12; avian, 1–2, 8, 10, 198; and biodiversity, 32; and biomedicine, 23, 26–27, 129, 240, 270, 298; conceptually reframing, 21; and culture, 16, 24, 27, 33–34, 36, 270, 274–76, 285; and culture of gender/sex, 34, 276; definitions of, 39–40; and disease, 235; and diversity, 215; and economics, 275; and evolution, 1–2, 22, 39, 114, 191, 219; and evolutionary biology, 32; evolutionary novelty in, 219; and feminism, 16, 21–22, 25, 29, 33, 78; feminist critiques of, 21; of fertilization, 340n6; and gender, 73; "geneticization" of, 258; and genetics, 29, 73, 257; and geology, 42, 246; and health, 270; as historical science, 42–43; human, 261; and human culture, 8, 27; and humanities, 210, 247, 306; and material feminism, 20; and medicine, 275; and metaphor, 25–27; and ontology, 42, 318n3; organismal, 1, 54, 118–19, 327n56; and ornithology, 10; and phenotype, 27, 275; philosophy of, 77; and politics, 22; and psychology, 48, 275; and queer feminism, 1, 245, 257; reframing, 21; and science, 12–13, 26, 29, 41, 276, 285; and self-organization, 284; and self-understanding, 276; and sex, 13, 19–20, 29, 44, 191, 277, 318n13; and sexual body, 22; and social science, 261; and sociology, 275. *See also* developmental biology; evolutionary biology; molecular biology; neurobiology; performative biology; queer: biology; systems biology
biomechanics, 125, 300, 309, 311–13

biomedicine: and biology, 23, 26–27, 129, 240, 270, 298; and culture, 162; and health care, 267; and pathology, 162; and performativity, 234; and research, 18–19, 153, 177, 182–83, 214, 234, 240, 255, 257, 264–69, 285, 298, 318n13; and science, 14, 48; and sex, 285, 318n13; sex and race categories in, 264–69; and sex differences, 163, 265; and sexual development, 162; and sexual variation, 161; and technology, 161

birds, 2–9; adaptive radiations in, 217; aesthetic evolution of, 9, 12–13; anus of, 145; biology of, 1–2, 8–9, 284; courtship display, 8–9; and gynandromorphs, 196; history of, 12; neognathous, 206, 335n44; neotropical birds, 3–4, 122, 250; origin of, 5; paleognathous, 197, 206, 335n44; and performance, 125; sexual development of, 196; sexual differentiation in, 197, 334n23, 335n44; sexual reproduction of, 194; songs of, 8, 250–52; and sperm temperature sensitivity, 331n38; structural coloration in feathers and skin, 6–7, 15, 102–3, 315n7; study of, 2–9, 12; subjective and aesthetic agency of, 9. *See also* avian; ducks; feathers; ornithology

birdsong, as sexual communication, 8, 250–51

birth order, and sexual preferences, 179, 331n46

bisexuality, 14, 17, 53

bisphenol-A (BPA), 180

Black feminism, 14

black holes (astronomy), 42, 189

blastocyst stage, 171, 222

BMP2 (bone morphogenetic protein 2), 90

Bodies That Matter (Butler), 66, 69, 74–76, 191

Bogaert, A. F., 331n46

Bohr, Niels, 66–68, 260

bone morphogenetic protein 2 (BMP 2), 90

Bono, James, 27

Borges, Jorge Luis, 289–90, 342n3 (app. 2)

Botelho, João Francisco, 104, 324n30

BPA (bisphenol-A), 180

brain: development of, 121, 136, 285; function, 121; and homeostasis, 107; and hormones, 285

breast cancer, 208, 267, 295, 335n47, 341n29

Brennan, Patricia, 10–11, 315n11

Brownmiller, Susan, 10

Butler, Judith, 22, 24, 28, 35, 58, 66–76, 78, 86, 111, 122, 125, 190–91, 248, 256, 274, 321n27

butterflies, 196, 322n7

cacti (Cactaceae), 302

caecilians, 41, 215

CAH. *See* congenital adrenal hyperplasia (CAH)

CAIS. *See* complete androgen insensitivity syndrome (CAIS)

Campbell, Polly, 205–6

canalization, 96–98, 119, 128, 158, 185–86, 201, 217, 224, 236, 253, 312

cancer, 28, 48, 111, 118–19, 181, 208, 266–67, 295, 312, 322n5, 325n41, 327n54, 331n29, 335n47, 338n11, 341n29

canonical, vs. performative pathways, 98–100

"Can There Be a Feminist Science?" (Longino), 22

Capel, Blanche, 241, 272, 336n50

capitalism, 51

Carroll, Lewis, 204, 211

Cartwright, Nancy, 239

cats, 171, 207

causality, 1, 99, 108–10, 236–41, 259,

269, 297, 308. *See also* genetic causality and causation
cells: avian, 196; and communication, 84, 91, 271; and cytoplasm, 8, 86, 88, 91, 96, 98, 103, 105, 136, 139, 152, 168, 200, 204, 323n14, 334n26; and cytoplasmic sterility mutations, 200; as docile citizens, 327n54; fetal, 222; and genes/genetics, 84, 91–95, 103, 271; genetically identical, 79, 113–14, 118, 249, 305–6; and genomes, 79; immune, 120, 222–23, 288–90, 292; Leydig, 140, 145, 172; mesenchymal, 101; and molecules, 84, 271; and morphology, 249; and organs, 59, 119; and parts of body, 119, 128; and phenotypes, 106; and proteins, 109; Sertoli, 140–42, 153; and sexual development, 6, 127; stem, 95, 180, 330n17; and tissues, 95, 100–105, 324n27
Century of the Gene, The (Keller), 283
Chang, Hasok, 26, 242, 319n14, 339n23
character identity mechanisms (ChIMs), 116–17, 326n51, 336n48
character identity mechanistic performativities (ChIMPs), 326n51
character identity networks (ChINs), 115–16, 141, 336n48
"Charles Darwin: Queer Theorist" (See), 273–74, 277
Chen, Yen-Shan, 203, 211, 330n23, 334n36
Cheney, Dick, 41–42
Cheng, Eugenia, 21
Cheverud, James, 312
chimerism: genetic, 166, 330n17; micro-, 180
ChIMPs. *See* character identity mechanistic performativities (ChIMPs)
ChIMs. *See* character identity mechanisms (ChIMs)
ChINs. *See* character identity networks (ChINs)

Cho, Sumi, 14
chromosomes: avian, 197; biological sexual binary determined by, 23; and differentiation, 201, 208; and genes, 23, 132, 166, 215, 318n2; and genomics, 285; and homology, 201–2; mammalian, variation in, 205–6, 215–16; and phenotype, 157; role of, 130–32; sex, 130–32, 196, 206, 229, 285, 334n28; sex determination by, 130; and sexual development, 127, 130–32, 166–67, 198; and sexual differentiation, 197. *See also* X chromosomes; Y chromosomes; Z chromosomes
Cipolla, Cyd, 321n26
citationality: and genealogy, 70; and homology, 114–17; and immunity, 291; and performativity, 69–72
Clark, A. G., 325n41, 338n10
Clayton, Janine A., 255, 317n2
climate change, 16, 18, 34, 198
clitoris, 44, 51, 117, 128, 147–48, 163, 172–77, 183, 186–87, 218–20, 230
cloaca and cloacal folds, 145–49, 214, 218, 230
clownfish (Amphiprioninae), 212–16
codons, 41, 82
coercion, sexual, 10–13, 123, 335n46
Cohn, Martin, 219, 279
colonialism, 19, 51, 60
communication, art as, 124. *See also* discourse; intra-actions and intra-activity; sexual communication
competence, and induction, 108
complete androgen insensitivity syndrome (CAIS), 173, 331n31
conflict, sexual. *See* sexual conflict
congenital adrenal hyperplasia (CAH), 175–76, 178–79, 254, 278
conservation: biodiversity, 16, 34; and ecology, 2
Couger, Matthew B., 205–6

COVID-19, 232, 234, 265, 287, 290–91, 292, 338n9
Crenshaw, Kimberlé, 14
Creutzfeldt-Jakob disease, 85
cryptorchidism, 165, 176, 179
cultural critique, 20, 28
cultural oppression. *See* oppression
culture: and biology, 16, 24, 27, 33–34, 36, 270, 274–76, 285; and biomedicine, 162; changes in, 17; and gender, 19, 22–23, 250; of human sex, 14; and human sexual possibilities, 14, 162; and language, 20–21, 69; and nature, 15, 19, 252, 283, 285; and politics, 15, 28, 33–34; and science, 1, 15–16, 18–19, 21–24, 27–29, 33–34, 38, 46–47, 152, 159, 249–50, 256–58, 270, 276, 283, 285, 329n3. *See also* human culture
culture of gender/sex: and biology, 276; and science, 33
culture of human sex, 14, 162; and biology, 34
cystic fibrosis, 108, 325n37
cytokines, 221–22, 292
cytoplasm. *See under* cells; genes

Darwin, Charles, 8, 30, 273–74, 277, 284, 285–86, 305, 308
Davis, Georgiann, 159, 161, 165, 186
Dawkins, Richard, 13, 31–32, 55–56, 112–13, 119, 250, 305, 309, 319n28, 326nn48–49
DAX1 (dosage-sensitive sex reversal, adrenal hypoplasia critical region, on chromosome X, gene 1), 177–78, 331n43
"Decade of the Gonad" (Fausto-Sterling), 317n15
Deleuze, Gilles, 284
deoxyribonucleic acid. *See* DNA (deoxyribonucleic acid)
de Quieroz, K., 318n3
Derrida, Jacques, 69
development: and anatomy, 161, 300; chemical communication in, 108; defined, 55; and differentiation, 5, 54–55, 100–101, 136, 167, 172, 255, 257, 299; of embryos, 197–98, 222, 329n32; and evolution, 24, 57, 114, 116, 125, 236, 239–40, 243, 272, 281; and genes/genetics, 1, 23, 35, 74, 78–84, 99, 105–6, 109, 119, 133, 176, 207, 211, 238, 240, 269–70, 325n39; and growth and space, 100–102; and materiality, 74, 78; molecular genetic mechanisms of, 6, 78, 297; organismal, 102, 107–8, 117–18; and phenotype, 170; and phylogenetics, 32; and phylogeny, 30–31, 243; physical forces in, 102–6; and physiology, 84, 86, 111, 113, 117–19, 121, 153–54, 182, 249, 268, 271, 298, 302, 306, 329n32; and psychology, 22, 121–22; and regeneration, 324–25n34. *See also* biological development; sexual development
developmental biology: agency in, 111–14; and behavior, 93; and discourse, 58; and evolution, 306, 312; and evolutionary biology, 30–31, 317n27; and feminism, 25; and feminist cause, 78; and gender performativity, 23; and genes/genetics, 35, 73, 99, 106, 110, 324n27; human, 13; and humanities, 91, 119; and material body, 22; and material feminism, 21, 101–2; and material-structural approach, 324n27; and modularity, 299; and molecular signals, 84; and New Synthesis, 306, 317n27; and organismal development, 108; and performative continuum, 24–25, 246; and performative theory, 35; and performativity, 98, 272, 332n48; and phylogenetics, 32; and phylogeny, 31; and physiology, 111, 113, 119; and queer bodies, 60–61; and queer feminism, 1, 13, 257; revolutionary

advances in, 33; and sexual body, 23; and signaling pathways, 98–99
developmental feminization, 334n26
developmental fields, 100–102, 105, 112–13, 116, 249, 255, 269
Developmental Plasticity and Evolution (West-Eberhard), 297
developmental systems theory (DST), 283–84, 296–97
DHT (dihydrotestosterone), 145, 147–49, 172–73, 176, 181–83, 229–30, 329n32, 331nn29–30
Dickkopf/bullhead. *See* DKK (Dickkopf/bullhead)
differentiation: and biological development, 54–55; chromosomal, 201–2; defined, 54–55; and development, 5, 54–55, 100–101, 107, 136, 167, 172, 255, 257, 299; and pattern formation, 54; and sexual development, 54–55. *See also* sex differentiation; sexual differentiation
DiFrisco, James, 116, 326n51
dihydrotestosterone. *See* DHT (dihydrotestosterone)
Dillon, Elizabeth, 316n22
dimorphism, 199, 206–8, 211, 335n46
dinosaurs, 3, 6, 115, 217
dioecy, 192, 200, 333n7
disability, 231–36, 291. *See also* disease; illness; impairment
discourse: and agency, 68; biological, 58, 74–75, 78, 117–18, 247, 323n12; cellular, 106–8; communicative, 247; and Continental philosophy, 284; cultural, 247; developmental, 91, 93, 181, 210, 300, 312; everyday, 274; and genetic action within cells, 91–94; intra-agential, 91; and language, 106; and material-discursive practices, 68; molecular, 78, 80, 83–89, 91, 95, 107, 118, 120, 122, 125, 155, 177, 209–11, 248–49, 272, 289, 300, 305, 312, 322–23n12; nonhuman, 322–23n12; and performativity, 68, 72, 115, 247–49; as persuasion, 84; power of, 72, 78; reiterative power of, 58, 78, 86, 248; as suasion, 84; theatrical, 126. *See also* communication, art as; intra-actions and intra-activity; language
discrimination, sexual, 17
disease: autoimmune, 234, 291–92; and biology, 235; complex, 235, 295, 325n41, 338n11; etiology of, 133, 232, 234, 325n41; and genes, 108, 325n37, 325n41; heritable risks of, 325n41; infectious, as controlled, 234; and medicine, 232; and performative phenotype, 232; and proteins, 85; single-gene, 108, 325n37. *See also* disability; health; illness; *and specific diseases*
diversity: and biology, 215; evolutionary, 114–15, 191, 204, 216; and evolutionary biology, 32; of feathers, 5; of individual becomings, 16; of mating types, 45; and phenotype, 35, 190, 264; and race, 263–64; sexual, 45, 123, 213, 315–16n15; and sexual difference, 263; of sexual differentiation, 200. *See also* biodiversity
DKK (Dickkopf/bullhead), 147–48, 329n30
DMRT1 (doublesex and mab-3 related transcription factor 1), 131–32, 141, 177, 196, 209, 328n6, 328n23
DNA (deoxyribonucleic acid), 85, 91–92, 99, 112, 116, 168, 181, 193, 195, 210, 247, 322n5, 323n22, 329n28; double helix, *81*; and gene expression, 94–95; and paracrine signaling, 88, 109; pre-DNA embryology, 102; sequences, 26, 41, 82, 130, 169, 201–2, 234–35, 322n11; structure of, 29, 80–82, 231. *See also* genes; RNA (ribonucleic acid)
Dobzhansky, Theodosius, 244
Donoghue, M. J., 318n3
dramaturgy, 65, 78

DST. *See* developmental systems theory (DST)
ducks: aesthetic evolution and sexual coercion in, 12–13; biology of, 284; Female, 10–11; penises of, 9–10, 132, 196; sexual autonomy of, 9–11; and sexual violence, 10–11
Dupré, J., 318n3, 322n27

echinoderms, 192, 213
Ecological Theater and the Evolutionary Play, The (Hutchinson), 126, 275
ecology: and conservation, 2; and eugenics, 286; and evolution, 126, 213, 218–19; and history of life, 190; invasive species, 286; and multispecies, 275; queer, 60, 275; terrestrial, 218–19
economic inequality, 16, 270
economics, 12, 260, 275, 305
eco-toxicology, 34
Edwards, Erika, 302
electrochemistry, 105, 242, 339n23
Ellis, Havelock, 51–52
embryos and embryology, 79, 102, 117, 162, 194; biomaterialization of, 69; development of, 126, 197–98, 222, 329n32; gendering of, 69; and genomes, 201; material developmental environment, 221; and morphology, 201; and pharmaceuticals' effect on, 181; pre- DNA, 102; sexual development of, 35, 151, 157; and sexual differentiation, 228; transplants, 108; twin, 179. *See also* performativity: placental
EMX2 (empty spiracles homeobox 2 gene), 173, 175, 331n33
endocrine signals and signaling, 86–98, *87–88*, 109, 116, 118, 123, 154, 155, 162, 170, 229–31, 255, 289, 337n6. *See also* paracrine signals and signaling

endocrinology, 151, 265
environmentalism, 60
environmental pollutants. *See* pollution and pollutants
epidemiology, 18, 34, 234
Epigenetic Landscapes (Squier), 96
epigenetics, 26, 96–98, *97*, 199, 240, 317n15, 323n22
Epistemology of the Closet, The (Sedgwick), 170
equal rights, 19
essence: biological, 258; and essentialism, 70; and gender, 321n17; individual, 45, 215–16, 258, 270; sexual, 58, 128, 215–16
essentialism, 215; anti-, 258; biological, 21, 271; and essences, 70; sex, 318n13
ethics: and feminism, 19, 286; and science, 260–62; and values, 260
ethology, and wild animal behavior, 3–4
etiology, of disease, 133, 232, 234, 325n41
etymology, 241, 323n17
eugenics: and ecology, 286; and genocide, 53; and normative sex difference, 53; and racial difference, 34, 262; and racism, 51–53, 263, 285–86; and statistical tools, 340n22
eukaryotes, 94, 193, 246, 248–49, 319n18, 326n46, 327n4, 333n4
eusociality, 112, 313, 339n24
Everything Flows (Nicholson and Dupré), 318n3
evolution: and adaptation, 31, 52, 267, 297; aesthetic, and sexual coercion in ducks, 12; and aesthetics, 8–9, 12–13; and anthropology, 189; and astronomy, 189; avian, 1–2, 12, 340n8; as construction, 297; defined, 189, 243; and development, 24, 57, 114, 116, 125, 236, 239–40, 243, 272, 281; and developmental

biology, 306; and ecology, 126, 213, 218–19; and gender performativity, 114; and genetics, 30–31, 264, 323n22; and genetic selection, 190, 308; of homologies, 4, 116; and humanities, 189; and individuality of sexes, 43; and innovations, 6, 119, 194, 217–19; of material body, 22, 272; of multicellular phenotypes, 32; and natural selection, 31; of organismal bodies, 1, 23–24, 75, 125, 236, 239–40, 242; and performativity, 325–26n45, 336n50; and phylogeny, 191, 317n27; and science, 33–34; selective genetic, and historical contiguity, 190; of sex, 191–95; of sexual body, 22, 37; and sexual development, 195–200, 217–18; and sexual differentiation, 211; and sexual essences, 215–16; and sexual reproduction, 224; and sexual variability, 189–225; and "survival of the fittest," 126; and systems biology, 298; and zoology, 189

evolutionary biology, 10, 22, 39, 77–78, 93, 112–15, 119, 121, 126, 145, 159, 191, 198, 215, 219, 223, 227, 237, 240–45, 281, 326n49; and agential realism, 76; avian, 2, 284; and biodiversity, 32; and biology, 32; and development, 313; and developmental biology, 30–31, 310, 317n27; and developmental constraints, 336nn55–56; and discourse, 58; and diversity, 32; and eugenics, 34; and evolutionary advantage of sexual reproduction, 333n5; and evolutionary genetics, 309; faux-synthesis in, 335n46; and feminism, 25; and feminist cause, 78; and gender performativity, 23; and genetic assimilation, 303; and genetics, 285–86; and genetic selection, 32, 305, 307–10, 313; historical ontology in, 318n3; and history, 308, 317n27; and humanities, 119; and internal selection, 313; and material feminism, 21, 46, 74; and modularity, 299; and morphology, 326n48; and natural selection, 335n46; and New Synthesis, 30–31, 242–44; ontology in, 318n3; and ornithology, 2, 315n4; and performative biology, 242–44; and performative continuum, 25, 246; and performativity, 190, 272; and phenotype, 30–32, 305; and phylogenetics, 32, 317n27; and phylogeny, 30, 317n27; and physics, 30, 32; and physiology, 119; and queer bodies, 60–61; and queer ecology, 275; and queer feminism, 1, 13, 257, 277; and queer theory, 317n28; and race, 278–79, 285–86; revolutionary advances in, 33; of sex, 60–61; sexist origins of concepts and models in active use in, 284; and sexual body, 23; and sexually disruptive selection, 336n46; and sexual reproduction, 333n5; stakes for, 30–32, 242. *See also* genotype-phenotype relationship

Evolution of Beauty, The (Prum), 123, 315–16n15

Evolution's Rainbow (Roughgarden), 273

fallopian tubes, 128, 142–44, 172–75, 178, 218, 230

Fausto-Sterling, Anne, 19, 23, 51, 129, 151, 159, 163, 165, 186, 255, 283–84, 317n15, 321n26, 334n37

feathers: and adaptation, 5–6, 217; and β-keratin, 7, 102–3; development and evolution of, 4–6, 8, 15, 47, 302; and dinosaurs, 3, 6; diversity of, 5; and evolutionary innovations in phenotype, 219; hierarchical com-

feathers (*continued*)
plexity of, 15; and modularity, 5–6; and natural selection, 9, 47; origin of, 5; performativity of, 303; and phenotype, 219, 313; and placodes, 102; structural coloration in, 6–7, 15, 102–3, 315n7

Feeling for the Organism, A (Keller), 283

Female, and Male. *See* Male, and Female

femininity, 52–53, 153, 215, 261–63. *See also* masculinity

feminism, 246; and biology, 16, 21–22, 25, 29, 33, 78; Black, 14; and ethics, 19, 286; gender-critical, 19; and genetics, 29; and philosophy, 1, 66, 286; political and ethical base for, 19; and politics, 19; and science, 11–13, 22, 283–85; second-wave, 21, 316n7; and traditional binary perspective, rejection of, 19. *See also* material feminism; queer feminism

feminist analyses, 11, 20, 25, 58, 68, 129–30, 257, 283–85

feminist cause, 78

feminist critique, 19–21, 285

feminist theory, 2, 13–16, 20, 25, 34, 66, 74, 270, 277

feminization, developmental, 334n26

Ferguson, Roderick, 60

fertilization, biology of, 340n6

FGF (fibroblast growth factor), 101, *135*, 136, 138, 168–70, 328n15

FGFR2 (fibroblast growth factor receptor 2), *135*

Fidelio (Beethoven), 63–64

Fisher, Ronald A., 307–8, 326n49, 340n22

fishes: flying, 106; genetic sexual differentiation systems in, 197; live-bearing, 337n61; osmoregulatory performance of, 125; parthenogenetic, 45–46, 337n64; and performance, 125; sexual reproduction of, 194, 218–19; spacer regions in, 322n7. *See also specific fishes*

5α-reductase, 148, 172–73, 175, 181–83, 329n32, 331n29

FOG2 (friend of GATA2), *135*, 138, 140, 328n17

food production, sustainable, 16

forced copulation (acts of sexual violence in nonhuman animals), 10–11

Forkhead box L2. *See* FOXL2 (Forkhead box L2)

Foucault, Michel, 70–72, 117–19, 157–59, 217, 247, 271–73, 318n7, 326–27n53, 327n54

FOXL2 (Forkhead box L2), *135*, 140–41, 214, 328n21

Framingham Heart Study, 208

freemartins (cows), 179

Freud, Sigmund, 22, 122, 159

friend of GATA2. *See* FOG2 (friend of GATA2)

Fringe, Lunatic, 133–34, 328n10

Frizzled (FZD), 140

From Darwin to Derrida (Haig), 305–6

fruit flies: early embryonic development of, 272; gene action in, 109–10, 133; and genotype-phenotype abstraction, 77; mutation in, 328n10; with red eyes, 293; and wing morphology, 301, 328n10

fungi, 40, 45; meiosis in, 193; multicellular, 79

FZD (Frizzled), 140

Galen, 158

gametes, 37–38, 43, 167, 193–94, 201, 213–14, 249, 323n22, 333n7, 334n30, 340n6; hetero-, 131, 196–97, 202, 205, 228–30, 334n32; homo-, 131, 196, 199, 207, 228–30

Garland, E. C., 340n8

Garland, Judy, 138

Gastaud, F., 332n59

GATA4 (GATA-binding protein 4), *135*, 138, 140, 328n17

gender: as becoming, 66; and biology, 73; construction, 66, 71, 73, 321n17; as culturally constructed, 66; and culture, 19, 22–23, 66, 250; of embryos, 69; and essence, 321n17; expression, 71, 321n17; as extended phenotype, 55–56; genealogy of, 252; historical critique of concepts of, 284; identity, 17, 70, 165–66, 183–84, 258, 321n17; in nature, 249–52; nonconforming people, 14; presentation, 22, 114–15, 321n17; as process, 66; -queer people, 14, 22, 32, 71; realization, 68–70, 321n17; reidentification, 184; revolution, 251; roles, 65; and science, 18; vs. sex, 18–19; in social world, 66; variations, 70–71, 216–17. *See also* gender performativity; gender/sex; gender theory

Gender Advertisements (Goffman), 320n2

gender performativity, 22–25, 63–76, 117, 126, 151, 178, 190, 216–17; and biological development, 114; defined, 64–67; elements of, 67–72; and humanities, 65–66; and language, 67; and literary criticism, 65, 67; and philosophy, 65; and politics, 70–71; and trans experience, 72–73

genderqueer people, 14, 22, 32, 71

gender/sex, 1–2, 14, 16, 22, 24, 28–29, 32–33, 35–36, 47, 51–53, 55–56, 60, 65, 68–70, 73–75, 115, 128–29, 149, 151, 154, 168, 179, 182–83, 187–88, 230, 234, 236, 246, 250–53, 256–59, 261, 265–66, 270–72, 276, 279, 283–85; defined, 17–20, 38. *See also* Male, and Female

gender theory, 15, 23, 28, 66, 72, 163, 178, 186, 258

Gender Trouble, Bodies That Matter (Butler), 66, 256

genealogy: of biological discourse, 247; and citationality, 70; of evolution, 191, 193; of gender, 252; of human reproductive biology, 193; of performative discourse, posthuman, 247–49; of performativity, 36, 70, 250; of sex, 246

gene expression, 24, 26, 50, 79, 82, 84–85, 89, 91–96, 100, 103, 105–7, 113, 119–20, 134–36, 140–41, 190, 205–6, 248, 297, 305, 309, 323n14, 323n22

genes: and anatomical homologs, 40; androgen receptor (AR), 331nn30–31; and body, 15; as causes, 108–11, 234; and cells, 84, 91–95, 103, 271; and chromosomes, 23, 132, 166, 215, 318n2; and communication, 322n11; and cytoplasm, 204; defined, 26, 82, 116; and development, 78–84, 105–6, 133, 176, 211, 270, 325n39; and developmental biology, 35, 73, 99, 106, 110, 324n27; and disease, 108, 325n37, 325n41; and environment, 295, 297; fringe, 134, 328n10; gene by environment (GxE) interactions, 27, 295–96; and genetic causality, 240; hedgehog, 134; and heritability, 24, 30, 82, 326n49, 338n11; immunoglobin, 287–90; and master switch, 138, 167, 173, 184, 203, 240–41, 336n52; and metaphor, 25–27; and molecular agents, 109; networks of, 116; nomenclature, 132–34; and performativity, 100, 113, 322n11; and phenotypes, 82; and proteins, 90, 134, 215, 328n13; regulation of, 59, 78, 80, 83–84, 92–99, 103, 106, 108–13, 116, 139–41, 182, 209–11, 219, 236, 248, 270–71, 325n39, 329n32; and reproductive tract development, 166; selfish, 31–32,

genes (*continued*)
56, 112–13, 119, 305–6, 309, 326n48, 335n46; and sexual development, 29, 35, 127, 132–34, 177; and transcriptomes, 321n14. *See also* DNA (deoxyribonucleic acid); gene expression; genetics; transcription factors

genetic assimilation, 36, 301–3

genetic causality and causation, 108–11, 234, 240, 257, 270, 308–9

genetic chimerism, 166, 330n17

genetics: and assimilation, 36, 301–3; and behavior, 338n11; behavioral, 285, 338n11; and biological development, 127; and biology, 29, 73, 257; and cells, 84, 91–95, 103, 271; developmental, 1–2, 23, 35, 74, 99, 109–10, 119, 207, 236–38, 240, 257, 269, 298, 320n3; and developmental biology, 35, 73, 99, 106, 110, 324n27; and evolution, 30–31, 264, 323n22; and evolutionary biology, 285–86; and feminism, 25, 29; and gender performativity, 23; and material feminism, 21; and mathematical models, 30; metaphor in, 322n12; and molecular biology, 43; molecular-developmental, 23; and molecular-developmental biology, 23; new synthesis of development, evolutionary biology, and, 297; and performance, 256–57, 260; and performative continuum, 25; performative theory of, 35; population, 30–31, 43, 77, 82, 242–44, 263–64, 284, 293–99, 303, 306–7, 311–13, 316n16, 326n49; and queer bodies, 60–61; and queer feminism, 1, 13, 257; racist and eugenic roots of common concepts in, 285–86; and sex, 23; and sexual body, 23; and sexual development, 29, 35, 177, 198–99; and sexual differentiation, 201; and sexual variability, 190; and transcription, 322n11. *See also* epigenetics; genes; molecular genetics; phylogenetics

genetic variation, 58, 77, 96, 110, 113, 139, 172, 189–90, 193, 199, 203–9, 217, 232, 235, 250, 267, 269, 295–97, 300, 307–8, 325n37, 325n41, 326n48, 335n46, 336n50, 338nn9–11; and biomedical research, 182; in gonad development, 167–71; among individuals, 288; noncoding, 177–78; and phenotypic diversity, 190. *See also* alleles (gene variants)

genitalia and genitals: ambiguous, 160, 165, 183, 186; cultural assumptions about, 186; development of, 132, 145–49, *146*, 154–55, 172–79, 229–30, 277, 329n32; differentiated, 147, 155; external, 147, 176; and genetics, 155; and hormones, 172; as inverted, 158; mammalian, 132, 277; reconstructions of, 187; undifferentiated in fishes, 214; undifferentiated in twins, 179; variable, 160, 167, 170, 182

genital tubercle, 147–48, 172–73, 176, 218–20, 230

genocide, 53, 263

genomes: as algorithm or developmental program, 83; binary, 43; as blueprint, 27, 82–85, 185, 247; and cells, 79, 83; and development, 57; and embryos, 201; as generational, 27, 43; and genes, 79–80; as highly structured, 82; human, 56–57, 130, 133, 234–35, 264, 266, 325n41, 328n11; as informational resource and lexicon, 27, 79, 84–85; and inheritance, 27, 43, 77; linguistic analogies of, 85, 322n12; and metaphor, 322n12; nuclear, 41–42; organismal, 4; and phylogenetics, 4; and proteins, 82; selfish genes in,

56; and sex determination, 23, 46, 56–58; and sexual development, 127; and sexually differentiated bodies, 201; variable, 288, 291
genome-wide association studies (GWAS), 295–96
genomics, 26, 28, 33, 79–85, 93–94, 106, 167, 177, 185, 190–91, 234–35, 264, 269, 279, 283, 285, 295, 306, 317n15, 322nn11–12, 325n41, 338nn10–11
genotype-phenotype relationship, 24–25, 27, 31, 35–36, 46, 77, 83, 207, 247, 274–75, 293–98, 300–303, 307
geology: and astronomy, 42; and biology, 42, 246; as historical science, 42–43; and ontology, 42
Ghiselin, M. T., 318n3
Ghost Stories for Darwin (Subramaniam), 285–86
Gill-Peterson, Jules, 129, 159, 165, 330n13
glands: endocrine, 86, 151, 172; mammary, 56, 192, 194, 216, 224; prostate, 48, 149, 175–76, 208. *See also* gonads; ovaries; testes
global warming. *See* climate change
Goffman, Erving, 65–66, 279, 320n2
Goldberg, Rube, 99
gonads: bipotent, 134–38, 140–42, 147, 154, 162, 167, 170, 172–73, 178, 209, 213; development of, 56, 132–42, *135*, 154–55, 166–71, 177–78, 209, 211; differentiated, 134–41, 155, 160, 336n52; and genetics, 155; genetic variations in, 167–71; and hormones, 184; mammalian, 132; molecular signaling pathways in, 134; SRY binding to SOX9 enhancer, 169; and SRY mRNA stabilization, 139; and SRY transcription factor function, 136
gonochory, 333n7. *See also* dioecy
Gould, Stephen Jay, 190, 336n55

Gowaty, Patty, 10, 283–84, 317n1
Gray, Russell, 296–97
green algae, 43, 193
Greene, Marjorie Taylor, 18
Griffiths, Paul, 296–97
Grosz, Elizabeth, 50–51, 284, 321n26
"Gut Feminism" (Wilson), 21, 22
GWAS. *See* genome-wide association studies (GWAS)
gynandromorphs (individuals that are bilaterally Male and Female), 196

Hacking, Ian, 318n7
Haig, David, 305–6, 308
Hall, J. M., 341n29
Halperin, David, 60–61, 273
Haraway, Donna, 12, 15, 28, 83–84, 101–2, 246, 269–70, 283, 321n26, 340n15
Harding, Sandra, 284
Hayward, Eva, 46
health: and biology, 270; disparities, 16; improved, 267; and medicine, 187, 232; public, 34; and social inequities, 266; and well-being, 235, 264; and wellness, 232, 255. *See also* disease; medicine
health care: access to, 16, 233, 266–67; and biomedicine, 267
Hekman, Susan, 321n26
Hennig, W., 318n3
hermaphrodites, 157; practicing, 164; pseudo-, 160; sequential, 213–14, 225; simultaneous, 213, 259
Herpin, Amaury, 211, 336n52
Herzig, R., 340n15
Hesse, Mary, 27, 317n19
heterochromatin, 94, 171
heteronormative, 1, 24
heterosexuality, 33, 53, 60, 68, 70, 159–60, 186, 188, 274, 285
Hird, Myra, 129, 159, 165, 284–85, 316n7

histones, 80, *81*, 94–95
historical individuality. *See* individuality, historical
history, study of, 12
history of life, 126, 190–91, 193, 247, 333n4
History of Sexuality, The (Foucault), 70
Hitt, Jack, 342n3 (app. 2)
"holon," 319n14
HomeoBox domain protein (HOX). *See* HOX (HomeoBox domain protein)
homeostasis, 24–25, 80, 107–8, 118–20, 153, 223, 236, 271; maternal hormonal, 329n32
homology: and anatomy, 4, 40, 44–45, 114–16, 128, 142, 189; and avian penis, 9, 132; and behavior, 4, 115; and chromosomes, 197, 201–2, 267, 334n32; and citationality, 114–17; and developmental genes, 133; embodied reproductive, 44; and evolution, 4, 116, 190, 218–19, 333n7; hierarchies of, 190; and individuality, 4, 318n3; and molecular modules, 219; and morphology, *220*; and phenotype, 117, 190, 219; and phylogenies, 191–93; and physiology, 128; reproductive, 44, 47, 130, 132, 142, 193, 213, 221, 252; sexual, 45, 147, 158–59; and sexual development, 154, 162, 195–96; and sexual organs, 9, 51, 132, 149, 164, 193, 213, 218–19, *220*, 225, 333n7; and "tree thinking," 4; vertebrate, 328n10. *See also* morphology; phylogeny
homosexuality, 179, 331n46. *See also* same-sex sexual attraction
hormones, *87*; and brain development, 285; and genitalia/genitals, 172; and gonads, 184; and molecules, 59, 93, 153; and phenotypes, 154; and proteins, 92; sex, 152–53, 179, 331n30; and sexual development, 35, 153–54, 179; and therapies, 18, 258
HOX (HomeoBox domain protein): HOXA1, 337n8; HOXA13, 175; HOXD13, 147, 219–20, 279, 329n28
Hull, David, 317n27, 318n3
human culture: and biology, 8, 27; gender/sex roles in, 65, 252–53; and historical ontology, 42; and metaphors, 27; patriarchal, 11; and performative discourse, 249
Human Genome Diversity Project, 264
Human Genome Names Consortium, 133
Human Genome Nomenclature Committee, 328n11
humanism, 321n20. *See also* posthumanism
humanities: and biology, 210, 247, 306; cultural discourses in, 247; and developmental biology, 91, 119; and evolution, 189; and evolutionary biology, 119; and gender performativity, 65–66; and performative theory, 35; and queer feminism, 66; reframed, 12; and science, 12–14, 51, 274–75, 305; and social sciences, 12, 27, 65, 249, 251–52
hummingbirds, 250
Hutchinson, Evelyn, 126, 275
Hutson, Matthew, 325n34
hypospadias (developmental condition), 163–65, 172–77, 180–83, 186–87, 218, 254

ideals: and monogamy, 285; sexual, 52–53
identity, sexual, 60, 160, 183, 185
IL (interleukin), IL1, IL9, IL17A, and IL19, 221–22, 336n60
illness: and performative phenotype, 232; performativity of, 231–36, 291. *See also* disability; disease

immune systems, 342n1 (app. 2)
immunity: and citationality, 291; and physiology, 119–20. *See also* acquired immunity
impairment, 235–36. *See also* disability
Imperato-McGinley, Julianne, 182–83, 331n29
individuality: biological, 4, 24, 124, 318n3; and homology, 4, 318n3; human, 29, 235; intellectual value of, 12; ontological, 39–44, 326n46, 338n14; of sexes, 43
individuality, historical, 39–44, 47, 215, 319n14, 326n46, 327n54. *See also* natural kinds; ontology
individuation, 90–91, 98, 128, 300
induction: and competence, 108; and developmental identity change, 108; and tissues, 108
inequality, economic, 16, 270
influenza viruses, 342n2 (app. 2)
injustice, social, 34
innovations: and behavioral novelty, 4; evolutionary, 6, 119, 194, 217–19; and historical analysis, 70; and norms, 216–19; and ornithology, 15; and performativity, 72; role of development in evolution of, 336n56; sexual, 193
interACT Advocates for Intersex Youth, 332n59
interferon regulatory factor 9 (IRF9). *See* IRF9 (interferon regulatory factor 9)
interleukins. *See* IL (interleukin), IL1, IL9, IL17A, and IL19
Internal Factors in Evolution (Whyte), 311
internal selection, 36, 243, 311–13, 343n2
intersexuality, 329n6, 332n59; defined, 160, 329n6; and identity, 61, 161; as marginalized, 38; and material feminism, 284–85; medical treatment of, 161, 165, 283, 285; pathological, 163; as sexual minority, 14
intra-actions and intra-activity, 9, 43, 49, 59, 67–72, 74, 78, 83–86, 89, 91–93, 99–101, 107–12, 116–18, 121–25, 141, 153–55, 167, 170, 179, 190, 196, 205, 210–11, 219, 221–23, 228, 235–39, 246–47, 249, 252–56, 259–60, 275–76, 290–91, 296–98, 300, 308–9, 312–13; aesthetic of art, 124; agential as causal enactments, 325n39; communicative, 271; defined, 67, 86; in developmental genetics, 320n3; meiosis as, 246, 339n3; sexual reproduction as, 45, 246. *See also* discourse
introns, 82
INUS conditions, 239
invertebrates: parthenogenetic, 45–46; sexual reproduction of, 194; terrestrial reproduction in, 194. *See also* vertebrate animals
ions, 339n23
IRF9 (interferon regulatory factor 9), 209, 336n49
Irigaray, Luce, 50–51, 284
Is Water H₂O? (Chang), 242

Jagose, Annamarie, 60
James, William, 309, 326n48
jellyfish, eusocial, 112
Joel, Daphna, 255
Johnson, Barbara, 51
Johnston, S. D., 333n14
Jordan-Young, Rebecca, 49, 129, 152–53, 285
Jung, Carl, 159

kangaroos, 202, 336–37n60
Karkazis, Katrina, 49, 129, 152–53, 159, 161, 165, 186, 285
Kashimada, Kenichi, 203
Kaur, R., 334n26
Kay, Lily, 26, 317n19, 317n21, 322n12

Keinan, A., 325n41, 338n10
Keller, Evelyn Fox, 78, 283, 321n26
Kessler, Suzanne, 159, 165, 186
Kirby, Vicki, 106, 284
Klinefelter's syndrome, 166, 171, 254
Klinge, Ineke, 265–66
Koestler, Arthur, 319n14
Koopman, Peter, 203
Krafft-Ebing, Richard von, 52–53, 319n23
Kuhn, Thomas, 184–85

labia, 44, 117, 128, 147–48, 149, 151, 163–64, 172–73, 176–77, 183, 230, 330n11
labiaplasty, 164, 330n11
Lacan, Jacques, 122
language: authority and power in, 70–72; and culture, 20–21, 69–70; and discourse, 106; and etymology, 241, 323n17; and "euphemism treadmill," 161; and gender, 251; and gender performativity, 67; and historical ontology, 42; and literary criticism, 67; and material body, 284; and materiality of body, 284; and metaphor, 26–27; and ontology, 42; and science, 26; and words, 323n17. *See also* analogies; discourse; metaphor
Laqueur, Thomas, 158–59, 188, 329n3
Latour, Bruno, 59
Lavoisier, Antoine, 242
LEF1 (lymphoid enhancer binding factor 1), 140, 328n20
Levin, Michael, 325n34
Lewontin, Richard, 77, 293–95
LGBTQ+ rights, 316n4
liberation: and politics, 60; women's, 19
"Library of Babel, The" (Borges), 289–90
lichen species, 42
life, origin of, 238, 247–50. *See also* history of life

Lillie, Frank, 159, 257
linguistics, 20–21, 29, 74, 80, 85, 210, 224, 322n12
Linnaean classification, and phylogeny, 2–3
literary criticism: feminist, 25, 51, 60; and gender performativity, 65, 67; and language, 67
literary theory, 20, 58, 60, 66, 69
Lives of the Cell, The (Thomas), 75
lizards: as "cold-blooded," 333n21; and genetic sexual differentiation, 197, 199, 215, 229; legless, 41; live-bearing, 229, 337n61; parthenogenetic, 45–46, 194–95, 337n64; penis of, 218; and performance, 125; physiology of, 333–34n21; sexual development of, 198–99; sexual reproduction of, 194
Lofton, Kathryn, 279, 320n2
Longino, Helen, 22, 284
Lorenz, Konrad, 4
Love, Alan, 116, 326n51
Lunatic Fringe, 133–34, 328n10
Lurtsema, Robert J., 339n1
lymphoid enhancer binding factor 1 (LEF1), 140, 328n20

Mackie, James, 239
Male, and Female, 18–19, 23, 43–49, 52, 122–23, 128–32, 142, 153–54, 158–64, 182, 196, 200, 205, 208, 212–16, 258–59, 265, 268, 275, 329n6, 333n7; defined, 1, 37–38, 43, 58, 166; as historical entities, 43. *See also* gender: vs. sex; gender/sex
mammals: chromosomes of, variation in, 205–6, 215–16; heterogametic, 131; homogametic, 131; live-bearing, 231, 337n61; penis of, 9, 194, 218; placental, 40, 57, 178–79, 192, 194, 197, 202–3, 206, 219, 221–24, 334n23, 337n61; red blood cells of, 327n4; reproductive

innovations and novelties of, 192, 194, 218; sex chromosomes of, 130; sexual development of, 139, 203, 231; sexual differentiation in, 197, 334n23; sexual reproduction of, 194; and sperm temperature sensitivity, 331n38; SRY genes in, 204; warm-blooded, 331n38
mammary glands, 56, 192, 194, 216, 224
manakins (Neotropical birds), 3–4
Mangold, Hilde, 108
manhood, 334n37. *See also* masculinity
Markowitz, Sally, 51–52
marriage equality, 17
marsupials, 192, 194, 197, 202, 209, 221–23, 336–37n60
Martin, E., 317n1
Marxism, 60
masculinity, 52–53, 152–53, 215, 261–63. *See also* femininity; manhood
masturbation, 53, 319n26
material body, 1, 20, 22, 25, 28–30, 35–36, 64, 67, 70, 73–76, 272; biology of, 246; and gender, 66; scientific representations of, 241; and sexual difference, 284. *See also* sexual body
material feminism, 1, 29, 36, 46, 73–76, 281, 283–86, 321n26; and developmental biology, 21, 101–2; history, boundaries, and goals of, 20; intellectual histories of, 316n7; and linguistic turn in feminist theory and analysis, 20–21; and medicine, 285; and new materialists, 269, 284–85; post-disciplinary, 269–72; and psychology, 285; richness and diversity of, 21; and science, 22
materiality, 14; and development, 74, 78; and gender/sex, 75; of sex vs. sex of materiality, 24, 67, 78. *See also* material body; material feminism
materialization, 55, 69, 74, 78, 98, 100, 190, 195, 270–71

mathematical models, 30, 298, 323n16
mating displays, 8–9, 123, 327n59. *See also* vocalizations, as sexual communication
Mayr, Ernst, 307
McCall, Leslie, 14
McClintock, Barbara, 283
McWhorter, Ladelle, 52–53
Mead, Margaret, 65–66, 279, 320n2
measles, 338n9, 342n2 (app. 2)
medicine: and biology, 275; and disease, 232; genomic, 234–35, 325n41, 338n11; and health, 187, 232; and material feminism, 285; and science, 159–60; Western, 160. *See also* biomedicine; health
meiosis (reduction division), 166, 170, 193, 197, 201–2, 246, 249, 339n3
meristems, 103
Merleau-Ponty, Maurice, 327n56
mesonephric ducts. *See* Wolffian ducts
metaphor: and biology, 25–27; and genes/genetics, 25–27, 322n12; and human culture, 27; and language, 26–27; linguistic in genetics, 322n12; and nature, 26; parliamentary for sexual development, 241; and science, 25–27, 317n19, 317n21; spatial in pre-DNA embryology, 102. *See also* analogies
methylation, 94–95, 98, 199, 323–24n22
mice: and behavioral genetics research, 285; embryonic, *220*; embryonic sexual development in, 173, *220*; gene action and regulation in, 82, 109–10; and genotype-phenotype abstraction, 77; homologous protein of, 134; queered strains of, 182; SRY gene in, 168, 204–5; transgenic, 182
microbes, and matter, 284
microchimerism, 180
Migeon, Barbara, 171
Miki, Y., 341n29
Mittwoch, Ursula, 196, 272, 337n6

modularity: biological, 36, 116, 299–300; of complex phenotypes, 312; and feathers, 5–6
modules: anatomical, 5–6, 300; defined, 299; developmental, 6, 300; and feather cells, 5–6; molecular, 219, 300; paracrine signaling, 93; and phenotypic functions, 300
molecular biology, 16; and agential realism, 76; and discourse, 58; and genetics, 43; genomic, 269; and material feminism, 21; and modularity, 299; and neurobiology, 121; and performativity, 272; and sex, 257–58; of sexual reproduction, 340n6; soft condensed matter in, 324n26
molecular-developmental biology, 6, 22–23, 25, 73, 99–101, 107, 180, 203, 210–11, 214, 245–46, 312
Molecular Feminisms (Roy), 286
molecular genetics: and animal development, 6; and development, 297, 329n3; and performative continuum, 246; revolutionary advances in, 33
molecules: bio-, 59, 82, 111–12, 120, 124, 271, 320n8; biological, 40, 109; and cells, 84, 271; and cell-to-cell communication, 84; and developmental biology, 98–99; and hormones, 59, 93, 153; inorganic, 108; and pathogens, 287–91; and proteins, 168, 322n11; signaling and receptors, 59, 84, 86; signaling and transcription, 210; signaling pathways of, 98–99
mole voles, chromosomes in, 205
monogamy, 53, 68, 285, 313
monophyly, 39–40
Moran, Jeff, 277
morphogenesis, 54, 79, 90, 97, 101, 105–6, 111, 119, 219–20, 324n31
morphology, 5–6, 8, 45, 51, 90–97, 100–102, 115, 127–28, 131–34, 139–40, 160–65, 175–77, 183, 190, 211, 214–19, 224, 231, 253, 258, 293, 300–302, 311, 336–37n60, 338–39n14; and behavior, 249, 326n48; cellular, 249; and embryos, 201; as feminist cause, 78; of hindlimbs, 336–37n60; and identity, 91; and physical science, 324n27; vaginal, in ducks, 10–11; of vertebrate digestive tracts, 324n31; of vulva, 323n14; of wings of fruit flies, 328n10; of zygodactyl grasping foot of parrots, 324n30. *See also* homology; phylogeny
mosaicism, 166–67, 171, 207, 268
Müllerian ducts, 142–43, *144*, 145, 148, 149, 153, 162, 168, 172–76, 178, 217–18, 230–31, 328n25, 337n8
Müllerian hormone, anti-. *See* anti-Müllerian hormone (AMH)
multispecies, 275, 283

nanostructures, 6–8, 102–3
nationalism, 60
natural kinds, 38–41, 43, 46, 318n7, 326n46. *See also* individuality, historical; ontology
natural selection, 6, 72, 113, 118, 124, 185, 193, 201–8, 219, 243, 248, 273, 276, 299–301, 305–7, 311–13, 326n49, 326n53, 335n46; and adaptation, 4, 8–9, 31–32, 190, 315n4, 319n28; and evolution, 31; and feathers for flight, 47; and sexual reproduction, 224, 253
nature: and culture, 15, 19, 252, 283, 285; gender in, 249–52; and metaphor, 26; and nurture, 297; and origin of consciousness, 238; and sexuality, 60
Nature, 272
Nature Reviews Genetics, 241, 272
neurobiology, 286; and behavior, 121; and psychology, 120–22

neuroligin proteins, 179, 332n46
neurons, 49, 55, 95, 100, 105, 107, 121–22
New Synthesis: and adaptation, 31–32, 317n27; and developmental biology, 306, 317n27; and evolutionary biology, 30–31, 242–44; and flatitudes, 31–32, 243; and genetics, 297; and phylogeny, 30, 307
New Yorker, 325n34
Nicholson, D. J., 318n3, 322n27
Nietzsche, Friedrich, 284
Noad, Michael, 251
Noë, Alva, 327n56
norms: animal gender, 251; cultural, 22, 28, 52, 66, 68, 70, 117, 216–17, 236, 251, 261, 273; and expectations, 66, 236; gender/sex, 32, 34–35, 56, 70, 182, 217, 236, 251, 274; and heteronormativity, 1, 24; and innovation, 216–19; queer gender, 274; reaction, 46; sexual, 163; social, 186
NR3C4 (nuclear receptor subfamily 3, group C, member 4), 331n30
NR5A1 (nuclear receptor subfamily 5, group A, member 1), *135*, 136, 140, 170, 328n14
nuclear weapons testing, and radioactivity, 34
nurture, and nature, 297
Nüsslein-Volhard, Christiane, 133, 272, 274

On Growth and Form (Thompson), 324n27
ontology: and astronomy, 42; and biology, 42, 318n3; and classes, 39–43, 47, 234; in evolutionary biology, 318n3; and genetics, 322n12; and geology, 42; historical, 38–46, 215, 263, 318n3, 318n5, 318n7, 319n14; and language, 42; and nonbiological things, 42. *See also* individuality, historical; natural kinds

opossums, 175, 218, 221
oppression: cultural and social, 14; resistance against, 19; of sexual minorities, 14, 60; of women, 14
orgasm, 151
ornithologists, 11–12
ornithology: and biology, 10; and evolutionary biology, 2, 315n4; and innovations, 15; and intersectionality, 13–16, 316n20, 316n22; and posthuman perspective in interdisciplinary research, 12. *See also* avian; birds
oscine songbirds, 250
osmoregulatory performance, 125
ovaries, 44, 128, 134–36, 147, 151, 160, 166–69, 172, 175, 178, 184, 194–97, 214, 216, 254, 333n17, 336n54
Owen, Richard, 114
Oyama, Susan, 296–97

PairedBox 2 (PAX2), 337n8
Pape, Madeleine, 255
paracrine signals and signaling, 86–101, *87–88*, 107–9, 116, 118, 121–23, 132–36, 140, 147–49, 155, 162, 168, 170, 184, 210, 221–23, 229–31, 255, 272, 289, 292, 323n14, 323n18, 337n6. *See also* endocrine signals and signaling
paramesonephric ducts. *See* Müllerian ducts
parrots, 8, 102, 104–5, 116–17, 250, 324n30; African Grey Parrot (*Psittacus erithacus*), *104*
parthenogenesis, 45–46, 194–95, 334n26, 337n64
Passeri. *See* oscine songbirds
paternalism, 163
pathogens, 120, 232, 265, 287–91, 338n9, 342n2 (app. 2)
pathology, 61, 162–66, 262
patriarchy, 1, 11, 28, 60, 70–71, 163
Pauling, Linus, 34

Pavlicev, M., 318n5, 338n14
PAX2 (PairedBox 2), 337n8
PCBs (polychlorinated biphenyls), 180
"Pelvic Politics" (Markowitz), 51–52
penis, 44, 333n14; avian, 9; development of, 148–49, 172–73, 175, 181, 218; of ducks, 9–10, 132, 196; of lizards, 218; mammalian, 9, 194, 218; reptilian, 194, 218; size of, 164, 330n11. *See also* hypospadias (developmental condition)
perception, as activity, 327n56
performance: all the way down, 17, 25, 78, 125, 129–30, 159, 196, 246; all the way up, 36, 196, 245–47, 249, 256; and birds, 125; defined, 125; and lizards, 125; and mating displays, 123; organismal, 125; osmoregulatory, 125; social, 65, 279, 320n2
performative, defined, 316n13
performative biology, 25, 27, 34–36, 67, 71–76, 112, 120–21, 124–26, 219, 274–76, 281, 320n8, 327n56; and evolutionary biology, 242–44; future of, 227–44; and queer biology, 34–35
performative continuum, 22–25, 29, 112, 189, 224, 246, 249, 256–58, 270, 272; and molecular-developmental biology, 22, 25; and sexual differentiation, 195
performative gender. *See* gender performativity
performative phenotypes. *See* phenotypes: performative
performative scientific hypotheses, 25–26, 172, 227–31
performative theory, 23, 25, 30, 35, 71–75, 126, 178
performativity: and agency, 66, 71–72, 112; and becoming, 66, 69; and behavior, 24, 64, 126; and biomedicine, 234; and bodies, 330n6; vs. canonical pathways, 98–100; of cellular discourse, 106–8; and citationality, 69–72; cultural, 36, 69, 247, 257, 320n8; defined, 58, 64–68, 78; and developmental biology, 98, 272; of disability, 231–36, 291; elements of, 67–72; and evolution, 325–26n45, 330n23, 336n50; and evolutionary biology, 272; genealogy and history of, 36, 70, 250; and genes, 100, 113, 322n11; of illness, 231–36, 291; and material, 66–68, 75; and mating displays, 123; and molecular biology, 272; of organisms, 74–75, 117–18, 190, 276; and philosophy, 321–22n27; and phylogeny, 247; and physiology, 117–18, 272; placental, 179, 219–23; posthuman, 66–67, 72, 320n8; of protein folding, 85–86; as reiterative power of discourse, 58, 78, 86, 248; and science, 227–31, 317n28; and sexual development, 35–36; and speech, 320n3; as useful language for scientific research, 26; variable, 334n37, 336n50. *See also* gender performativity
perversion, sexual, 52–53
PGD2 (prostaglandin D2), *135*, 136, 138, 168–69, 328n15
PGE2 (prostaglandin E2), 221–22, 336–37n60
PGF2A (prostaglandin F2α), 222, 337n62
phenotypes: and behavior, 152–53, 302; and biology, 27, 275; and cells, 106; complex, 106, 118, 121, 190, 244, 295–97, 300, 306, 309, 311–12, 325n41; and culture, 319n28; and development, 102–6, 170; and diversity, 35, 190, 264; and evolutionary biology, 30–31, 305; evolution of, 31–32, 35–36, 191, 243–44, 281; extended, 55–56, 274–75, 319n28;

and feathers, 219, 313; and genes, 82; and hormones, 154; multicellular, 32; organismal, 46, 67, 82, 110, 122, 125, 190, 242, 298, 305, 313; as performance, or enactment, of self, 24; performative, 26–27, 30–32, 36, 78, 83, 96–98, 102–6, 126, 128, 131, 162, 190–91, 218, 232–36, 242–44, 247, 271, 287, 294, 296, 303; and physiology, 125, 235, 237, 296; and pluralism, 242; and reproductive possibilities, 46; robustness of, 96; sexual, 55–57, 73, 98, 117, 126–29, 139, 154, 157–62, 173, 177–79, 182–85, 189, 195, 200–201, 207–8, 216, 243–44, 249–53, 255, 275, 281, 329n32, 335n46; and variation, 46, 56, 157–59, 162, 217–18, 227, 239, 268, 270, 294–95, 297–98, 329n32. *See also* genotype-phenotype relationship

philosophy: aesthetic, 11; of biology, 77; Continental, 20, 51, 58, 60, 70, 283–84; and cultural critique, 20; of ethics, 286; feminist, 1, 66, 286; and feminist theory, 66; and gender performativity, 65; and literary theory, 66; and material feminism, 20; and performativity, 321n27; of science, 1, 21–22, 46, 116, 242, 283, 309, 317n28; of sex, 70; and sociology of science, 309

phlogiston, 242

phosphorylation, 93–95, 98, 136, 168, 203–5, 210, 330n23

photosynthesis, 302, 333n4

phylogenetics, 3–4, 31–32, 178, 191, 195–98, 218, 250, 302, 317n27, 333n4

phylogeny: avian, 2–4, 12, 104; and development, 30–31, 243; and developmental biology, 31; and evolution, 191, 195–96, 317n27; and evolutionary biology, 30, 317n27; and feminist cause, 78; and homology, 191–93; and Linnaean classification, 2–3; and New Synthesis, 30, 307; and performativity, 247; reconstruction of, 307, 333n4; and reproductive traits, 192; and selection, 243; and sexual innovations, 193; of vertebrate animals, 192. *See also* homology; morphology

physics, 102–3; and engineering, 69, 298; and evolutionary biology, 30, 32; and genetics, 106; and material science, 4–5, 30; Newtonian, 238; optical, 340n22; scientific study of, 68; soft condensed matter, 102, 324n26. *See also* quantum physics

physiology: and anatomy, 18, 44–45, 51, 54, 118, 121, 128, 130, 158, 161, 167, 181, 207, 253–54, 302, 329n6; and behavior, 153–54; and biological performativity, 117–18; and biomechanics, 125; and cellular development, 121; "cold-blooded," 333n21; and development, 84, 86, 111, 113, 117–19, 121, 153–54, 182, 249, 268, 271, 298, 302, 306, 329n32; and developmental biology, 111, 113, 119; and development of complex body, 54, 83; and evolutionary biology, 119; and feminist cause, 78; fetal, 221, 228; hormonal, 83, 152, 258; human reproductive, 153; and immunity, 119–20; intercellular discourse in, 86; of lizards, 333–34n21; and modules of body, 299; and molecular communication, 84; of multicellular organisms, 58; and neurobiology, 121; and organs of body, 271; and performative continuum, 29; performative theory of, 35; and performativity, 64, 117–18, 272; and phenotype, 125, 235, 237, 296; reproductive, 18, 128, 154, 161, 253–54, 329n6; sexual, 51, 54, 130, 158, 160; and sexual difference, 50

Pickering, Andrew, 317n28, 320n8
Pipridae. *See* manakins (Neotropical birds)
placenta, 179–80, 192, 194, 219–23, 228–29, 329n32, 330n17, 332n48, 337n61
placodes, 47, 102, 116, 305, 320n34
planetary science (historical science), 43
plants: anisogamy in, 333n7; and cytoplasmic sterility mutations, 200; dioecy in, 333n7; evolutionary origin of, 200; evolutionary success and diversity of, 333n8; flowering, 40, 213, 217, 333n7; and invasive species ecology, 286; meiosis in, 193; and meristems, 103; multicellular, 79, 103; and roots, 78; seed, 194, 333n8; sexual development in, 200; sexually transitioning, 258–59; succulent, 302; terrestrial reproduction in, 194; vascular, 103. *See also specific plants*
pluralism, 239, 242, 285, 318n13
Polanco, Juan Carlos, 203
politics: and biology, 22; and culture, 15, 28, 33–34; and feminism, 19; and gender performativity, 70–71; and joint material presence for invisible people, 71; and liberation, 60; queer, 73; and science, 11, 28, 33–34; sexual, 20; and sexual identity, 160; and sociology, 31
pollution and pollutants, 34, 180–81, 335–36n47
polychlorinated biphenyls (PCBs), 180
polygyny, 3
Popeil, Ron, 333n9
pornography, 163, 319n26
porpoises, songs of, 250
posthumanism, 247–49, 321n20; and biodiversity, 75; and performativity, 66–67, 72, 320n8; perspective, 12, 322n12; and power, 117–19. *See also* humanism

post-structuralist analysis and critique, 58, 284
Pradeau, T., 318n3
predators, and stotting, 125, 327n62
Presentation of Self in Everyday Life, The (Goffman), 65
Previn, Andre, 245, 339n1
primates and primatology, 114, 168, 203, 210, 252, 283, 330n23
Prosser, Jay, 72–73, 258
prostaglandin D2. *See* PGD2 (prostaglandin D2)
prostaglandin E2. *See* PGE2 (prostaglandin E2)
prostaglandin-endoperoxidase synthase 2. *See* PTGS2 (prostaglandin-endoperoxidase synthase 2)
prostaglandin F2α. *See* PGF2A (prostaglandin F2α)
prostate cancer, 48, 181, 208, 266–67, 335n47
prostate glands, 149, 175–76
prostates, 214, 331n29
protandry, 213
proteins, 85–86, 94, 101; biological, 108; in bird feathers, 8; and cells, 109; and disease, 85; and genes, 90, 134, 215, 328n13; and hormones, 92; and molecules, 168, 322n11; neuroligin, 179, 332n46; new, 39; transcription factor, 92
protogyny, 213
Prum, Richard O., 123, 315n15
Psittaciformes. *See* parrots
psychology, 260; and biology, 48, 275; developmental, 22, 121–22, 258; and gender, 66, 117; and gender realization, 321n17; and gender/sex development, 151; and material feminism, 20, 285; and neurobiology, 120–22; and performative continuum, 29; and performativity, 64, 66, 117
Psychopathia Sexualis (Krafft-Ebing), 52

Ptashne, Mark, 323n22
PTGS2 (prostaglandin-endoperoxidase synthase 2), 221, 336–37n60
puberty, 18, 166, 168, 173, 180, 182

Quammen, D., 326n46, 333n4
quantitative trait loci (QTL) mapping, 295–96
quantum mechanics, 111, 238–39
quantum physics, 66–67, 238, 259–60
queer: biology, 34–35, 190, 275–76; defined, 60–61, 273; ecology, 60, 275; evolution as, 201, 273–74; and gay closet, 274; gender-, 14, 15, 22, 32, 71, 274; identity, 32, 61; as marginalized, 38; politics, 73; and science, 272–76, 285; and sexual development, 209; and sexual differentiation, 201–7, 312; studies, 13, 246; systems biology, 298; vulnerabilities of individuals, 32, 35; and Y chromosome, 203
queer bodies, 61, 159–60, 181–82, 189, 201, 204, 257, 272
queer feminism: and biological diversity, 284–85; and biology, 1, 33, 245, 256–57; and developmental biology, 1, 13, 257; and evolutionary biology, 1, 13, 257, 277; and gender performativity, 23–24; and genetics, 1, 13, 257; and humanities, 66; and monogamy, 285; and performative continuum, 25; and performativity, 23–24, 29; in science, 15
queer feminist: analyses, 15, 257, 284–85; culture and comprehension, 73; inquiry, 22; science studies, 13, 29; studies, 1, 24, 67, 317n28; theory, 2, 13–16, 25, 270, 277; thought, 1, 66–67, 272; vision, 285
queer gender: norms, 274; theory, 15
queering, 60–61, 71, 209, 312; biology, 35, 272–76; evolution as generatively, 201–4, 207

"Queer of Color Critique" (Ferguson), 60
queer performative: biology, 275; phenotype, 293
queer science, 2, 15, 25, 35, 166, 181–88, 272–76, 285. *See also* queer: biology
queer theory, 11–13, 15, 34–35, 60, 245, 277, 317n28
quorum sensing (coordinated group behavior), 248

race: biological concept of, 34; and diversity, 263–64; scientific fact of, 263; self-identified, 263; and sex, 14, 51–54, 256, 259–69, 294, 308–9, 330n13
Race to the Finish (Reardon), 264
racism, 16, 60; and eugenics, 51–53, 285–86; and genocide, 53, 263; and normative sex difference, 53; and white supremacy, 263
radioactivity, and nuclear weapons testing, 34
rape, in nonhuman biology. *See* forced copulation (acts of sexual violence in nonhuman animals)
rats: chromosomes in, 205, 215–16; eusocial, 112; osmoregulatory performance of, 125
realism: and intra-action, 260; and reality, 26–27. *See also* agential realism
Reardon, Jenny, 264
reductionism, 9, 12–13, 15, 29, 98, 101, 113, 121, 238, 243, 298, 326n49
regeneration, and development, 325n34
Reich, David, 266–69
reproduction, sexual. *See* sexual reproduction
reproductive bottleneck, 44, 185–86, 224–25, 318n9
reproductive rights and autonomy, 16
reproductive tract development, 142–48, *143–44*, *146*, *150*, 155, *172–77*, *174*

reptiles, 145; non-avian, 197; penis of, 194, 218; reproductive innovations of, 192, 194, 218. *See also* amphibians

ribonucleic acid. *See* RNA (ribonucleic acid)

Richardson, Sarah, 23, 29, 46, 129, 132, 159, 215, 256, 258, 285, 317n15, 318n13, 321n26, 334n37, 335n41, 335n43, 340n13

Ritz, Stacey, 255, 279

RNA (ribonucleic acid), 26, 85, 112, 138–39, 170, 204, 247–48, 321n14; and universal genetic code, 41. *See also* DNA (deoxyribonucleic acid)

rodents, chromosomes in, 205

Rooney, Mickey, 138, 169, 328n16

Roughgarden, Joan, 273

roundworm: development of, 89, 323n14; as model organism, 109–10

Roux, Wilhelm, 311

Roy, Deboleena, 286, 316n7, 321n26

RSPO1 (R-spondin 1), *135*, 141, 170, 328n22

RuPaul's Drag Race, 69

Salamon, Gayle, 73

same-sex sexual attraction, 170, 331n46. *See also* homosexuality

Sandilands, Catriona, 60

SARS-CoV-2, 234, 291, 338n9

Schartl, Martin, 211, 336n52

Schiebinger, Londa, 265–66, 284

Schwenk, Kurt, 311–13, 343n2

science: anti-, 70; and biology, 12–13, 26, 29, 41, 256, 276, 285; and biomedicine, 14, 48; boundaries of, blurring, 12–13, 15, 22; and culture, 1, 15–16, 18–19, 21–24, 27–29, 33–34, 38, 46–47, 152, 159, 249–50, 256–58, 270, 274, 276, 283, 285, 329n3; and engineering, 265; and ethics, 260–62; and evolution, 33–34; feminist, 11–13, 22; feminist critiques of, 19–20; and gender, 18; historical, 42–43; history of, 21–22, 283–84; and humanities, 12–14, 51, 274–75, 305; intellectually queer space in, 272–76; as interdisciplinary, 12–13; and intersectionality, 13–16, 316n20, 316n22; and language, 26; and material body, 29; as material-discursive practice, 67; and material feminism, 22; and medicine, 159–60; and metaphor, 25–27, 317n19, 317n21; and objectivity, 12; and performativity, 227–31, 317n28; philosophy of, 21–22, 46, 116, 242, 283, 309, 317n28; and politics, 11, 28, 33–34; queer critique of, 285; queer feminism in, 15; queer space in, 272–76; and reality, 27; of sex, 18–20, 33, 283; sexual, 275; situated style of, 12, 15, 269–70, 283; and sociology, 264, 268–69; sociology of, 21, 309–10; and subjectivity, 24; women in, 283–84. *See also* queer science; social sciences; *and specific sciences*

scrotum, 44, 117, 128, 147–51, 160, 165, 172–77, 181, 183, 187, 230, 331n38

Second Skins (Prosser), 72–73

Sedgwick, Eve Kosofsky, 22, 25, 60, 66, 170, 270, 274

See, Sam, 272–74, 277, 341n1

selection and selectionism: and adaptation, 6; and agency, 112; and behavior, 271; disruptive, 336n46; gene-level, 30, 32, 36, 112–13, 229, 231, 242–43, 305–13, 326nn48–49, 334n36, 335n46; intrasexual and intersexual, 122–23; and phylogeny, 243; sexual, 3, 8–9, 72, 112, 122–24, 185, 201, 207–8, 211, 217, 224, 253, 271, 273, 277, 309, 315n4; sexually disruptive, 207–8, 336n46. *See also* internal selection

selective serotonin uptake inhibitors (SSRIs), 122
self-becoming, 23, 24, 34
self-enactment, 24, 58, 77–126, 195
Selfish Gene, The (Dawkins), 112–13, 305
selfish genes, 31–32, 56, 112–13, 119, 305–6, 309, 326n48, 335n46
self-realization, 58, 158, 256, 258
self-understanding: and biodiversity, 32, 191, 249–50; and biology, 276
Serano, Julia, 73, 258
sex: as binary fact established at birth, 18–19; and biology, 13, 19–20, 29, 44, 191, 277, 318n13; and biomedicine, 318n13; contextual account of, 318n13; cultural concepts of, 19–20; defined, 1, 17–20, 38, 43, 46–47, 129, 179, 186, 252–56, 263, 284–85, 318n3, 318n13; as dynamic dyadic kind, 46; embodied, 73, 196, 214, 225, 284, 329n3; essentialism, 318n13; evolution of, 191–95; vs. gender, 18–19; genealogy of, 246; genetic and identity, 17–18; "geneticization" of, 159; as history, 43–47, 128, 256, 263, 318n3; objective science of, 19; philosophy of, 70; and race, 14, 51–54, 256, 259–69, 294, 308–9, 330n13; science of, 18–20, 33, 283; and social behavior, 19. *See also* gender/sex
"Sex and Gender Analysis Improves Science and Engineering" (Tannenbaum et al.), 265
Sex and Temperament (Mead), 65
sex determination, 19, 23, 28, 56–58, 127, 130, 132, 138–39, 159, 185, 196, 231, 240–41, 317n15, 327n1. *See also* sexual determination
sex-determining region on the Y. *See* SRY (sex-determining region on the Y)
sex difference, 48–54, 152–54, 158–59, 161, 265–69; and behavior, 262; binary, 262; essential, 285; genetic, 267–68; hormonal research on, 154; normative, 53; phenotypic, 261; and race, 51–54; science of, 52; vs. sexual difference, 48–53, 128–29, 154, 284. *See also* sex differentiation; sexual difference
sex differentiation, 197, 203, 207, 228–29. *See also* sex difference; sexual differentiation
sex/gender. *See* gender/sex
Sexing the Body (Fausto-Sterling), 283–84
Sex Itself (Richardson), 29, 46
sex reversal, 56–58, 127, 170, 173, 185, 319n29, 331n43
sex/sexual roles, 18, 52, 65
sexual anatomy. *See* anatomy: sexual
sexual autonomy, 11, 123, 315–16nn15–16, 331n46
sexual becoming, 6, 73, 189, 248–49, 252, 258
sexual body, 25, 36, 73; and biology, 22, 54; development of, 127–55; evolution of, 22, 37; molecular developmental of, 22, 329n3; scientific views of, 284; and sexual differentiation, 200. *See also* material body; queer bodies
sexual coercion, 10–13, 123, 335n46
sexual communication, 3, 8, 103, 250–51
sexual conflict, 10–11, 207, 315n15, 331n46, 335n46
sexual contextualism, 318n13
sexual determination, 167, 170, 173, 184. *See also* sex determination
sexual development, 19–20, 154–55; and behavior, 153–54; of birds, 196; of butterflies, 196; and cells, 127; and chromosomes, 166–67, 198; differences in (DSD), 161, 165–67, 173, 240, 258, 329n6, 330n13, 330n15,

sexual development (*continued*)
  332n54; and differentiation, 54–55; disorders of (DSD), 161, 240; of embryos, 35; and environment, 178–81, 198–99; and environmental pollutants, human-created, 180–81; and evolution, 195–200, 217–18; evolutionary variability of, 195–200; and genes/genetics, 29, 35, 177; and genetics, 29, 35, 177, 198–99; haplodiploid, 199–200; hormonal signaling in, 35, 229–31; and molecular discourse, 209–11; nonnormative, 332n54; novel performative scientific hypotheses about mechanisms of, 25–26, 226–31; and pathology, 162–66; performative, 35–36, 129, 155; phallocentric, 140; in plants, 200; post-embryonic, 151–54, *150*; and queer biology, 190; variations in, 57, 155, 157–88, 196, 332n59
sexual difference, 158, 161, 202, 228, 255–56, 284, 332n54; and anatomical homogeneity, 207; and behavior, 50, 207; and biomedicine, 265; development of, 128; and diversity, 263; embodied, 182; genetic, 131; and individual sexual bodies, 54; and material body, 284; and pathology, 61; in phenotype, 195; vs. sex difference, 48–53, 128–29, 154, 284; and sexual development, 129; and sexual reproduction, 253; and sexual variation, 53–54, 129, 266. *See also* sex difference; sexual differentiation
sexual differentiation: in birds, 197, 334n23, 335n44; cell-autonomous, 196; diversity of, 200; and embryos, 228; and evolution, 211; and genetics, 201; initiation mechanisms, 198; in mammals, 207; and performative continuum, 195; and performative variability, 334n37; and queering, 201–7, 312; science of, 272; and sibling effects, 332n48. *See also* sex differentiation; sexual difference
sexual discrimination, 17
sexual essences, 58, 128, 215–16
sexual identity, 160, 183, 185
sexuality: and biology, 19; history of, 246; and identity, 60; and nature, 60; and social order, 188
sexual perversion, 52–53
sexual phenotype. *See* phenotypes: sexual
sexual reproduction, 38, 43–45, 48, 55, 119, 128, 185, 193–95, 201, 207, 216–17, 224, 246–47, 249, 253, 333n5, 335n46, 340n6
sexual selection. *See under* selection and selectionism
sexual transition, 212–15, 258–59
sexual variation, 19–20; embodied, 159–61, 211; and evolution, 189–225; as marginalized, 38; and sexual difference, 53–54, 129
sexual violence, 10–11, 335n46
SHH. *See* Sonic Hedgehog (SHH)
Shubin, Neil, 145
sickle-cell anemia, 108, 325n37
"Situated Knowledges" (Haraway), 12, 15, 269–70, 283
snakes, 41, 192, 198, 215, 218, 229
Snow, Sam, 316n16
Sober, Eliot, 308
social environment, 65, 68, 121, 159, 178, 200, 214, 217, 236, 321n17
social inequities, and health, 266
social injustice, 34
social oppression. *See* oppression
social sciences: and biology, 261; and gender performativity, 65; and humanities, 12, 27, 65, 249, 251–52; reframed, 12; and testosterone, 285
sociology: and biology, 275; and communicative discourses, 247; and

politics, 31; and/of science, 21, 264, 268–69, 309–10
Sonic Hedgehog (SHH), 90, 101, 112, 133–34, 147, 148, 219
SOX (SRY-related high-mobility group box): SOX3, 178, 184, 202, 209, 254, 331n44, 336nn50–51; SOX9, *135*, 136–41, 168–69, *169*, 172, 177–78, 182, 184, 203, 209–11, 214, 254, 328n13, 336n51. *See also* SRY (sex-determining region on the Y)
Spemann, Hans, 108
sperm ejaculation, 128
Squier, Susan, 96
SRY (sex-determining region on the Y), 131, *135*, 136–41, 167–73, 192, 202–6, 215, 254–55, 328nn12–13, 330n23, 331n44, 334n36; binding to SOX9 enhancer, *169*; mRNA stabilization by WT1+KTS, *135*, 138–39, *139*, 184, 204, 254; origin of from SOX3, 202; transcription factor, 136, *137*. *See also* SOX (SRY-related high-mobility group box)
SSRIs (selective serotonin uptake inhibitors), 122
starfish, 78, 192, 194
Stearns, Stephen, 208, 335n46
stem cells, 95, 180, 330n17
stotting, 125, 327n62
Structure of Scientific Revolutions, The (Kuhn), 184–85
Subramaniam, Banu, 285–86, 316n7
sustainable food production, 16
symbiosis, 42, 275–76; endo-, 333n4
synapses, 121–22, 179
syngamy (fertilization), 193, 246, 249
syringes (manakin family vocal organs), 3–4
systems biology, 240, 297–98, 299

Tannenbaum, Cara, 265
Taylor, Peter, 77, 293–94

tenascin R (TNR). *See* TNR (tenascin R)
termites, eusocial, 112
testes, 44, 128, 132–41, 149–51, *150*, 168–70, 172, 178, 184, 203, 209–11, 329n32, 329n34, 331n38, 333n17, 336n51, 336n54
testosterone, 49, 140, 142, 145, 147–53, 172–84, 227–30, 254, 285, 329n32, 331n30, 334n36
Testosterone (Jordan-Young and Karkazis), 49, 152–53
tetrapods, 40–41, 44, 114, 128, 213–19, 299
thermodynamics, 30, 43, 326n49
theropods, 3
Thomas, Lewis, 75
Thompson, D'Arcy, 324n27
thrips, haplodiploid sexual development of, 199–200
Through the Looking Glass (Carroll), 204, 211
Tinbergen, Niko, 4
tissues: and cells, 95, 100–105, 324n27; and induction, 108
TNF (tumor necrosis factor), 222, 337n62
TNR (tenascin R), 221, 336–37n60
Tolstoy, Leo, 59, 231–32
toxicology, eco-, 34
transcription factors, 85, 88, 91–95, 98–99, 107, 109, 132–33, 136–41, 168, 175, 177–78, 204, 210, *220*, 248, 254, 272, 322n11, 325n39, 328n6, 328n23, 330n23
transcriptomes, and genetics, 70, 157, 298, 317n15, 321n14
transgender people, 14; and gender performativity, 72–73, 258; and identity, 61, 73, 258; medical treatment for, 165; medicine for, 214; recognition and rights of, 17–18; and sports participation, 18; vulnerabilities of, 32

transition, sexual, 212–15, 258–59
transsexual people: and gender performativity, 72–73; identity, and material body, 72–73; and intersex experience, 284–85; and material feminism, 284–85; nonperformative, seeking to be, 73; and performative continuum, 258–59; and performative perspectives, 258–59; and queer feminism, 73, 284–85; rights of, 284–85
Tree of Life, 3–4, 191, 249
Tresky, Shari, 236
Trochilidae. See hummingbirds
Trump, Donald, 17, 19
tumor necrosis factor (TNF), 222, 337n62
tunica albuginea and tunica vaginalis, of testes, 329n34
Turing, Alan, 90–91, 274, 323n16
Turner's syndrome, 166
turtles, 198, 201, 215, 218, 256, 333n17, 340n15
twins, 171–72, 179–80, 288, 291, 322n4
Tyrannosaurus rex, 115

Undoing Monogamy (Willey), 285
universality, 15, 270
urethra, 51, 56, 146–49, 163, 173–76, 187, 218–19, 254
urogenital sinus, 143–49, 172–76, 230–31, 337n8
uterus, 48, 128, 142–44, 151, 171–75, 178–79, 214–24, 230, 254

vagina, 10–11, 53, 128, 142–51, 158, 163–64, 172–78, 181, 183, 187, 218, 230–31, 254, 337n8
van Anders, Sari, 19
Velocci, Beans, 159
vertebrate animals, 40–41; acquired immune system of, 287–92; anatomy of, 277; evolution of, 195–96; homologous protein of, 134; legs of, 44; and phylogeny of reproductive traits, *192*; sex determination of, 241; sexual bodies of, 189; and sexual differentiation, 272. See also invertebrates
"vicious abstractionism," 309, 326n48
violence, sexual. See sexual violence
vocalizations, as sexual communication, 3, 15, 250–51, 340n8. See also mating displays
vulva, 163–64, 254, 323n14

Waddington, Conrad, 96–98, 108, 159, 217, 240, 257, 274, 301, 311–12, 323n22
Wade, Michael, 307–9, 311, 313, 325n45
Wagner, Günter P., 93, 115–16, 141, 221–22, 278, 311–13, 318n3, 325n40, 326n51, 336n48, 336n56, 337n1, 343n2
War and Peace (Tolstoy), 59
W chromosomes, 192, 196–97, 199, 202, 206
Weismann, August, 311
well-being, 32, 235, 264
wellness, 232, 255, 332n59
West-Eberhard, Mary Jane, 297, 301, 303
whales, 115–16, 250–52, 340n8
Who Wrote the Book of Life (Kay), 317n21, 322n12
Whyte, L. L., 311, 343n2
Wieschaus, Eric, 133, 272
Wiley, E. O., 318n3
Willey, Angela, 25, 129, 285, 316n7, 321n26
Wilms' tumor protein 1. See WT1 (Wilms' tumor protein 1)
Wilson, E. O., 339n24
Wilson, Elizabeth A., 21–22, 129, 284–85
Wilson, Tony, 279
Wingless-related integration site pro-

tein. *See* WNT (Wingless-related integration site protein)

Winther, Ramus, 309, 326n48

WNT (Wingless-related integration site protein), 148, 184; WNT4, *135*, 140–41, 328n19; WNT7A, 101; WNT7B, 170

Wolffian ducts, 142–43, 145, 162, 176, 230, 328n25

women's liberation, 19

World War II, 34

Wright, Sewall, 326n49

WT1 (Wilms' tumor protein 1), 170, 328n17; SRY stabilization by, *135*, 138–39, 184, 204, 254; WT1−KTS, *135*, 138; WT1+KTS, *135*, 138–39, 184, 204, 254

X0, 166, 205, 215–16

$X_1X_1X_2X_2X_3X_3X_4X_4X_5X_5$, 197, 216

$X_1X_2X_3X_4X_5Y_1Y_2Y_3Y_4Y_5$, 197, 216

X chromosomes, 50, 130–31, 166, 170, 184, 192, 197, 202, 205–7, 215–16, 239, 254, 267–68, 331n31, 331n43, 334n30; inactivation, 171–72

XMXP, 205–6, 216

XX, 23, 50, 130–31, 140, 153, 166–67, 170–73, 175–76, 178–80, 186, 196–99, 205–7, 209, 216, 227–31, 254, 268, 294, 329n32, 331n31, 334n23, 336n50, 337n6

XXX, 166

XXY, 166, 171, 254

XY, 23, 50, 130–31, 136–40, 153, 166–68, 170–73, 175, 177–82, 196–98, 202–3, 206–7, 209, 215–16, 227–31, 254, 268, 294, 329n32, 331n31, 332n46, 334n23, 335n41, 337n6

XYY, 166

Y chromosomes, 50, 57, 130–32, 136, 139–40, 166, 170, 179, 184, 192, 197, 202–6, 215–16, 228, 254–55, 267–68, 332n46, 334n37, 335n41, 335n43

Young, Hugh H., 164–65, 330n27, 336n54

Your Inner Fish (Shubin), 145

Z chromosomes, 131–32, 192, 196–97, 334n26

zebrafish: gain-of-function gene mutations in, 106; gene action in, 106, 109–10, 197, 199; multigene genetic variation in, 199; as one of best-studied model organisms, 197; sexual differentiation in, 197, 229

zone of polarizing activity (ZPA), 101, 219

zoology: and evolution, 189; and sequential hermaphrodites, 213

ZPA. *See* zone of polarizing activity (ZPA)

ZW chromosomes, 131, 196–99, 229, 334n23, 335n44

zygodactyly, 104, 116–17, 324n30. *See also* anisodactyly

zygotes (fertilized eggs), 23–25, 43, 54, 57–58, 79, 89, 114, 128, 130, 158, 166, 180, 185, 217, 224, 252–53, 256, 322n4, 323n14, 334n26

ZZ chromosomes, 131, 196–99, 206, 215, 229, 334n23, 334n26, 335n44

Milton Keynes UK
Ingram Content Group UK Ltd.
UKHW011935221223
434869UK00003B/91